The Geography of Agriculture
in Developed Market Economies

The Geography of Agriculture in Developed Market Economies

Edited by Ian R. Bowler

Copublished in the United States with
John Wiley & Sons, Inc., New York

Longman Scientific & Technical,
Longman Group UK Limited,
Longman House, Burnt Mill, Harlow,
Essex CM20 2JE, England
and Associated Companies throughout the world.

Copublished in the United States with
John Wiley & Sons, Inc., 605 Third Avenue, New York, NY 10158

First published 1992

British Library Cataloguing in Publication Data
A catalogue record for this book is available from the British Library

ISBN 0-582-30161-0

Library of Congress Cataloging-in-Publication Data
The Geography of agriculture in developed market economics / edited by
  Ian R. Bowler.
   p.   cm.
  Includes bibliographical references and index.
  ISBN 0-470-21869-X (USA only)
   1. Agricultural industries – Location.   2. Economic geography.
  I. Bowler, Ian R.
  HD9000.5.G377   1992
  338.6'042 – dc20   92-12994
                                        CIP

Set by 9 in Bembo
Printed in Malaysia by PA

To our students of agricultural geography
past, present and future

Contents

Contents

List of figures

List of tables

List of contributors

Professor John Aitchison, Department of Geography, University College of Wales, Aberystwyth, Wales

Dr Ian Bowler, Department of Geography, University of Leicester, England

Professor Christopher Bryant, Département de Géographie, Université de Montréal, Montréal, Canada

Dr Gordon Clark, Department of Geography, University of Lancaster, England

Mr Christopher Edwards, Department of Environmental Studies, University of Ulster, Northern Ireland

Dr Patrick Hart, formerly Department of Geography, College of Ripon and York St John, England

Professor Richard Munton, Department of Geography, University College London, England

Professor Martin Parry, School of Geography, University of Oxford, England

Dr John Tarrant, School of Environmental Sciences, University of East Anglia, England

Acknowledgements

We are grateful to the following for permission to reproduce copyright material:

Akademia Kiado and the author, Prof. D. Gillmor for Fig. 6.1 (Gillmor, 1977); Butterworth-Heinemann Ltd. for Figs 9.8a & b (Robinson, 1991); Cambridge University Press and the author, Dr T. Bayliss-Smith for Figs 8.1 & 8.2 (Bayliss-Smith, 1982); Central Statistical Office for Table 7.2 (Department of Trade & Industry Business Statistics Office, 1988); Paul Chapman Publishing Ltd. for Fig. 4.3 (Clark, 1991); Croom Helm for Table 1.1 (Troughton, 1986); Office for Official Publications of the European Communities for Figs 4.2, 4.4 (Commission of the EC, 1991) & 9.6 (Henry, 1981); Elsevier Science Publishers BV and the author, Prof. C. R. Bryant for Fig. 10.2 (Bryant, 1984); Farming Press Ltd. for Table 6.6 (Slee, 1989); Food and Agriculture Organization of the United Nations for Fig. 7.4 (FAO, 1990), Tables 7.1 (FAO, 1986a), 7.4 & 7.5 (FAO, 1989); the editor, *Food from Britain* for Fig. 7.1 (Tanbaum, 1981); The Free Press, a Division of Macmillan, Inc. for Figs 6.3a–d (Rogers & Shoemaker, 1971) Copyright © 1971 The Free Press; Geographical Society of Finland for Fig. 5.2 (Varjo, 1984); Geographical Society of Ireland for Tables 6.1 & 6.2 (Edwards, 1974); the author, Prof. H. Gregor for Fig. 5.5, Tables 5.6 & 5.7 (Gregor, 1982); the Controller of Her Majesty's Stationery Office for Fig. 7.2 (National Food Survey Committee); the author, Dr K. Hoggart for Fig. 7.3 (Hoggart, 1979); the editor, Dr B. Ilbery for Tables 1.2 & 1.3 (Bowler, 1985); *'Information Géographique* for Figs 7.6a & b (Guellec, 1983); The Institute of British Geographers for Fig. 8.8 & Table 8.9 (Parry, 1991); the editor, *Journal of Agricultural Economics* for Table 6.3 (Gasson, 1973); Kluwer Academic Publishers for Figs 10.3 & 10.4 (Bryant & Russwurm, 1982); Longman Group UK Ltd. for Figs 8.3, 8.4, Tables 8.2 & 8.5 (Briggs & Courtenay, 1985); Methuen & Co. for Fig. 6.4 (Brown, 1981); Ministry of Agriculture, Fisheries and Food for Figs 3.2 & 3.3 (MAFF/IR) © Crown Copyright; Nielsen for Table 7.3 (Nielsen Marketing Research, 1982, 1987); The Open University for Figs 6.2a & b (Jones, 1967); Oxford University Press for Table 8.3 (World Commission on Environment and

Development, 1987); the editor, Dr M. Pacione for Fig. 5.1 (Aitchison, 1986); Pergamon Press Ltd. for Tables 4.3 (Munton & Marsden, 1991), 4.8 (Marsden *et al.*, 1987), 8.7 (Myers, 1979) & 10.3 (Fuller, 1990) Copyright 1979, 1987, 1990 & 1991 Pergamon Press Ltd.; Polish Scientific Publishers PWN Ltd. for Fig. 5.3 (Scott, 1983); Royal Dutch Geographical Society for Figs 4.1 (Windhorst, 1989) & 7.11 (Hart, 1978); Wadsworth Publishing Co. for Fig. 8.7 (Miller, 1982); Worldwatch Institute Publications for Figs 8.6a & b, Tables 8.4 & 8.6 (Worldwatch Institute, 1985).

Whilst every effort has been made to trace the owners of copyright material, in a few cases this has proved impossible and we take this opportunity to offer our apologies to any copyright holders whose rights we may have unwittingly infringed.

List of abbreviations

AWU	Annual work unit
BSU	business size unit
CAP	Common Agricultural Policy
CGIAR	Consultative Group on International Agricultural Research
CPE	centrally planned economy
EAGGF	European Agricultural Guidance and Guarantee Fund
EC	European Community
ECU	European currency unit
ESA	Environmentally Sensitive Area
ESRC	Economic and Social Research Council
ESU	European size unit
FAO	Food and Agriculture Organization
FCGS	Farm and Conservation Grant Scheme
FNVA	farm net value-added
FWS	Farm Woodland Scheme
GATT	General Agreement on Tariffs and Trade
GDP	gross domestic product
GIS	Geographical information systems
GNP	gross national product
HLCA	Hill Livestock Compensatory Allowance
LFA	Less Favoured Area
MAFFA	Ministry of Agriculture, Fisheries and Food
MMB	Milk Marketing Board
MSS	Multi-Spectral scanner
NASA	National Aeronautical and Space Administration
NFU	National Farmers' Union
NOMIS	National On-line Manpower Information Service
NSA	Nitrate Sensitive Area
OECD	Organization for Economic Cooperation and Development
OFW	off-farm work
PIK	payment in kind
PYO	pick your own

SCP	simple commodity production
SGM	standard gross margin
SICA	Sociétés d'Intérêt Collectif Agricole
SMD	standard man-days
SPOT	Système Probatoire de l'Observation de la Terre
UK	United Kingdom
UNCTAD	UN Conference on Trade and Development
US	United States (of America)
USDA	United States Department of Agriculture
USGPO	United States Government Printing Office

Introduction

The broad objective of this book is to describe and explain the contemporary geography of agriculture in developed market economies. The objective has been approached by a team of agricultural geographers, each writer contributing an analysis of a particular topic. The contributors have been given freedom in interpreting their topics with the deliberate intention of developing a variety of theoretical and methodological approaches to the overall objective of the book. However, so as to retain a coherence to the various chapters, two themes run through the ten essays: one is the recent transformation of farming, seen by some contributors as the industrialization of agriculture; the other theme is the place of farming in the evolving food supply system.

Looking first at the recent transformation of agriculture, over the last four decades farming in most developed countries has been changed at a speed and to an extent never previously experienced. Employment in farming has declined drastically, together with the number of independent farm businesses; intensive farming practices have raised agricultural output significantly but have become dependent on high levels of capital expenditure, especially as regards mechanical and biological technology; most farm production now passes through food processors and manufacturers before reaching the consumer. When taken together with other changes, these developments have been described variously as comprising the third agricultural revolution, the modernization of farming, the restructuring of agriculture, and the industrialization of agriculture (Healey and Ilbery 1985: 83). Not surprisingly, such significant developments have attracted widespread attention, although most writers have treated agriculture as if it were an homogeneous entity. A distorted picture of modern agriculture can result, for while the processes of change are international in scope, their effect on different countries, regions and individual farms has varied enormously. One contribution of this book, therefore, is to redress the balance of analysis by describing and explaining the geography of the characteristics, extent and implications of the recent transformation of agriculture. So as to limit the scope of the book, attention has been focused on countries with developed market economies; even so, similar transformations are apparent within many state socialist (Enyedi and Volgyes 1982) and developing countries (Arnon 1987).

Turning now to the food supply system, as recently as 10 years ago an analysis of the transformation of agriculture would have been largely confined to the farm or production sector. Indeed the most cursory reference to recent books on the geography of agriculture will confirm this focus of attention (see e.g. Morgan and Munton 1971; Tarrant 1974; Ilbery 1985; Pacione 1986; Robinson 1988). In recent years, however, the analytical base has been broadened to place agriculture in the context of (1) those industries which provide (upstream) inputs, for example fertilizers and machinery, to (2) the production sector itself, together with (3) industries (downstream) which process farm outputs and (4) distribute food products. Lang and Wiggins (1985) make a strong case for also including (5) the nature of consumer demand (consumption). These five sectors together comprise the 'food chain' (Bowler and Ilbery 1987), with the large business organizations that increasingly dominate one or more sectors of the chain termed 'agribusinesses' (Wallace 1985).

Nevertheless the concept of the 'food chain' does not capture the full complexity of the structures which support and promote a transformed agriculture. For this we must turn to a model of the 'food supply system': that is, the inputs–production–outputs–distribution–consumption sectors of the 'food chain' as connected to the supporting structures comprised by state farm policies, international food trade (imports and exports), the physical environment and credit/financial markets. This approach to studying the contemporary geography of agriculture in developed economies is summarized in Fig. 1.1. The model displays both the flows of capital and materials, as well as the direction of the dominant power relations within the food supply system. Previous writers on modern agriculture have tended to concentrate on limited parts of the model, for example trade patterns (Hine 1985), state–agriculture relations (Bowler 1979), or agriculture–environment interactions (Briggs and Courtenay 1985). In this book, however, elements from within the whole food supply system have been drawn together, with attention given to their varied spatial characteristics. Even so, the farm or production sector has been retained as the central focus of the book, with no attempt at a comprehensive, integrated analysis of the whole food supply system; that must remain the objective of another book. Rather than elaborating on the structure of the food supply system here, the model is introduced at relevant points in the following chapters.

This book has been written at a particularly interesting time in so far as the development of agriculture is concerned. Increasing doubts are being expressed over the long-term sustainability of contemporary agriculture in developed market economies, now summarized in the expression 'the international farm crisis' (Goodman and Redclift 1989). There are three elements to this concern. First, the recent transformation of agriculture has been sustained by increasingly expensive, state-financed, farm support programmes. Governments around the world have found their budgets strained by expenditure on the farm sector, while agricultural production itself has grown to such an extent that, in most developed countries, domestic supply now exceeds demand. Attempts to disperse agricultural surpluses through subsidized exports have served only to distort international food trade and create international trade disputes. Indeed food surpluses have become both politically and economically embarrassing to many governments. One consequence has been a world-wide movement to reduce the level of protective state subsidies to agriculture, as evinced by the Uruguay round of negotiations under the General Agreement on Tariffs and Trade (GATT). But falling levels of farm price support have reduced farm incomes, raised the level of farm bankruptcies and thrown into question the economic sustainability of many modern farming systems. With international pressure to reduce state intervention further, agriculture is faced by a period of continuing adjustment to new economic conditions.

A second concern involves the rising dependency of a transformed agriculture on purchased energy, particularly petroleum (Pimentel and Pimentel 1979). This dependency is found not just in heating glasshouses and the motive power for the plant and machinery used in farm production, but also in the consumption of petroleum-based agrichemicals, including inorganic (artificial) fertilizers. In addition, the distribution network of food products beyond the farm gate, through processors/manufacturers, wholesalers and retailers, has increased the energy dependency of modern food supply systems. With petroleum recognized as a finite, as well as increasingly expensive commodity in the long term, the energy dependency of modern agriculture has become a cause for concern.

Thirdly, the long-term environmental sustainability of a transformed agriculture has also been questioned. Beginning in the 1960s with the damaging ecological consequences of using pesticides such as DDT (Carson 1963), concern has been extended to the landscape impacts of removing farm hedgerows and woodlands (Hamilton and Woolcock 1984), the consequences for wildlife of destroying habitats such as wetlands and moorlands (Adams 1984), the degradation and pollution of soil and water resources (Brown and Wolfe 1984) and, most recently, the declining health standards of the food we eat (Lang and Wiggins 1985). In the last case attention has been focused on pesticide residues in crops, and on bacterial and viral infections in livestock products. Thus, at the same time as the long-term sustainability of a transformed agriculture has been questioned in terms of its economic viability and energy dependency, so parallel doubts have been expressed about its environmental consequences. Together these concerns comprise the contemporary 'international farm crisis', although again agriculture is not homogeneous: the dimensions of the 'crisis' vary by country and region.

Looking now at the organization of the book, at the outset the main characteristics of recent developments in the agricultural sector are described in terms of the 'industrialization of agriculture'. The organizing concepts of 'intensification', 'concentration' and 'specialization' are introduced so as to emphasize the primary processes of transformation in farming and their

3

secondary consequences in relation to the food chain. Attention is then turned to a number of competing theories or conceptualizations of the transformation of agriculture: they direct our analysis to the interpretation of the role of non-farm capitals in transforming the food chain. Chapter 2 considers the sources of information that can be used to study modern agriculture; just as the theoretical base of agricultural geography has broadened, so the number of sources of information has been increased to include remote sensing, farm accountancy data and 'humanistic' material, such as diaries, photographs and tape recordings.

The book then turns in successive chapters to analyse the factors of production in agriculture (inputs) (Chapter 3), the farm or production sector itself (Chapters 4–6), and food processing/manufacturing and distribution (outputs) (Chapter 7). In Chapter 3, attention is given to the changing relationships between land, labour and capital in the recent transformation of agriculture, and the critical role played by farm indebtedness in an era of falling product prices and relatively high interest rates. Chapter 4 examines the agricultural significance of farm size and land tenure, drawing attention to the relationships with farm income, scale economies, the spatial organization of farmland and the social structure of farming communities. Chapter 5 then turns to the problem of how types of production in agriculture can be classified so as to categorize their variation over space. Farm types and farming regions are complex, dynamic systems, and this chapter needs to be read in conjunction with Chapter 6 in which the decision-making basis of enterprise choice is examined. How farm produce is marketed also has significance for the stability, profitability and characteristics of change in individual farming types, and this theme is explored in Chapter 7.

Then two of the structures supporting the food chain are examined in detail so as to reveal their relationships with developments within the farm sector: the physical environment (Chapter 8) and state farm policies (Chapter 9). In the discussion about the physical environment, agriculture is defined as a managed ecosystem; this permits the inherent instability of modern farming systems to be analysed, including the dependency on purchased energy and agrichemicals. The theme of instability also emerges from the discussion on state intervention in farming, an instability that has ramifications for world, as well as national, food supply systems.

Finally, one specific geographical context is examined within which elements of the food supply system interact, namely the urban fringe (Chapter 10). In this chapter attention is drawn to the different contexts for agricultural development that can be found in different sectors of the fringe around a single metropolitan area, as well as between different metropolitan areas; a spatially varied agricultural response is generated by differences in the resource base, rate of urban expansion and market opportunity in particular localities.

The Conclusion to this book turns from the immediate past to the future of agriculture in developed economies. Whether defined in terms of a search for 'sustainable agriculture', as in North America, or an 'alternative agriculture', as in Western Europe, new agricultural structures are likely to emerge. As in the recent transformation of agriculture, a spatially uneven pattern of development can be anticipated.

Finally, the book has been written mainly with students of agricultural geography in mind, although those based in agricultural economics, political science, rural sociology, countryside management or other cognate disciplines should find value in the treatment of the topics. We trust the reader will be stimulated into further thought and enquiry by the variety of approaches that are presented by the different contributors to this book.

References

Adams, W. M. (1984) Sites of Special Scientific Interest and habitat protection. *Area* **16**: 273–80

Arnon, I. (1987) *Modernization of agriculture in developing countries*. Wiley, New York

Bowler, I. R. (1979) *Government and agriculture: a spatial perspective*. Longman, London

Bowler, I. R. and **Ilbery, B. W.** (1987) Redefining agricultural geography. *Area* **19**: 327–32

Briggs, D. J. and **Courtenay, F. M.** (1985) *Agriculture and environment. The physical geography of temperate agricultural systems*. Longman, London

Brown, L. R. and **Wolfe, C. W.** (1984) *Soil erosion: quiet crisis in the world economy*. Worldwatch Paper 60, Worldwatch Institute, Washington DC

Carson, R. (1963) *Silent spring*. Hamish Hamilton, London

Enyedi, G. and **Volgyes, I.** (eds) (1982) *The effect of modern agriculture on rural development*. Pergamon, New York

Goodman, D. and **Redclift, M.** (eds) (1989) *The international farm crisis*. Macmillan, Basingstoke

Hamilton, P. and **Woolcock, J.** (1984) *Agricultural landscapes: an approach to their improvement*. CCP 169, Countryside Commission, Cheltenham

Healey, M. J. and **Ilbery, B. W.** (1985) *The industrialization of the countryside*. Geo Books, Norwich

Hine, R. C. (1985) *The political economy of European trade*. Wheatsheaf, Brighton

Ilbery, B. W. (1985) *Agricultural geography: a social and economic analysis*. Oxford University Press, Oxford

Lang, T. and **Wiggins, P.** (1985) The industrialization of the UK food system: from production to consumption. In Healey, M. J. and Ilbery, B. W. (eds) *The industrialization of the countryside*. Geo Books, Norwich, pp 45–56

Morgan, W. B. and **Munton, R.** (1971) *Agricultural geography*. Methuen, London

Pacione, M. (ed) (1986) *Progress in agricultural geography*. Croom Helm, London

Pimentel, D. and **Pimentel, M.** (1979) *Food, energy and society*. Edward Arnold, London

Robinson, G. M. (1988) *Agricultural change: geographical studies of British agriculture*. North British Publishing, Edinburgh

Tarrant, J. R. (1974) *Agricultural geography*. David and Charles, Newton Abbot

Wallace, I. (1985) Towards a geography of agribusiness. *Progress in Human Geography* **9**: 491–514

1

The industrialization of agriculture

Ian Bowler

The main purpose of this chapter is to establish the general characteristics of agriculture under advanced capitalism. The theme concerns the industrialization of agriculture: its interpretation as the 'third agricultural revolution', its main features and the ongoing debate on how to theorize its recent development. Later chapters will examine the geography of contemporary agriculture as revealed by different countries, regions and farms; here the discussion centres on gaining a broad overview.

Taking this broad view, the history of world agriculture can be interpreted as long periods of slow, incremental, evolutionary change punctuated by relatively short periods of acccelerated, radical transformation. Troughton (1986) argues that the term 'revolution' can be used to describe three periods of particularly rapid change in agriculture, even though the 'revolutions' have occurred at various times in different parts of the world. Two comprehensive reviews of the available literature on the first and second of these agricultural revolutions have been completed by Grigg (1974, 1982), while Table 1.1 summarizes and compares the essential features of all three 'turning points' in world agricultural development.

Three agricultural revolutions

The first agricultural revolution

Beginning more than 10 000 years ago, this revolution is now associated with the development of seed agriculture, the plough and draught animals. This emphasis tends to devalue the role of vegeculture (plants reproduced by vegetative propagation) in early farming, especially in the tropics, for there is partial evidence that vegeculture preceded seed agriculture in some regions. However, early agriculture is usually associated with the successful selection and domestication of crops and animals such as wheat, barley, millet, rice, maize, cattle, sheep, goats and pigs, while the archaeological evidence suggests

Table 1.1 The three agricultural revolutions

	1. Beginnings and spread	2. Subsistence to market	3. Industrialization
Time	Pre-10 000 BP to 20th century	*c.* AD 1 650 to present	1928 to present
Key periods	Neolithic. Medieval Europe	18th century England. 19–20th century in 'European' settlement areas	Present day
Key areas	Europe and South East Asia	Western Europe and North America	USSR and Eastern Europe. North America and Western Europe
Major goal	Domestic food supply and survival	Surplus production and financial return	Lower unit cost of production
Characteristics	Initial selection and domestication of key species Farming replaces hunting and gathering as way of life and basis of rural settlement and society Agrarian societies proliferate and support population growth Subsistence agriculture: labour intensive, low technology, communal tenure	Critical improvements, mercantilistic outlook, and food demands of Industrial Revolution, replace subsistence with market orientation Agriculture part of sectoral division of labour: individual family farm becomes 'ideal' for way of life and for getting a living Commercial agriculture develops growing reliance on technological inputs and infrastructure	Collective (socialist) and corporate (capitalist) ideologies and common agrotechnology favour integration of agricultural production into total food-industry system Emphasis on productivity and production for profit, replace agrarian structure and farm way of life Collective/corporate production utilizes economies of scale, capital intensity, labour substitution and specialized production on fewer, larger units

Source: Troughton 1986.

that only a few regions were involved in originating these developments. A broad area of South-west Asia, stretching from Greece and Crete in the west to the foothills of the Hindu Kush south of the Caspian Sea in the east, appears to have been particularly influential. Village communities, based on a settled way of life, replaced the earlier nomadic hunting and gathering societies, while flood-plain locations, for example along the Tigris, Euphrates and Nile rivers, became so favoured as to form the basis of complex civilizations.

Other 'hearths' for the domestication of crops and animals, and their subsequent diffusion across the continents, have been identified in parts of Central and South America, northern China, the north-east of India and East Africa. Debate continues over the location, timing and independence of origin for many of these 'hearths', although it is clear that North America, northern Europe and Australasia were receiving areas for crops and livestock domesticated elsewhere. For example, agriculture did not reach Spain, southern France and the shores of the North Sea until approximately 4000 BC.

The second agricultural revolution

The origins of this revolution have been ascribed by most historians to Western Europe, although again a debate continues over the exact date. For example, some writers favour the period of accelerated urban and economic growth in Western Europe between AD 1000 and 1340; these developments were preceded by a number of innovations in agriculture such as an improved yoke for oxen, three-course rotations and the replacement of the ox by the horse, which together can be interpreted as forming an agricultural revolution. Other writers, especially for English agriculture, favour a later period for the second agricultural revolution and point up the agricultural improvements of the late-sixteenth to mid-eighteenth centuries as marking a more significant period of accelerated development. A range of new farming practices was introduced, particularly new crop rotations (convertible husbandry) which included grass leys containing legumes (clover, lucerne, sainfoin) and improved varieties of mangel, turnip and swede. Convertible husbandry reduced the need for bare fallows, released land for cultivation, provided pasture and winter fodder for more livestock, renewed soil fertility through the nitrogen-fixing leguminous crops, and increased the supply of animal manure. Other writers have favoured those rapid agricultural changes later in the nineteenth century as comprising the second agricultural revolution. Thompson (1968), for example, writing on British agriculture, cites the 1820s as one of three 'revolutionary' phases during which the large-scale purchase of inputs from off the farm began, especially fertilizers (bone meal and imported guano) and animal feed (oilcake), but including field drainage and the construction of new farm buildings.

On a broader scale, and using measures such as crop and livestock yields, labour input and capital formation, it is possible to show that the eighteenth and early nineteenth centuries comprised the major period of rising agricultural productivity throughout Western Europe (Goodman et al. 1987). Placing the second agricultural revolution in the early decades of the nineteenth century also enables a linkage to be formed with parallel developments in the manufacturing sector which followed the Industrial Revolution. Indeed, the industrialization of the broader economy and society in the nineteenth century had a wide range of consequences for agriculture throughout Western Europe, especially through the urbanization of the workforce, the rising demand for food from that population, the improvement of transport systems, as well as the development of new agricultural technologies, including horse-drawn farm machinery. From this perspective, therefore, the creation of a commercial market for food among a growing urban–industrial population was an essential feature of the second agricultural revolution. Even so, Bairoch (1973) maintains the argument that the second agricultural revolution occurred earlier in the seventeenth century, and was subsequently absorbed into the broader social and economic developmental processes that swept Western Europe following the Industrial Revolution.

The focus on agricultural innovations to define the location and timing of a second agricultural revolution can have a number of unfortunate consequences. For example, there is always a time-lag between the initial adoption of an innovation, its spatial diffusion, and consequently its impact on the economic and social organization of a country or region. Thus, while historians such as Ernle and Toynbee may be correct in placing the origins of the second agricultural revolution in eastern England between approximately 1760 and 1815, the wider impact of these developments on other regions of the United Kingdom (UK) came in later years. In addition, innovations tend to focus attention on the inputs to farming rather than the outputs, whereas the latter, especially the productivity of the factors of production (land, labour, capital) can be of equal importance in defining a period of accelerated change. Also, by focusing on technical innovations in farming, attention is drawn away from equally important features of the second agricultural revolution. For example, the 'revolutionary' developments in West European farming occurred at a time when the feudal landholding system was being replaced by private property rights. The medieval open fields, with their strip-farming, communal farming practices and common lands, were progressively replaced by enclosed, consolidated, individually owned or tenanted farms, with land given a price in its own right. As Tracy (1982: 8) observes: 'the [new] structure of agriculture, with large farms and a class of wealthy landowners ready to invest in their estates, made . . . farming particularly receptive to technological progress'. The new agrarian system transformed an essentially subsistence or peasant agriculture which had been only partially integrated into a market economy.

While debate continues over the most appropriate date for the second agricultural revolution, few would argue with Grigg (1982: 91) that 'the nineteenth century marked the most profound break in the long history of agriculture, when industrialization and urbanization transformed the rural world and revolutionized agriculture'. From its origins in Western Europe, the new, commercialized system of farming was diffused by European colonization during the nineteenth and twentieth centuries to other parts of the world (Peet 1969). A dominant agrarian model of commercial, capitalist farming was established, based on a structure of numerous, relatively small family farms, and from this period can be traced both the dependence of agriculture on manufacturing industry for many farm inputs, and the increasing productivity of farm labour which released large numbers of workers from the land to swell the ranks of factory workers and city dwellers. Moreover, the production of food surplus to domestic demand enabled international patterns of agricultural trade to be established.

The third agricultural revolution

Like its predecessor, this revolution can be subdivided into a number of developmental phases. The terms 'mechanization', 'chemical farming' and 'food manufacturing' are used to describe the rapid agricultural changes that have successively swept through agriculture in developed countries over the last 50 years. Each phase appears to have had its origins in North America, later being diffused to other economies; for example, the first petrol-driven tractor was built in the United States (US) in 1892, the first manufacturing plant was opened in 1907, while the tractor progressively replaced the horse in the 1920s and 1930s. In Europe, by comparison, the widespread adoption of the tractor was delayed until the years following the Second World War. Chemical farming – the use of inorganic fertilizers, herbicides, fungicides and pesticides (agrichemicals) – developed in the US in the 1950s and in Europe in the 1960s, although pesticides based on derris and pyrethrum were in use in the late nineteenth century. The origins of the third phase – food manufacturing – can be traced to the 1960s in North America and the 1970s in Europe, so that today almost all foodstuffs are subjected to some 'value-added' treatment off the farm before reaching the consumer. Whereas the first two developmental phases were associated with mechanical and biological inputs to farming, and brought about the internal 'modernization' of farm businesses, the third phase is more concerned with the outputs from farming, and the external relations between farm businesses and firms involved with the processing and manufacturing of food. Taken together, all three phases have brought about the 'industrialization of agriculture' and they are described in more detail in the following section.

The meaning of the 'industrialization of agriculture'

The term 'industrialization of agriculture' more accurately describes the processes of change in the most recent transformation of agriculture, rather than the actual physical condition of agricultural production itself. The composite of features normally associated with manufacturing industry are found in only a few farming systems, for example horticulture and intensive livestock, but the processes leading to the acquisition of those characteristics are present in many sectors of capitalist agriculture. Indeed Troughton (1986) draws attention to the acquisition of similar characteristics by agriculture in the USSR and Eastern Europe, even suggesting that tendencies towards agricultural industrialization were evident in socialist before capitalist farming systems. In developed market economies, agriculture has been absorbed into an industrial model of food production, as described in Fig. 1.1, within the 'food supply system'.

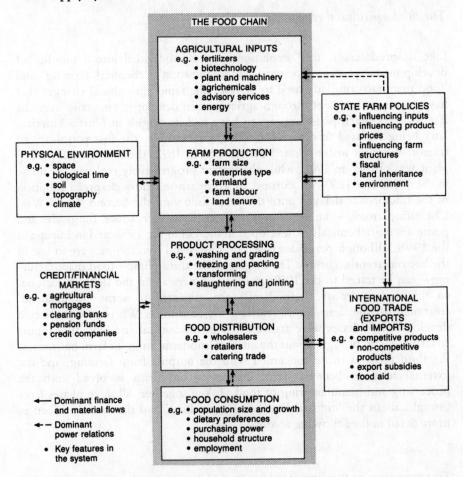

Fig. 1.1 The food supply system

The characteristics of agricultural industrialization include the creation of scale economies at the farm level (larger farms), the increased reliance on purchased inputs from other sectors of the economy (machinery, fertilizers, feed, agrichemicals), resource substitution (capital for land and labour), the implementation of organizational features associated with the concept of the 'firm', specialization of the labour function within the farm business and mechanization of the production process (Symes and Marsden 1985). These processes are most pronounced in two types of agriculture: the glasshouse sector of horticultural production and intensive livestock production (pigs, poultry and feed-lot beef). In the former case, food crops such as tomatoes and salad crops, flowers and pot plants are grown in rigorously controlled environments within glasshouses. Temperature, moisture and soil conditions are carefully regulated, there is a rapid and constant throughput of materials from seeds, fertilizers, agrichemicals and the growing medium, to the final product, while plant and machinery are manipulated by a workforce with specialist skills. The glasshouse regions of the Netherlands – in the Hook of Holland (Westland) and between Leiden/Haarlem/Amsterdam – are probably the most advanced examples of this form of agricultural industrialization. In the case of intensive livestock, the permanent housing of pigs and poultry, with beef cattle confined either in covered yards or in open feed-lots, creates an economically optimum environment by removing constraints on an animal's energy metabolism, especially variations in temperature and humidity. Feed is taken to the livestock; watering, faeces removal and ventilation are carried out mechanically; the livestock are usually constrained by cages or cubicles, so as to maintain high-density conditions without the problems of aggressive social behaviour; while antibiotics are employed to contain health risks to the stock. With shortened breeding cycles, for example, a broiler chicken can be reared for poultrymeat from the egg stage to slaughter within 47 days; together with conveyor-belt production methods, intensive livestock production has attracted the term 'factory farming'. The poultry sector in western Ontario (Canada), located mainly in the counties of Huron, Lambton, Wellington and Oxford, exemplifies this type of agricultural industrialization, especially with its linkages to adjacent feed manufacturers and poultry processors.

Similar tendencies, but at lower levels of development, are evident in other farming systems. For example, crop hybrids developed since the 1920s have permitted the complete mechanization of the major crops, including post-harvest handling, washing, drying, grading and packing, while plant breeders have developed plant characteristics to suit the needs of food processors as regards the standardized conformity of the product. Projecting these developments, visions of the next agricultural revolution place most food production, both crops and livestock, within buildings under controlled, factory-like environments; by that time the term 'industrialized agriculture' may be most appropriate (Busch et al. 1989).

This analysis of the meaning of the 'industrialization of agriculture' can be carried further by identifying three structural dimensions in the recent transformation of farming: intensification, concentration and specialization (Bowler 1985). The term 'intensification' is used to describe the rising level of purchased non-farm inputs in agriculture (capitalization) and the resulting increases in output per hectare of farmland (Table 1.2). Increased purchases of

Table 1.2   Primary process response in the industrialization of agriculture

Structural dimension	Primary process response
Intensification	Purchased inputs (capital) replace labour and substitute for land – increasing dependence on agro-inputs industries
	Mechanization and automation of production processes
	Application of developments in biotechnology
Concentration	Fewer but larger farming units
	Production of most crops and livestock concentrated on fewer farms, regions and countries
	Sale of farm produce to food-processing industries – increasing dependence on contract farming
Specialization	Labour specialization, including the management function
	Fewer farm products from each farm, region and country

Source: Bowler 1985.

plant and machinery have displaced farm labour while fertilizers, land drainage and agrichemicals have enabled crop yields to rise consistently over several decades. In addition, developments in biotechnology have created crop varieties and livestock strains that permit an increased output of agriculture per hectare of land. The process of intensification has been described as a 'treadmill' for those working in agriculture (Cochrane 1958): costs of production tend to rise at a faster rate than the prices obtained for farm produce, thereby creating a cost–price squeeze; innovative farmers gain a short-term financial advantage by reducing their production costs or increasing the output from each hectare of land using new farm technology; as output increases and product prices fall, other farmers are forced to apply the new farming methods in order to survive; further downward pressure is exerted on agricultural prices thus causing the cycle to be repeated. But when governments intervene to support farm prices and incomes, they also tend to reward the process of intensification. The Common Agricultural Policy (CAP) of the European Community (EC), for example, has maintained many agricultural prices at such favourable levels as to encourage and sustain excess production for the domestic market. In addition, the intervention system within the EC has provided, in effect, a limitless market for all the major agricultural products (see Ch. 9).

'Concentration' describes the process whereby productive resources, and the output of particular products, have become confined to fewer but larger farm businesses, and to fewer regions and countries (Table 1.2). Concentration can be measured as the proportion, or share, of total productive resources or output located in a defined production unit (farm, region, country). At the farm level, the process of concentration is implemented through the land market by farm amalgamation: the land of small unprofitable (marginalized) farms is absorbed by purchase into larger farming units. This process of 'structural change' in agriculture, leading to fewer but larger farms, is an international phenomenon, while Gregor (1982) has identified the large farm, albeit measured by the quantitiy of capital employed, as the principal indicator of the industrialization of agriculture in the US. One motive force behind the process of concentration is the search for economies of scale by individual producers. Most studies reveal an increasing efficiency in the use of resources in moving from a small to a medium-sized scale of operation in farming. Just what 'small' and 'medium' imply varies by type of farm, while in the longer term the minimum-size threshold of efficient economic activity tends to move upwards. Nevertheless, there is a limit to the economies of scale that can be obtained in agriculture; in the context of the UK, for example, this appears to operate at approximately 1 200–1 800 standard man-days (SMDs) per year, or the equivalent of a business giving full employment to four to six people (Britton and Hill 1975). Thereafter, few scale economies appear to operate (see p. 92), a condition broadly supported by research in other developed countries.

Associated with 'concentration' in agriculture is an increasing linkage with the food processing industries. Larger farm businesses, on the one hand, require an assured market for their produce, particularly when such businesses are highly capitalized and economically vulnerable to changes in product prices. Food processors, on the other hand, require assured supplies for the continuous operation of their production facilities. Since the returns to capital investment are less risky and remain higher in industry compared with agriculture, the food processing industries have been cautious in promoting vertical integration by purchasing the farms that provide their raw materials. A greater benefit has been obtained through a system of annually renewable contracts signed between individual producers and processors with the risks of production passing to the farmer. Some contracts concern only farm produce, others encompass the inputs used in the farm system, while some specify the production and management techniques that are to be employed. A number of observers in the US have concluded that farmers have become a class of 'propertied labourers' because of these ties with agribusiness (Vogeler 1981). The situation in Western Europe is probably not as extreme as this, but further discussion is deferred to Chapter 7.

'Specialization' is measured as the proportion of the total output of a farm, region or country accounted for by a particular product. For crops, hectares offer a convenient base but there can be no statistical integration with live-stock, which are usually measured by the number of animals or birds. Consequently the value of output of each type of crop and livestock is often used as a common measure. Specialization begins at the farm level when individual producers focus their resources of land, labour, capital and management expertise on a narrow range of products, thereby withdrawing from the production of less profitable, usually minor, enterprises (Table 1.2). Once again, economies of scale can be obtained by this strategy. On a wider spatial scale, an increasing regional or national specialization is obtained because farmers at the extensive margins of production (Chisholm 1966: 51–2) of each enterprise are the first to cease or decrease their production. Any disequilibrium in the

Table 1.3   Secondary consequences in the industrialization of agriculture

Structural dimension	Secondary consequences
Intensification	Development of supply (requisites) co-operatives
	Rising agricultural indebtedness
	Increasing energy intensity and dependence on fossil fuels
	Overproduction for the domestic market
	Destruction of environment and agro-ecosystems
Concentration	Development of marketing co-operatives
	New social relations in rural communities
	Inability of young to enter farming
	Polarization of the farm-size structure
	Corporate ownership of the land
	Increasing inequalities in farm incomes between farm sizes, types and locations
	State agricultural policies favouring large farms and certain regions
Specialization	Food consumed outside region where it was produced
	Increased risk of system failure
	Changing composition of the workforce
	Structural rigidity in farm production

Source: Bowler 1985.

market is absorbed at the intensive margins of those producers who continue in production. Comparative advantage in unit costs of production are usually claimed for those areas where a product becomes a more specialized element in agriculture, even though this condition is extremely difficult to establish using empirical data. One of the spatial effects of specialization is the regional concentration of agricultural production.

An extensive set of secondary consequences is associated with the three structural dimensions noted above (Table 1.3) and together they help define the meaning of the 'industrialization of agriculture'. Looking first at the intensification of agriculture, a consequence attracting the most adverse criticism is the destruction of the rural environment and certain agro-ecosystems. At issue is the pollution caused by the use, some would argue over-use, of agrichemicals in crop production, including artificial fertilizers. The indiscriminate application of aerial and field sprays has damaging effects on flora and fauna in rural environments while there is some concern about chemical residues in food crops. Also, leakages from slurry pits and tanks can have devastating local effects on plants and wildlife in watercourses, and long-term damage to water quality can be caused when excess fertilizers, especially nitrogen, are leached off fields into streams and rivers. All of these problems emanate from the rising volume of purchased, non-farm inputs in agriculture. At the same time the amenity and conservation value of hedgerows, small woodlands, heaths and moorlands have been reduced by field enlargement and land improvement. These developments in part are a function of the increasing farm size and the use of larger machinery; but also at issue is the desire to maximize output from every available hectare of agricultural land.

Another aspect of the intensification of agriculture is the increasing dependence of contemporary farming on purchased energy. Not only is an industrialized agriculture an inefficient user of energy, but an agriculture dependent either directly or indirectly on fossil fuels, such as petroleum, appears to be vulnerable in the long term. The first point is indisputable: the average ratio of energy inputs to energy outputs (production) in contemporary agriculture is low (1:1.1) and falling. There is some variation by country and product. Using data for Western Europe, the ratio is 1:2.4 for potatoes in the Netherlands but 1:1.6 in the UK; 1:4.7 for sugar-beet in Belgium and 1:0.01 for tomatoes in the UK (Commission of the European Communities 1982). On the other hand, relatively few gains can be made in terms of the use of energy by importing food. For example, once the energy used to transport and handle food imports to the EC is accounted for, only beef, sheepmeat, pigmeat, barley and tomatoes can be imported with lower total energy inputs. The evidence is equally equivocal on the vulnerability of agriculture in the long term. Again drawing on available evidence for the EC, farming consumes only 5 per cent of the total energy used by the economy of the Community. Of this, one-third is used directly as power for machinery and plant, or to heat buildings and glasshouses, about three-quarters of the energy being derived from petroleum. Two-thirds of agriculture's energy is consumed indirectly through purchased inputs of fertilizers (44 per cent), feeding-stuffs (31 per cent), machinery (14 per cent) and agrichemicals (4 per cent).

It can be argued that in the long term society will give precedence to agriculture in the allocation of limited petroleum resources and make savings more readily elsewhere in the economy. But the more immediate issue is the cost of such resources rather than their availability. In the EC, intermediate consumption (purchased inputs) in agriculture is approximately 44 per cent of the value of final production. With farm gate prices being allowed to fall in comparison with the prices of inputs, the economic vulnerability of agriculture is increasing. Thus the rising energy intensity of agriculture, with its dependence on fossil fuels, causes concern for the economic costs of food production. Nevertheless, there are major savings to be made in the conservation of energy elsewhere in the food chain, especially in food processing and distribution. To date insufficient attention has been given to the energy used in freezing, canning, packaging, transporting, displaying (freezer cabinets) and retailing food, while the energy consumed in cooking food, either in the home or by catering establishments, has yet to be fully emphasized. Available estimates indicate a far greater scope for energy conservation 'downstream' of agriculture than in farming itself.

The process of concentration in agriculture has also had several secondary consequences (Table 1.3). In order to combat the market power of the food processing industries, for example, farmers have increasingly organized themselves into marketing co-operatives. Initially these grew out of requisites co-operatives that were set up to purchase farm inputs and so combat the economic power of the agro-inputs industries, but more recently an increasing emphasis has been given to specialist marketing co-operatives. Indeed most countries provide technical and financial assistance to farmers who wish to set up a co-operative. A strong regional network of co-operatives can bring great benefits to an area: on the one hand, co-operatives can be an agent for development, as seen in their role in the evolution of intensive livestock, as well as vegetable production in Brittany; on the other hand, co-operatives can act to defend an agricultural system, as exemplified by wine production in the Midi (southern France). In the latter case, the regional co-operative network has been mobilized on several occasions to counteract the threat posed by the importation of cheaper Italian wine, to protest against measures from the EC to limit or curtail wine production, and, in the late 1970s, to put pressure on French politicians concerned with negotiations on the entry of Spain and Portugal into the EC.

Another secondary consequence of concentration in agriculture is the growing inequality in incomes between different farm sizes and locations. Large farms, because of the sheer magnitude of the resources involved and the economies of scale that can be realized, yield their occupiers incomes comparable with those earned elsewhere in the economy. This is despite the returns per hectare of land being inferior to those obtained on more intensively worked small farms. The gap between small and large farms has widened as, under the process of farm amalgamation, the farm-size structure has tended to polarize. In so far as there is a relationship between farm type and farm size, so income differentials by farm type have also widened. A spatial dimension is introduced when regions vary in their farm type and size.

The secondary consequences associated with specialization in agriculture have received relatively little attention (Table 1.3). It is clear that specialized farming systems are open to greater risks from the spread of crop and livestock diseases, the impact of climatic hazards such as drought and frost, and the often unforeseen side effects of particular agricultural practices such as soil erosion or depletion. The traditional checks and balances created by integrated (mixed) crop-with-livestock systems have been lost, thus exacerbating the environmental problems caused by the intensification of agriculture. In cereal-growing areas, for example, straw is often burned, whereas in a mixed economy it would be fed to livestock or used for their bedding. In livestock areas, particularly those with intensive livestock, farmyard manure or slurry is often spread intensively on fields merely as a means of disposing of problematical animal wastes. Thus, in the Netherlands, leaching of nutrients from animal wastes poses a major pollution threat to water resources; traditionally such waste products would have been recycled through crop enterprises on the farm. A further inefficiency caused by specialization is the need to transport food longer distances. Regions have become decreasingly self-sufficient in their own food supplies and 'import' from increasing distances. The increased transport (energy) costs are passed on to the consumer in the price of food but the scale of this diseconomy has yet to be fully researched.

Regional specialization has also caused problems for agricultural policy-making. Recently attempts have been made in many countries to reorientate agricultural production away from products in oversupply on the domestic market (see Ch. 9). The rigidity of farming systems under specialization makes it most difficult for farmers to change their production in the face of policy measures such as quotas on milk production, or grant aid to grub up orchards or leave dairying and viticulture altogether. Regional specialization hinders rather than helps the solution of overproduction in agriculture.

Competing theorizations on the dynamics of agriculture

There can be little disagreement over the empirical characteristics of the industrialization of agriculture, but theorizing or conceptualizing the relevant processes of agricultural change, especially in a spatial dimension, has proved more contentious. Among the complex and voluminous literature on the dynamics of agriculture, three broad schools of thought can be discerned, although even this simplifying categorization will provoke argument among the protagonists and mask variations of interpretation within each school. For convenience the terms 'commercialization', 'commoditization' and 'industrialization' are used to summarize the competing theorizations. The first two schools of thought have been concerned mainly with the transformation of peasant societies, especially in Third World countries, although, like the 'industrialization' school, their concepts are now being applied to historical analyses of the transformation of agriculture in developed economies, as well as to more recent agriculture trends. The following summary turns towards this last interpretation.

Commercialization

Following the argument of Vandergeest (1988), the 'commercialization of agriculture' school developed from the work of post-Second World War modernization theorists, although Newby (1987) has traced the origins of their concepts to classical economists, such as Smith and Ricardo, and nineteenth-century social theorists, especially Max Weber. The main proposition of this school of thought is that traditional, peasant societies are transformed by the introduction of market relations, with the degree of commercialization measured by the proportion of farm production sold in the market. Commercialization is supported by research and development to produce the appropriate new farm technologies, the development of industries to manufacture the inputs, and education and promotion work to give farmers the ability to apply the new inputs. From this viewpoint, the integration of farm households into a market economy forms an integral part of the broader processes of social and economic development, including benefits for the rural economy and society. Indeed, in Rostow's (1960) 'stages of economic growth', the transformation of agriculture becomes a prerequisite for 'take off' in the whole economy. Within this context, capitalist farms are assumed to have a higher technical efficiency, a more rational form of organization and a greater sensitivity to the demands of the market than the peasant sector. A similar logic of 'economic efficiency', 'profit maximization' and 'competitiveness' is used to explain the recent linkages between the farm sector and other parts of the food chain, for example food processors and food retailers. Nevertheless, the commercialization literature recognizes that a competitive capitalist structure of individual farm businesses will tend to marginalize the smaller (peasant) farms, leading to a dual farming economy. In this structure, a technically superior capitalist sector (larger farms) coexists with a transformed peasant sector (smaller family farms), the survival of the latter being dependent on an ability to develop types of farm enterprise in which there are few economies of scale, production is less capital intensive and high-quality work is rewarded in the market.

In some contemporary interpretations within the 'commercialization' school, a 'traditional–modern' continuum replaces the concept of a dualistic farm-size structure (e.g. Hoselitz 1960), with the cost–price squeeze (see p. 14) explaining the continued economic rationalization of farm numbers. The capacity of farms to remain in business depends on adopting new farm technology, with the competitive structure of capitalist agriculture marginalizing those who fail to 'modernize' their farms. Interpreted in this way, technology is not a means but a cause of agricultural modernization and economic development. However, any explanation of the type of technology developed for, and adopted by, modern agriculture requires a further step in the argument. Here a theory of induced innovation has been constructed, whereby new technology is developed to substitute the more by the less expensive factors of production (Hayami and Ruttan 1985: 84). Since factors of production are classically divided into land, labour and capital, and historically labour has been the most expensive factor of production, with land the factor most fixed in supply, in the long term capital has been substituted for both land and labour. In practice farmers have used capital to purchase output-increasing (land-saving) technology, such as fertilizers, genetically improved seeds and livestock breeds, as well as labour-saving technology, for example farm plant and machinery, and agrichemicals.

Until recently, most agricultural geographers and agricultural economists tended to draw on the 'commercialization' theory of agricultural change. For example, the 'commercialization' literature helps to explain why the process of agricultural transformation has occurred earlier, and to a greater extent, in some farming locations compared with others. Using the early theoretical work of von Thünen (1783–1850), Schultz (1953) developed an 'urban–industrial impact' hypothesis. Schultz took as his starting-point the empirical observation that agricultural transformation often occurs first adjacent to large, urban–industrial concentrations, with a declining intensity of transformation at increasing distances. Further from urban–industrial areas, both farm and farmer incomes, output per hectare and the value of farmland all tend to be lower. Schultz contended that factor and product markets function more efficiently in areas of rapid urban–industrial development, especially where the urban–industrial sector is the source of new and more productive farm inputs. Savings on transport costs for marketing farm produce, as well as acquiring farm inputs, provide additional cost advantages and investment incentives to farmers in urban–industrial locations.

The spatial diffusion model of technological change (pp. 143–49) offers an alternative interpretation for the uneven transformation of agriculture. Characteristically, new farming practices, including production for the market, are first adopted in a particular region and then spread with a distance decay function into adjacent regions. Locations further from the source of innovation not only begin the process of transformation later, but ultimately achieve lower levels of adoption of the new farming practice. Empirical studies have shown that urban–industrial regions can be, but are not always, the first locations to be transformed by market relations or new farming practices. Rather three interacting processes, discussed more fully in Chapter 6, appear to operate at the regional level. Here it is sufficient to note that farmers in regions characterized by large, capitalist farms can be the most innovative over a wide range of new agricultural practices; farmers in those regional agricultural systems which gain the highest economic benefit from a particular technological development can be the first to adopt; the regional marketing strategy implemented by the manufacturing firm producing a new farm technology, or the agency responsible for supplying investment capital or relevant information, can also determine the spatial pattern of transformation.

Commoditization

The 'commoditization of production' school grew out of the dependency theory literature of the 1960s and its emphasis on the persistence of underdevelopment in Third World countries. This concern had itself emerged from a neo-Marxist critique of modernization theory and capitalist development in general. In fact Newby (1987), in the context of this discussion, has traced the intellectual roots of much recent writing within the 'commoditization' school back to the original ideas of Marx and Lenin on the processes of agrarian development. Consequently writers within the 'commoditization' tradition have tended to adopt a political economy (or historical–structural) perspective in their work. As described by Redclift (1984: 5), this approach 'locates

21

economic analysis within specific social formations and explains the development processes in terms of the benefits and costs they carry for different social classes'. Compared with the 'commercialization' school, therefore, writers in the 'commoditization' tradition have tended to emphasize social over economic structures and relations (Marsden et al. 1986a).

Summarizing the argument of Vandergeest (1988), commoditization occurs when non-market forms of production are transformed, with the factors of production and subsistence increasingly obtained through the market. As farm households are drawn into a dependency on goods obtained in the market, so they are compelled to produce agricultural commodities with an exchange value in order to obtain a cash income and thereby purchase inputs for the farm. However, compared with the 'commercialization' school, emphasis is placed on those inputs to farming (land, labour, capital) which are *purchased* in the market, rather than on products *sold* in the market. This perspective on the transformation of agriculture throws two theoretical issues into relief: the trajectory of the farm-size structure in capitalist agriculture, including the relationship with farm-family households, and how control is exercised over the means of production (farm inputs).

From a Marxist perspective, the 'differentiation' model of agrarian development assumes that, like industry, capitalist accumulation 'tends' ultimately to produce a polarized class structure. The trajectory of development is towards, on the one hand, an agrarian bourgeoisie, comprised by a capitalist landowning class, occupying or owning large-scale, wage-labour farms. On the other hand lies the development of an agricultural (or rural) proletariat, comprised by the marginalized and ultimately landless peasant class that supplies the wage-labour for the capitalist farms. At any point in time there exists a continuum of 'transitional' states between pure 'peasant' farming, at one extreme, and large-scale capitalist (wage-labour) farming at the other. Some writers have proposed an analytically separate concept of 'simple (or petty) commodity production' (SCP) within this structure, to describe a stratum of small farms based exclusively on the use of family labour and with no economic pressures for expanded reproduction, for instance by increasing the farm size. Simple reproduction is adequate for each farm household, with any expanded reproduction motivated by demographic and cultural factors (Friedmann 1986). This interpretation of SCP has been criticized by other writers as forming a 'historically contingent phenomenon' (Goodman and Redclift 1986) and not a theoretical concept. For example, SCP assumes no class relations within the farm enterprise, whereas empirical research has shown that even family farms periodically turn to hired workers at early and late stages in the farm-family demographic cycle. This hiring of workers on to small and medium-sized 'family farms' blurs the distinction between family-labour and capitalist, wage-labour farms, while, in addition, recent research has shown that many large-scale, capital-intensive, multiple-unit family businesses can be owned and operated exclusively by family labour (Marsden and Symes 1984). Also a debate continues over the empirical evidence for the 'disappearing middle' of medium-sized farms and the development of a polarized (dualist) farm-size structure (see p. 89).

While some neo-Marxists still explain the survival of non-capitalist farms as 'arrested marginalization', an increasing number of writers within the 'commoditization' school are prepared to recognize the persistence of family labour-based farms as a structural feature of agriculture under advanced capitalism (Rheinhardt and Barlett 1989). The survival of small-scale (transformed peasant) farming was first confronted by Kautsky in 1899. Recognition was given to the role of off-farm work by the farmer, including that provided for capitalist farms, in generating alternative sources of income for the farm family; worker-peasants, or part-time farmers, enable a stratum of family labour-based farms to survive. This perspective has been developed more recently under the term 'pluriactivity', as applied to the whole farm family. The farm household is not viewed as a passive receptor of market forces, but rather as a more active unit entering into a variety of relations with external, non-farm capitals in order to ensure its survival (reproduction). Either, or both, of the spouses can be engaged in off-farm employment, as well as the children or other members of the extended family. In addition, non-commoditized social relations within the farm family have been emphasized, including non-wage work, inter-household exchange and women's labour (Friedmann 1986). While pluriactivity can be interpreted as a form of proletarianization of the rural workforce, it enables ownership of the means of production to remain in the hands of farm families.

Pluriactivity is not the only path of development which has enabled a stratum of family farms to survive under advanced capitalism. Hobby farming and diversification, for example, provide two further examples of the complex forms of family farm structures that ensure their survival. Marsden et al. (1986b) have formalized these strategies into six categories:

1. Full-time, family-owned agricultural businesses;
2. Hobby and 'retired part-time' farm businesses;
3. Sub-marginal farms reliant on pensions, savings and insurance schemes (survival);
4. Farm households with diversified activities on and off the farm (survival);
5. Farm households with merged capitals on and off the farm (accumulation);
6. Corporate businesses (accumulation).

While an imperfectly determined combination of criteria is used to define the different categories – for example, mixing motivations with business structure and income from non-farming activities – the typology points up the variety of farming strategies that accompany the industrialization of agriculture under advanced capitalism, and the absence of a unitary pattern of development.

A second but related theoretical issue addressed by the 'commoditization' literature concerns how control is exerted over the means of production. This issue gains importance because control over the means of production constitutes the basis of power: under the process of commoditization the peasant class suffers a differential impoverishment by a progressive loss of control (C. Smith 1984). A distinction between the 'formal' and 'real' subsumption of labour by capital has been theorized to explain this process. Conventionally, capital can be subdivided into 'fractions', as represented, for example, by production, merchant, financial and industrial interests. Capital fractions seek to valorize (establish value in monetary terms) all productive activities in farming and extract surplus value. With the labour process acting as the main mechanism for creating surplus value, capitals can seek the 'real' subsumption of labour by direct ownership of the means of production (horizontal and vertical integration). The agro-industrial complex, for example, is a well-known feature of economies under state socialism, with elements of the whole food chain organized into one business structure, including food shops and hotels. Gregor (1982) has identified similar tendencies among corporately owned farms in California. In general, however, capitalist agriculture has not followed this linear theory of agrarian development as theorized by neo-Marxists, and consequently the concept of the 'formal' subsumption of the labour process has been developed. In this perspective, and as described by Marsden et al. (1986a: 276–7): 'the "subsumption" of the family farm [takes] place . . . through the expropriation of surplus value from the labour process by capital but without directly revolutionizing the labour process on the farm'. More explicitly, 'the legal ownership of the farm business and land remains [with the farm family] whilst they become separated from effective control, as management depends on external technological and economic factors governed by monopoly industrial and finance capitals' (Marsden et al. 1986a: 512).

In addition, an increasing number of tasks have been separated from the farm labour process and appropriated by external agencies. For example, the processing of milk into butter and cheese, once a farm-based activity, has been appropriated by dairies within the food-processing industry. In this interpretation, agriculture retains only those residual activities which have resisted transformation into industrial processes. Mann and Dickinson (1978) offer the concept of 'production' and 'labour time' to explain the survival of such 'residual activities' on family farms. By 'production time' is meant the time during which capital is employed in commodity production; 'labour time', however, is the proportion of production time during which labour is needed in creating value. Since labour time is less than production time, for example in annual crop and livestock farming, the average rate of profit is reduced and external capitals find direct investment in all agricultural production phases inefficient. For example, food processors prefer forward contracts with farmers to obtain their raw materials, rather than the direct ownership of farmland. This practice is less prevalent in those farming systems, such as intensive pig and poultry production, where labour time can be synchronized with biological processes, and vertical integration becomes an economically efficient business structure. From this viewpoint, therefore, family farms will be preserved only when they are functional for the maintenance of some capitalistic production relations.

From a 'commoditization' perspective, the uneven spatial and temporal transformation (differentiation) of agriculture can be explained as a function of regional variations in the externalization process. External capitals penetrate those regional farming systems where the greatest financial returns can be obtained, as mediated by farm type and farm size. As limitations are placed on the maximizing of surplus value ('accumulation crises'), so external capitals switch their interests between regions to produce phases or waves of capitalization (Marsden et al. 1987). Some writers have reversed this argument by claiming that differential degrees of commoditization partly result from the resistances to transformation (survival strategies) devised by specific categories of family-farm households. In this view, local and regional responses are reduced to a matter of empirical circumstance, including cultural and historical diversity, enabling some writers to claim them as 'historically superfluous' (Long et al. 1986: 40). In general, though, both 'land' (its productive capacity) and 'space' are viewed as constraints/opportunities in the process of commoditization, helping to explain both the differential transformation and durability of the family farm under advanced capitalism.

Industrialization

Recent developments in the relationship between farming and other elements in the food supply system have exposed the limitations of both the commercialization and commoditization theorizations of the transformation of agriculture. Writers in the former tradition, for example, have had to acknowledge that the unequal access to the means of production, especially capital, and the closer integration of agriculture with various fractions of capital, increases income disparities within agriculture and permits many farm families to remain relatively impoverished despite being integrated into a market economy. The 'commoditization' school, on the other hand, has had to concede a greater degree of manœuvre by farm families as they are confronted by the process of commoditization, and the varied paths of development open to them. Both traditions have had to come to terms with the increasing role of the state in determining the trajectory of agricultural development, the internationalization of the food supply system, and the growing influence of the retail and catering (fast food) trades within the food chain.

Writers developing more recent theorizations of agricultural change can be grouped into an 'industrialization' school. While their ideas are still being refined, they draw upon a number of earlier concepts in both the 'commercialization' and 'commoditization' literatures, and adopt the organizing framework of the food chain (Fig. 1.1). For example, Goodman et al. (1987) emphasize the 'uniqueness' of agriculture in terms of its natural production process – nature as the biological conservation of energy, as biological time in plant growth and animal gestation, and as space in land-based rural activities ('physical environment' in Fig. 1.1). In their view, these features have prevented industrial capitals from imposing a 'unified transformation' upon agriculture, and they offer 'appropriationism' and 'substitutionism' as two theoretical concepts.

The term 'appropriationism' is drawn from the 'commoditization' literature to describe the discontinuous but persistent transformation into industrial activities of certain parts of the agricultural production process, and their subsequent reintroduction in the form of purchased farm inputs. The appropriations are 'partial' in the sense that not all of the production processes on farms have been affected, 'discrete' in that quite specific parts of agricultural production have been appropriated, and 'discontinuous' in that innovatory phases are introduced from time to time. 'Substitutionism', in contrast, concerns the outputs from the farm sector. Agriculture is viewed as a source of raw materials for industrial processing and marketing, with industrial capital applying mass-production methods to create 'value-added' food products. A trend is under way within the food-processing industry which is substituting 'natural' farm produce by chemical and synthetic raw materials and so eliminating the rural base of agriculture. This process has already occurred in the production of fibres, with farm-produced wool and cotton being substituted by 'artificial' fibres such as rayon and nylon. With food increasingly 'fabricated' from reconstituted generic food components (e.g. starch, glucose and vegetable protein) and chemical additives, the constraints of land, space and biological reproduction on both the capitalist transformation of the production process and the social division of labour can be removed (Goodman et al. 1987: 58).

The growing industrial control of food production is reflected in the interest of the 'industrialization' school in theorizing the role of 'agribusiness' in the capitalist mode of production. Wallace (1985), for example, has exposed the varying use of the term 'agribusiness' – for specific enterprises, a range of industries, a sector and a system – while directing attention to those capitals that control the majority of food production, including their capitalist structures and processes. This argument has been taken up by Le Heron (1988): monopoly capital, basically organized within national boundaries, has yielded to global capital in agribusiness as elsewhere in the manufacturing sector; not only are agribusiness firms merging into large, vertically integrated, international corporations, often with multiple non-farming interests, but they are also seeking out those lowest-cost agricultural raw materials on the international market for their inputs ('international food trade' in Fig. 1.1). These behaviours by agribusiness are increasingly influential throughout the whole food chain, both nationally and internationally, and form a focus for recent research interest (Marsden and Little 1990). Indeed the search has begun for a theory of agricultural differentiation (location) that operates at an international and not just a national level of explanation; writers such as Marsden et al. (1986a), Bowler and Ilbery (1987) and Le Heron (1988), for example, have begun to sketch out the dimensions of the problem. Following the lead given by industrial geographers, the focus on agribusiness is likely to result in 'agricultural space' being viewed as a context within which successive international as well as national territorial divisions of capital and labour are superimposed and then destroyed (restructured) as capitalism itself develops.

A third area of theoretical concern for the 'industrialization' school lies in the relations between the state and agriculture in the transformation of food production ('state farm policies' in Fig. 1.1). Bowler (1989) has cast the relations in a 'policy process model': a linear but recursive relationship that exists between societal values – beliefs about reality – policy objectives – policy instruments and agencies – policy impacts, all set within a pluralist model of the state. Producers, consumers, taxpayers, political parties, civil servants and industrial capital are all assumed to be able to compete in policy formation. Similar assumptions underlie W. Smith's (1984) 'vortex model' of agricultural change: he argues that agricultural producers and consumers can 'exit' from market relationships ('linkages') with which they are dissatisfied, while expressing 'voice' or political action in shaping and developing those relationships. Both processes can have the effect of structural adjustment in the food supply system. A structuralist theory of state–agriculture relations, by contrast, allows for unequal power among interest groups in shaping agricultural policy. For example, in most developed countries farm groups and industrial capital appear to have enjoyed more influence than consumers and taxpayers in recent years. In addition, a more pro-active role is ascribed to the state, especially in serving the interests of capital accumulation, while the state is also able to prioritize certain areas for research and farm extension (advisory) work through its control of funding. Thus Newby (1982: 138) concluded that 'the technological transformation of agriculture can be viewed no longer as the product of the hidden hand of the market, rather it is the outcome of deliberate policy decisions, consciously pursued and publicly encouraged'. A 'corporatist' framework has also been proposed as a description of the way in which the state, farm groups and agro-industrial capital appear to have worked in coalition to support the development of agricultural industrialization. In this structure, an interest organization, such as a farm union or farm product group, can undertake an intermediary role between the state and individual producers: the organization negotiates with agencies of the state on matters such as price subsidies, but it also ensures that individual members of the organization comply with the agreement that has been reached with the state (Bowler 1989). More recently, interest has shifted to negotiations between individual states, or groups of states, at the international level, in forums such as the General Agreement on Tarrifs and Trade (GATT). These international negotiations have caused states to reappraise their farm policies with a view to reducing levels of price protection and subsidy. Any liberating effect on international trade, and capitalist accumulation, will have significant implications for agricultural adjustments in the international division of capital, labour and product specialization.

Conclusion

The central concern of this chapter has been the transformation of agriculture in developed market economies over the last four decades (see the summary in Table 1.4). The discussion has been set in the context of the industrialization of agriculture or, more formally, 'the competitive movement of industrial capitals to create sectors of valorization by restructuring the inherited "pre-industrial" rural labour process' (Goodman and Redclift 1986). Until very recently this process of transformation was viewed as unilinear in character, although acting unevenly from region to region, but now the limitations of the 'industrial model' of agriculture are being recognized. The Introduction has described the three dimensions of the 'international farm crisis' (economy, environment and energy) which appear to be forcing two main and divergent paths of development on agriculture. On the one hand lies the 'industrial model' of intensive, high input–high output farming, with a continuing emphasis on the quantity of food produced, the further development of biotechnology and international capital. On the other hand is a 'post-industrial model' of low input–low output farming, with an emphasis on the 'sustainability' of farming systems and the quality rather than the quantity of food production. It is possible for these two models to coexist, with some farming systems producing high-volume, low-value, mass-produced food, and others concentrating on low-volume, high-value, quality foods for those consumers able to pay the higher prices. Whether spatial differentiation occurs on a national, regional or farm-by-farm basis, however, will depend to a considerable extent on the agricultural policies devised by individual countries, or groups of countries such as the EC.

Table 1.4   Key words and terms used in the competing theorizations of agricultural change

Theorization	Key word or term
Commercialization	Modernization theory; produce (output) sold in the market; technical efficiency; profit maximization; competition; traditional-modern continuum; cost-price squeeze; induced innovation; diffusion of technology
Commoditization	Dependency theory; political economy; factors of production (inputs) purchased in the market; farm size dualism; capitalist accumulation and concentration; marginalization; labour process; family farm; survival strategy; formal and real subsumption; production and labour time
Industrialization	Food chain; appropriationism; substitutionism; agribusiness; global capital; international corporation; territorial division of capital and labour; theories of the state (pluralist, structuralist, corporatist)

The 'international farm crisis' has also emphasized a third path of development with in capitalist agriculture. As food surpluses have driven market prices downward, and caused governments to reduce agricultural subsidies, so the levels of farm bankruptcies and farm indebtedness have risen (see p. 76). In the latter case, the role of banking and credit relations in agriculture has come to the fore as farmers have been forced into new financial relationships in order to buy farmland, new farm technology or even their recurrent farm inputs ('credit/financial markets' in Fig. 1.1). Le Heron (1988) is of the opinion that the capacity to obtain credit, and the policies of the organizations supplying the credit, will exert a major differentiating influence over agriculture in the immediate future. Of equal significance, however, is the pressure on farm businesses to diversify their sources of income by investing in new enterprises both on and off the farm. On the farm, there has been a recent movement in Western Europe towards enterprises such as sport fishing, forestry, accommodation, deer farming and farm shops. Off the farm, labour and capital have been diverted to a range of alternative sources of income under the heading of 'pluriactivity'. In both cases the farm family has reduced its dependence on traditional agricultural production for an income.

The most recent phase in the third agricultural revolution, therefore, consists of divergence rather than convergence in the types of survival and accumulation strategies being adopted by farm businesses, a process that is varying in its outcome between countries, regions and individual farms. Central to an understanding of this diversity is the interaction between social and economic relations within, and political and economic relations external to, the farm business. This interaction is of great importance in understanding the contemporary geography of agriculture, as subsequent chapters will show.

References

Bairoch, P. (1973) Agriculture and the industrial revolution. In Cipolla C. M. (ed) *The Industrial Revolution*. The Fontana Economic History of Europe, Vol 3, London, pp 457–60

Bowler, I. R. (1985) Some consequences of the industrialization of agriculture in the European Community. In Healey, M. J. and Ilbery B. W. (eds) *The industrialization of the countryside*. Geo Books, Norwich, pp 75–98

Bowler, I. R. (1989) Revising the research agenda on agricultural policy in developed market economies. *Journal of Rural Studies* 5: 385–94

Bowler, I. R. and **Ilbery, B.** (1987) Redefining agricultural geography. *Area* 19: 327–32

Britton, D. K. and **Hill, B.** (1975) *Size and efficiency in farming*. Saxon House, Farnborough

Busch, L., Bonanno, A. and **Lacy, W. B.** (1989) Science, technology and the restructuring of agriculture. *Sociologia Ruralis* **29**: 118–30

Chisholm, M. (1966) *Geography and economics*. Bell, London

Cochrane, W. (1958) *Farm prices, myth and reality*. University of Minnesota Press, Minneapolis

Commission of the European Communities (1982) Energy consumption per tonne of competing agricultural products available to the EC. *Information on Agriculture* **85**. The Commission, Brussels

Friedmann, H. (1986) Family enterprises in agriculture: structural limits and political possibilities. In Cox, G., Lowe, P. and Winter, M. (eds) *Agriculture: people and policies*. Allen and Unwin, London, pp 41–60

Goodman, D. and **Redclift, M.** (1986) Capitalism, petty commodity production and the farm enterprise. In Cox, G., Lowe, P. and Winter, M. (eds) *Agriculture: people and policies*. Allen and Unwin, London, pp 20–40

Goodman, D., Sorj, B. and **Wilkinson, J.** (1987) *From farming to biotechnology: a theory of agro-industrial development*. Basil Blackwell, London

Goss, K. F., Rodefeld, R. D. and **Buttel, F. H.** (1980) The political economy of class structure in US agriculture: a theoretical outline. In Buttel, F. H. and Newby, H. (eds) *The rural sociology of the advanced societies: critical pespectives*. Croom Helm, London, pp 83–132

Gregor, H. F. (1982) *Industrialization of US agriculture: an interpretative atlas*. Westview, Colorado

Grigg, D. (1974) *The agricultural systems of the world: an evolutionary approach*. Cambridge University Press, London

Grigg, D. (1982) *The dynamics of agricultural change: the historical experience*. Hutchinson, London

Hayami, Y. and **Ruttan, V. W.** (1985) *Agricultural development: an international perspective*. Johns Hopkins, Baltimore

Hoselitz, B. F. (1960) *Sociological factors in economic development*. The Free Press, Chicago

Le Heron, R. (1988) Food and fibre production under capitalism: a conceptual agenda. *Progress in Human Geography* **12**: 409–30

Long, N., Van der Ploeg, J. D., Curtin, C. and **Box, L.** (1986) *The commoditization debate: labour process, strategy and social network*. Paper of the Department of Sociology 17, Agricultural University, Wageningen

Mann, S. and **Dickinson, J.** (1978) Obstacles to the development of a capitalist agriculture. *The Journal of Peasant Studies* **5**: 466–81

Marsden, T. and **Little, J.** (eds) (1990) *Political, social and economic perspectives on the international food system*. Avebury, Aldershot

Marsden, T., Munton, R., Whatmore, S. and **Little, J.** (1986a) Towards a political economy of capitalist agriculture: a British perspective. *International Journal of Urban and Regional Research* **10**: 489–521

Marsden, T. K. and **Symes, D. G.** (1984) Land ownership and farm organization: evolution and change in capitalist agriculture. *International Journal of Urban and Regional Research* **8**: 388–401

Marsden, T., Whatmore, S. and **Munton, R.** (1987) Uneven development and the restructuring process in British agriculture: a preliminary exploration. *Journal of Rural Studies* **3**: 297–308

Marsden, T. K., Whatmore, S., Munton, R. and **Little, J.** (1986b) The restructuring process and economic centrality in capitalist agriculture. *Journal of Rural Studies* **2**: 271–80

Newby, H. (1982) Rural sociology and its relevance to the agricultural economist, a review. *Journal of Agricultural Economics* **33**: 125–65

Newby, H. (1987) *Emergent issues in theories of agrarian development.* Occasional paper Series 2, Arkleton Trust, Langholm

Peet, J. R. (1969) The spatial expansion of commercial agriculture in the nineteenth century: a Von Thünen interpretation. *Economic Geography* **45**: 283–301

Redclift, M. (1984) *Development and the environmental crisis: red or green alternatives?* Methuen, London

Rheinhardt, N. and **Barlett, P.** (1989) The persistence of family farms in US agriculture. *Sociologia Ruralis* **29**: 203–26

Rostow, W. W. (1960) *Stages of economic growth: a non-communist manifesto.* Cambridge University Press, New York

Schultz, T. W. (1953) *Economic organization of agriculture.* McGraw-Hill, New York

Smith, C. (1984) Does a commodity economy enrich the few while ruining the masses? *Journal of Peasant Studies* **11**: 60–95

Smith, W. (1984) 'The vortex model' and the changing agricultural landscape of Quebec. *Canadian Geographer* **28**: 358–72

Symes, D. and **Marsden, J.** (1985) Industrialization of agriculture: intensive livestock farming in Humberside. In Healey, M. J. and Ilbery, B. W. (eds) *The industrialization of the countryside.* Geo Books, Norwich, pp 99–120

Thompson, F. M. L. (1968) The second agricultural revolution, 1815–1880. *Economic History Review* **21**: 62–77

Tracy, M. (1982) *Agriculture in Western Europe: challenge and response 1880–1980.* Granada, London

Troughton, M. J. (1986) Farming systems in the modern world. In Pacione, M. (ed) *Progress in agricultural geography.* Croom Helm, London, pp 93–123

Vandergeest, P. (1988) Commercialization and commoditization: a dialogue between perspectives. *Sociologia Ruralis* **28**: 7–12

Vogeler, I. (1981) *The myth of the family farm: agribusiness dominance of US agriculture.* Westview, Colorado

Wallace, I. (1985) Towards a geography of agribusiness. *Progress in Human Geography* **9**: 491–514

2

Data sources for studying agriculture

Gordon Clark

The argument developed in Chapter 1 showed how studies of the contemporary transformation of agriculture have had to embrace concepts such as 'the food chain' and 'the food supply system'. With the widening range of studies conducted by geographers, traditional concerns, such as the distribution of farm enterprises and the agricultural classification of regions, have been complemented by studies of landscape changes, marketing, decision-making by farmers, the effects of policy, and radical perspectives on power and social relations in food production. This extension of interests has been paralleled by a considerable increase in the number of sources of information which are relevant to the geography of agriculture. Information is now needed at many scales and on new aspects of farming. Advances in technology have provided new sources (remote sensing, for example), while traditional sources are becoming available in new formats such as on-line databases (e.g. the 'Farmlink' system in the UK). However, even this wealth of information leaves some questions unanswered while still presenting problems for the geographer who wishes to keep abreast of the new sources of information.

This survey cannot possibly include all the kinds of information which have been used by agricultural geographers; it will confine itself to the major sources and to those new ones whose potential is considerable (Table 2.1). The features of each source will be described along with details of what

Table 2.1 Types of information for the geography of agriculture

Maps (see Table 2.2)
Aerial photographs and remote sensing
Agricultural census
Other statistical sources (government and agency publications)
Geographic information systems
Bibliographic sources
Humanistic sources – photographs and film
                – sound recordings and oral history
                – novels and diaries
Fieldwork

information it can provide and how it can be used. Stress will be laid on the difficulties likely to be encountered when interpreting the information. The weaknesses of certain sources need to be borne in mind, since their true meaning is sometimes rather different from one's first impression of them. This chapter concentrates on the general principles for the interpretation of data in agricultural geography – it is not a list of sources. Further details can be found in the list of references at the end of this chapter and, for British sources, in Boddington's contribution in Edwards and Rogers (1974), in Coppock (1978), in Bowler (1980) and, more recently, in Peters (1988).

Cartographic sources

Maps are the most distinctively geographical source of information and they have been used in three overlapping ways. First, the map may present data also available in other forms. Climatic maps are of this type since they are derived from meteorological records which have been converted into cartographic form. Second, maps may be used to present original data which to all intents are only available in map form (information on topography and land use, for example). Third, maps may be used to present the results of studies which involve considerable judgement and interpretation rather than the recording of raw data; maps of farming regions are of this synthetic character. These types of map are illustrated in Table 2.2 which shows the great range of cartographic material available to researchers.

The first and most obvious point to bear in mind about any map is the date

Table 2.2   Types of maps for agricultural geography

Category of map	Examples of maps
Physical context	Geology Soils Topography Climate
Evaluation of environment	Land quality (capability)
Economic context	Landownership Farm boundaries
Production	Land use Type of farming Maps of enterprises, labour, machinery, crops and livestock
Processing and marketing	Livestock movements Creamery catchments Distribution of processing works

of the survey on which it is based. Maps generally take a long time to produce and it is often expensive to revise them. This is not of great importance for a stable subject like topography, or even climate, where any long-run changes in climatic averages are likely to be slow to appear. The date of survey for soil maps is more important, since the amount of first-hand data on soil types is increasing annually and so the latest edition will have fewer blank areas and less crude extrapolation. Recently produced maps are most important in the more economic aspects of agriculture where variations in production are frequent and can occur rapidly. In a few years new crops may be introduced, the distribution of an enterprise may alter and farms may become more or less specialized. Maps of enterprises, farming systems and types of farms are particularly susceptible to obsolescence.

Maps can also become out of date by virtue of the purpose behind them changing. Agricultural land classification exemplifies this. It is not so much that the basic conditions of the environment change, rather it is the priorities and purposes embodied in the process of classification which are revised. The purpose of one particular land classification map might be to show the areas of best quality farmland, where quality is defined as the ability to produce high yields of a wide variety of products with the least environmental constraint. This definition of quality stresses flexibility of land use and minimum investment given current technology. If public policy required that quality be defined in some other way (e.g. maximum yields of just one specific crop after considerable investment) or if technology improved, then a new system of classification would be needed and different maps would result. Therefore, even a land classification dealing with stable features of the environment, such as soils and slopes, can soon become misleading.

The question of scale is also important when using maps. Nearly all maps of relevance to the geography of agriculture are produced at a single scale and that scale may not be the one at which the agricultural geographer wishes to work. There is usually very little difficulty in reducing a map at one scale, say 1:50 000, to a smaller scale, say 1:100 000. However, more substantial reductions from say, 1:50 000 to 1:250 000 pose particular problems concerned with generalization. If the reduction in scale is considerable, then there will not be room on the smaller-scale map for all the information. Generalizing lines (e.g. boundaries between farming regions) is relatively easy since a sinuous boundary will become smoother as scale reduces. Generalizing areas, however, poses the question of when the area should disappear because it is no longer legible, given normal eyesight and printing technology. The amount of detail left on the map is important because too much detail reduces the clarity of the map and too little implies a misleading simplicity to the distribution of the object being measured. This may tempt the map reader into inferring simplistic correlations between, for example, type of farming and altitude – the so-called ecological correlations. Whereas reducing a map's scale loses detail, enlarging a map rarely enhances detail. Many maps are limited by the basic scale of their data – the farm, the county or the network of meteorological stations. No higher level of resolution is possible: maps can lose data but they cannot create information.

Maps must also be assessed for their accuracy, since a map is only as good as the data on which it is based. The quality of the survey will obviously affect the reliability of the map and this has to be borne in mind in Third World countries particularly. Accuracy may also be affected by extrapolation: many maps purport to provide a complete record of the spatial variation of the variable in question, yet they are based on measurements taken at discrete places which have been applied to the surrounding area. Rainfall and soil maps of large areas are usually based on only a small number of measuring points, while topographic surveys may also use extrapolation.

The limitations of printing and human comprehension usually require some classification of the many varieties of information available before they are presented cartographically. Most maps place considerable weight on clarity, which leads to simple classifications. Land-use maps, such as the First and Second Land Use Surveys of Great Britain, classify land into types and usually only one type per place. Multiple land uses (e.g. recreation on farmland) are common but technically difficult to illustrate. Maps showing the classification of farming (e.g. the UK Ministry of Agriculture's Types of Farming series) can act as a mental straight-jacket, camouflaging complexity, diversity and different uses of the same patch of ground. Classification is also a scale-dependent exercise. Many classifications use roughly the same number of groups irrespective of the size of the area to be mapped. Thus Whittlesey classified the agriculture of the whole world into 16 groups, while Coppock used 19 enterprise types just for England and Wales. Consequently, the larger the area to be classified, the greater the degree of generalization and so the less informative are the very broad categories which are produced. It must also be remembered that the method used to create the classification is likely to affect, perhaps radically, the map of farming regions which is produced (Clark 1984). Useful sources of maps on agriculture are found in national atlases (e.g. those for Finland and Canada) and in agricultural atlases by Coppock (1976a, b) for the UK, Gregor (1982) for the US and Keddie and Mage (1985) for southern Ontario.

Finally, one has to remember that maps may be subject to legislation for the protection of copyright. This may restrict their use within agricultural geography or require that permission be sought from the copyright holder for their reproduction.

Aerial photography

Like the map, the aerial photograph looks down on the world from above but without the intervention of the classification, selection and symbols which are the language of the cartographer. The aerial photograph records what can be seen and that is both its strength and weakness (Campbell 1983). The emphasis on the visible allows more details to be extracted than from a map, although the nearly vertical viewpoint may make the identification of crops or buildings more difficult. Complex equipment may also be needed to rectify the distortions of size and shape which are found on many photographs. Sometimes the use of colour photography will resolve the uncertainties inherent in the much more extensive panchromatic (i.e. black and white) coverage; the distinction between different crops or types of vegetation may be clearer in colour or by using infra-red film. The latter is particularly useful for detecting water, soil moisture and different states of vegetation – its age and health, for example.

The photograph records the scene at one instant and, when compared with other photographs, has great potential for studying agricultural changes over various lengths of time. Seasonal changes in cropping can be followed, the movements of outdoor livestock can be traced, and the evolution of the farmed landscape in terms of field boundaries and buildings can be recorded. Parry's (1975) work on the changing altitudinal limits to cultivation was a notable example of such a study using aerial photographic evidence. Finally, longer-term changes in the area of farmland can be seen by comparing aerial photographs taken in different decades. The ability to measure seasonal and historical changes depends on the frequency of photographic coverage which is very variable. In West Germany, the whole country has been photographed every 2 years, while in the UK cloud-free coverage is less frequent for many areas; in remoter parts of the world, aerial photography may be patchy and infrequent. Other problems include different flight paths and photography at different times of year. In practice building a sequence of comparable photographs is difficult and expensive. None the less a decade of aerial photography allowed the agricultural potential of 1 550 000 km^2 of Australia's Northern Territory to be mapped, which demonstrates the potential of the medium when conditions are favourable.

There are usually a number of sources of aerial photographs and it is costly, but not wholly impossible, for the agricultural geographer to supplement these with his/her own. The national mapping agency often maintains a collection acquired for the purpose of revising its maps: the Canadian National Air Photographic Library, for example, has over 3 million prints. In the UK, the Ordnance Survey maintains the register of official aerial photographs. Some commercial surveying companies can also supply photographs, while the extensive university collections at Cambridge and Keele may also be of interest for agricultural geographers.

Aerial photographs are taken at a variety of scales and photographs can be rescaled by printing them at different sizes. It is unusual, however, to find photographs at a smaller scale then about 1:50 000; consequently they are excellent for studies of restricted areas where cloud cover is minimal. However, the larger the area to be studied, the more tiresome and expensive it becomes to work with many photographs and assemble them into mosaics corresponding to map sheets. A photographic type of coverage of large areas is better provided by remote sensing. None the less, notable recent examples of aerial photography used in agricultural contexts include the study by Langdale-Brown (1980) of agricultural habitats, the work of Budd (1988) in Cumbria as part of the National Countryside Monitoring Scheme, and Ilbery and Evans' (1989) estimates of farmland loss in the urban fringe.

Remote sensing

Remote sensing is the name now given to the coverage of large areas of the earth's surface mainly from imaging sensors mounted in orbiting or geo-stationary satellites (Lintz and Simonett 1976). Compared with cameras mounted in aircraft, satellite systems yield images with relatively low resolutions – that is, the ability to detect objects of a given size. For example, the early orbiting Landsat satellites carried sensors with a ground resolution of only 79 m, although this improved to 30 m on later satellites in the series. The French SPOT satellite, by comparison, carries sensors capable of resolutions of 10 and 20 m (Chisholm 1986). These are all civilian satellite systems: military satellites are believed to have already achieved a resolution of under 0.5 m. Geostationary satellites, which tend to operate at altitudes in excess of 32 000 km, have such low resolution capabilities that they are best suited to monitoring meteorological conditions rather than land-use features.

The sensors carried by satellite platforms measure the intensity of radiation reflected or emitted from the earth's surface. The radiation is measured within closely defined wavelengths or 'bands'. Different satellite systems operate varying numbers and definitions of bands, but Table 2.3 shows the band widths (spectral ranges) employed by the Landsat 5 sensors. The radiation is measured for a multitude of small areas of the earth's surface termed 'pixels' (picture elements), their size defining the resolution capability of the sensing system. The sensing device transforms the measured radiation into a numerical value and transmits the information in digital form to a ground receiving station. There the information is stored by computer on magnetic tape. Such is the volume of data generated – over 30 million observations for each image in a Landsat system, for example – that computer storage and manipulation are required to reconstruct an image of the earth's surface.

Recording data for a number of different wavelengths has decided advantages. For example, when cloud obscures measurement of the earth's surface in the visible light sector of the electromagnetic spectrum, radar wavelengths permit imaging to take place. Also, the infra-red wavelengths are particularly sensitive to differences between land and water, between different tree and crop varieties, and between diseased and healthy vegetation. Moreover, with data available in several bands, statistical methods of analysis can be employed to classify land surface features and to create 'colour composite' images of land use. In this technique, the simultaneous display of one band in shades of (say) red and others in green and blue can yield either normal colour or false colour images. Up to 127 intensity levels (shades) are possible in each band using the multi-spectral scanner (MSS) data from Landsat, while 255 levels are available on the SPOT system, although over a smaller number of bands.

As with aerial photographs, agricultural land use can only be interpreted by establishing the 'ground truth' for at least a part of the image. In the early 1970s, the correct identification of land use by satellite systems was possible in 50–70 per cent of cases under ideal conditions, although by the late 1970s figures of over 80 per cent were being reported in the literature. These are maximum figures which vary for each land use – for example, wheat is easier to identify than hay – and depend on ideal conditions such as large, single land uses in flat fields, the time of day, the maturity of the crop and the degree of contrast with the surroundings. The 'automatic', computerized classification of land use is improving all the time, as the unique 'spectral signatures' of each land use are recorded and refined. The military and commercial value of predicting the yields of key crops, especially wheat, has been a major spur to this work (Rhind and Hudson 1980).

Remote sensing has two further characteristics which differentiate it from aerial photography. One is the much larger area that can be measured in any one image or scene. Each Landsat scene, for example, covers 34 000 km². The other feature is the ability of the satellite to 'revisit' an area of the earth's surface to make repeated observations. The Landsat satellite now revisits every 16 days, and the SPOT satellite on successive days if its 'side-scan' facility is employed. Thus in any one year, the Landsat satellite system is capable of yielding over 20 images of one farming area, subject of course to cloud-cover conditions.

Table 2.3 sets out some of the applications of remote sensing for each of the wavelengths (bands) commonly measured by satellite systems. While each

Table 2.3 Landsat 5 Thematic Mapper system

Band	Spectral range (µm) (band width)	Main applications
1	0.45–0.52 (blue–green)	Differentiates coastal and estuarine waters, and deciduous and coniferous woodland
2	0.52–0.60 (green)	Vegetation mapping and showing urban area
3	0.63–0.69 (red)	Vegetation mapping, biomass estimation and soil-type boundaries
4	0.76–0.90 ('photographic' near infra-red)	Estimation of biomass and edges of rivers and lakes
5	1.55–1.75 (near-middle infra-red)	Detects wilting in plants and differentiates between snow and clouds
6	2.08–2.35 (middle infra-red)	Geological and soil mapping
7	10.4–12.5* (thermal infra-red)	Mapping surface temperatures

* The spatial resolution of this band is 120 m compared with 30 m for the other bands.

band tends to have particular applications, the use of bands in combination rather than singly is now a common feature of digital image analysis. For studies in agriculture, the principal source of imagery in the UK is the National Remote Sensing Centre at Farnborough, while the supply of coverage in the US was transferred in 1985 from NASA to a private company (Eosat). Summarizing, remote sensing provides a very distinctive kind of information for agricultural geographers which Thaman (1974) has reviewed: it is best for large areas, for simple broad land-use categories and for the repeated coverage of areas. An example of such work can be found in Martin's (1986) study of land-use changes on the urban fringe using Landsat imagery. On the other hand, satellite images lack the resolution of aerial photographs, they tell us nothing about farms or other economic units, they are still rather suspect for detailed crop identification, and they give little information about livestock. For information on these more economic aspects of agriculture, it is best to turn to official statistical sources of information such as censuses.

Agricultural censuses

The central problem with photographs and digital images is the need to infer what the visual information means in terms of practical farming. This need for inference is also important when one is trying to use an agricultural census. Today, most developed countries attempt to survey their farms regularly: this provides information on food production which is invaluable in peacetime and war for shaping agricultural policy. The cost of a complete agricultural census prevents many Third World countries from having one and some developed countries conduct a full census only every 10 years (e.g. Canada). In the UK and Sweden, annual censuses are publicly available which have provided good quality agricultural data for well over a century. The main US census is now quinquennial while the Australian one is annual. In some countries, censuses and samples are intermingled. In the Republic of Ireland about 60 per cent of farms have been surveyed each year since 1953 when a full annual enumeration ceased; a 5-year full census is used to check this sample.

The exact information provided by the agricultural census varies between countries, but in most cases one can find out about the size of farms, what crops and animals they produce, and how much labour they employ. Sometimes information is available on farm incomes (US), investment (Poland), farm buildings (Sweden), turf bog (Ireland) and land tenure (UK). Generally the details provided have increased over time: the first US census (1840–41), and the first British one (1866), were about one-third the size of the current censuses in these countries. Thus agricultural censuses provide some information on inputs to farming, on the organization of agriculture, on production and on commercial outputs.

In order to ensure comparability of results, nearly all censuses are nominally taken on a single day: for example, 4 June in the UK and 31 March in Australia. The census is, therefore, a snapshot taken at one season of the year and this can mean that farming activities in other seasons are not fully recorded. Census day in the US, for instance, has varied considerably throughout the year, sometimes being held in spring, early summer or autumn. A census can also be incomplete if it is viewed with hostility by farmers. They might regard it as an invasion of privacy, or as likely to lead to higher tax bills. In these cases farmers may not respond or may give false information. The first 30 years of the British agricultural census in the nineteenth century were marked by such non-response, particularly in south and east England.

The ideal agricultural census would allow historical comparisons of agriculture farm by farm. Unfortunately this is not always possible for a number of reasons. Often the census is not as accurate as it might be: very large-scale farmers sometimes do not know exactly how many hectares of land they have; the number of hens in a major egg-production battery may be recorded roughly rather than correct to the last bird; statistics on land tenure may be misleading where farms have to be recorded either as tenanted or owner-occupied, with no scope for recording information for farms comprising a mixture of land tenure; figures for crop yields are notoriously inaccurate. Small farms are particularly difficult to survey, so that data for them, and for the enterprises in which they specialize, are rarely of as high a quality as the information on large farms. In England and Wales, for example, the census discovered, as late as 1941, that 18 000 small farms had been omitted from the census, even though the census had been compulsory by law since 1925 and had been in operation since 1866.

Another major obstacle to the ideal of farm-by-farm studies is that the data on individual farms are usually confidential and are not available to the public. A growing problem for the agricultural geographer is the desire to protect information on individuals either for reasons of commercial advantage or privacy. The Data Protection Act (1984) in the UK has made even the unofficial disclosure of farm-level data less likely; indeed parish summaries have not been published since 1989.

Agricultural census data are published for farming in defined areas. Often there is a hierarchical arrangement so that data are available for the whole country, for regions and districts, and perhaps even for small areas such as parishes. This is not as useful as individual farm results, particularly given the arbitrary nature of these areas. They are usually administrative units, often of some antiquity, which are convenient for the collection of data; they do not necessarily reflect the pattern of farming and may conceal considerable variations in farming within each area, for example by combining upland and lowland areas. Administrative areas are also subject to periodic reorganization which means that historical comparisons of farming are not so straight-forward. In Poland, census data are available for voivodships (roughly counties) which were increased in number from 22 to 49 in 1975; this artificially obscured the way agriculture evolved in Poland in the 1970s. In the UK, county boundaries have changed (1974/75) and so have parish boundaries in a more piecemeal fashion. There are now 26 per cent fewer parishes in England and Wales than in 1870.

Generally, the larger, more important administrative areas, such as the states of Australia and the US, and the provinces of Canada, are more stable and so more suitable for historical studies than the smaller statistical units. However, they are also less interesting to the geographer because they conceal so much of the spatial variety of farming. But while small statistical units are more informative than large ones, there is the difficulty of interpreting the census data when farms are expanding in size. This process may lead to the larger farms operating land in more than one administrative unit, although census data are usually recorded as though each farm were wholly in a single unit. To confuse matters further, the criteria for allocating farms to statistical areas are not constant. The location of the farmhouse or the majority of the land, or even farmers' preferences, have all been used. The result is that the agricultural statistics for a parish may include some land outside the parish, while excluding some land within it which is worked by farmers in other parishes. Thus one must exercise caution with agricultural statistics because the same name may be used for an administrative unit covering a particular area, and for a group of farms cultivating a rather different area of land. Consequently, as farms expand and take over extra land, it becomes increasingly difficult to map agricultural statistics for small areas.

The historical value of a census is also impaired by changes in the way the census is structured. The census organization may alter the questions it asks: new questions are included and old ones discarded. This is inevitable as new farming practices are taken up and old ones fall out of favour. During the Second World War, for example, British farmers were asked how many prisoners of war were working on their farms: in peacetime this question became redundant. It is always difficult to study the origins of any new kind of farming: census organizations are loath to alter the census form unless this is really necessary, since changes make it more difficult for farmers to fill in. Extra questions are only added once a new type of farming has expanded to such an extent as to show that it is going to continue on a significant scale, for example maize and oilseed rape in the UK in the 1970s.

Much more fundamental, however, are changes in the central definitions of the 'farm', 'farmland' and the 'farmer'. These are not self-evident terms – if someone keeps three hens, is that person a farmer? – but their definition is critical to an agricultural census. The difficulty arises because the definitions of these terms vary from country to country, which makes international comparisons less easy, and they also vary over time within a single country, which impairs historical studies.

The farm

The definition of a 'farm' is the most critical of the terms. In the UK the definition of a 'farm' has been widened and narrowed at various times by altering the minimum size of unit deemed large enough to merit inclusion in the census. In 1892, for example, the minimum size of farm included in the British agricultural census was raised from 0.25 to 1 acre. Units smaller than the prevailing minimum were originally excluded because they produced little food. More recently, small units have been excluded in order to lessen the cost of running the census. Smallness is increasingly defined in relation to the size of the farm as a food-producing business rather than just in terms of land area. In the UK, size has also been measured in terms of the estimated amount of labour needed to run the farm ('standard man–days' (SMDs) – see p. 106), while in the US a combination of farm area and the value of produce sold is used.

A farm is also less easy to define where there are large farm businesses with several blocks of land in different districts of the country. Are these blocks of land separate farms or just part of one big farm? The census organization is drawn in two directions. If the separate blocks are treated as one large farm, there are administrative savings in the cost of running the census and the census will record what that farm does as an economic unit. On the other hand, the mapping of the statistics and their presentation for small areas become increasingly unrealistic. The easiest guideline to use is the distance between the blocks of land. However, the maximum feasible separation between blocks which are still run as a single farm will vary both between countries and over time, depending on the terrain and the prevalence of motor transport. In the UK, a rule of thumb of 8 km (5 miles) has been replaced by one of 24 km (15 miles), although the farmer's preferences are also taken into account.

Farmland

The definition of 'farmland' is also an important one, since it affects both how many farms are covered by the census and how much of each farm's land is recorded. The trend in the UK has been for the definition of 'land' to become more all-embracing, while the definition of 'farms' has been narrowing. For instance, the extensive upland areas have been progressively included more fully in the census. First, rough grazings were recorded (1892), then unfenced grazing land in the uplands (1911), followed by the extensive deer forests (open moorland usually with a few sheep) of Scotland, principally in 1932 and 1959. These changes have been of major importance, particularly in affecting the total agricultural area of the UK and in the debate over how much agricultural land is being lost to forestry and urban development.

The farmer

The definition of a 'farmer' is also complicated because of the increasing frequency with which farming is combined with other jobs. Usually the agricultural census records a rather higher number of farmers than found in other sources of employment data (e.g. the population census) – they record part-time farmers under their principal, often non-agricultural, occupation. The family structure of agriculture also has to be remembered – spouses, children, parents and siblings often play some role in managing, financing and running the 'family farm': such activities need not be the sole responsibility of a single person – the farmer. Hence, in some cases there may be more farmers than the official statistics show. In some countries farm managers may perform many of the functions of the farmer but be excluded from the statistics on the number of farmers. In contrast, if many farmers run several nominally separate farms as a single business, then the number of farmers could be less than the number of farms. Generally the data on how many farmers operate in an area, and what sort of people they are, tend to be much poorer in quality than our knowledge of their farms and the farmland.

This section has tried to highlight some of the difficulties with using census data. However, this should not leave the impression that the agricultural census should be avoided. It is not as helpful as geographers would wish, particularly for historical studies, international comparisons and for detailed information on spatial variations in agriculture. Nevertheless, it does provide much useful material for the agricultural geographer. The agricultural census is usually produced by the Ministry of Agriculture of the country concerned. However, in the US the Bureau of the Census is responsible, in New Zealand it is the Department of Statistics, in Australia the Bureau of Statistics and in Ireland the Central Statistical Office. The national *Statistical yearbook* is a good starting-point for sources of agricultural data, while the British censuses are described in more detail in publications by Coppock (1978) and Clark (1982).

It is sometimes tempting to make international comparisons of agriculture and some sources are ideal for this. The numerous publications of the European Community (EC) (the Commission or Eurostat – the Statistical Office), which allow international studies, include *Agricultural Statistics Yearbook, The agricultural situation in the Community, Earnings in agriculture, Agricultural price indices* and *Farm structure 1985 – main results*. The Organization for Economic Co-operation and Development (OECD) produces *Food consumption statistics*, while the United Nations publishes a *Statistical yearbook* and *Current world food situation*. The Food and Agriculture Organization (FAO) in Rome issues a *Trade yearbook* and a *Production yearbook*.

Publications such as these can perform a useful function but they have to be treated with great caution. Each country compiles its agricultural statistics in its own way, although the EC is trying to impose some standardization.

Hence international comparisons can be very misleading, since words and concepts are not defined in the same way in each country. African systems of land tenure, for example, cannot be fitted meaningfully into European categories of owned and rented land. Best (1981) provides some good examples of the limitations in comparing land–use statistics between European countries.

Other statistical sources

The agricultural census provides much information on the nature of farming and how it varies throughout a country. However, in no country does it provide a complete picture of the industry. It is necessary to be aware of other official sources of data which can complement the census. What is available varies from country to country but sectoral studies are particularly useful.

Whereas the census gives an overview of agriculture, sectoral studies are concerned to provide an in–depth analysis of a specific aspect of the industry. This may be a particular enterprise such as milk or potatoes: for instance, much valuable information is available in the UK from the Milk and Potato Marketing Boards. The sector might also be an area of the country or a special type of farm, such as crofts in Scotland (the annual reports of the Crofters' Commission) or smallholdings (Wise Committee 1967). Alternatively the sector could be a theme running through farming, like structural change (USDA 1981; OECD 1972), landownership (Northfield Committee 1979), or tenure (Gasson and Hill 1984).

The second major area of data concerns farmers' incomes. In most censuses this topic is not well described and so information from colleges of agriculture, university departments and the taxation service is sometimes used (Hill 1984). None of these sources is perfect. Farmers may not reply to such enquiries or, at least, not truthfully. Where the responding farmers have volunteered their services, there is always doubt about how representative they are of the farms generally in the area (e.g. MAFF *Farm incomes in the United Kingdom*). The data provided by these sources are usually confined to the farmer's agricultural income. This underestimates the incomes of two particular groups of farmers. Small-scale farmers who have other jobs would appear to have lower incomes than in fact they do; also the less important farming is to them, the more misleading the farm income data will be (Gasson 1983). The other group with higher incomes than will be apparent are very large-scale farmers with an extensive income from investments, farm rents or other rural occupations, such as forestry or tourism. These wealthy farmers often form their business ventures into companies or trusts as a means of minimizing their tax liability and so conceal their wealth. Recently the EC has attempted to standardize information on farm incomes, but the incompatibility between income data and the census remains considerable even within one country, let alone between countries. If information on farmers' incomes is patchy, details of their wealth are even more obscure, particularly in countries, such as the UK, where there is no annual tax on wealth.

Most countries have tried to classify their farms into types, usually on the basis of what the farm produces but sometimes also involving farm size and tenure. These classifications are a valuable guide to how farming varies geographically: cereal farms can be distinguished from dairy units, and upland farming from horticulture (Tarrant 1974). The results may also be mapped to show farming regions, as was described in an earlier section (MAFF 1974). The classification of farming has the disadvantage that it is heavily influenced by the scale at which it is conducted and the statistical methods used. There are many ways the farming of an area could be classified and so it is dangerous to put too much faith on just one classification (Clark 1984) (see also Ch. 5).

Other sources of information which have been used by geographers can be classified into two groups: those concerned with inputs to the agricultural process and those dealing with its outputs. Inputs include, of course, items already discussed, such as data on land quality, soil type and climate. Equally farm labour is a major input and the agricultural census usually provides such statistics. Separate, non-agricultural censuses of employment will often give different figures for the size of the farm labour force, depending on how workers combining employment on and off the farm are classified. Data on farm labour are tending to become less reliable as fewer full-time workers are employed and an increasing proportion of farm labour comes from the farmer, his/her family, agricultural contractors and casual, seasonal or part-time workers. Such workers tend to be seriously under-enumerated by agricultural censuses taken on a single day during the year. In addition to these measures of the volume of inputs to agriculture, there are usually figures available on the costs of inputs (e.g. wage rates, land prices). The former usually derive from an official survey, while the latter may include government data, statistics from tax authorities or from agricultural banks. Mortgage companies (such as the Agricultural Mortgage Corporation in the UK) may also provide data. However, the confidentiality of commercially sensitive or private information may limit access to these sources.

Data on outputs differ from census figures (which deal with production) because some produce will be used on the farm for subsistence, seed or fodder. In the US, the Department of Agriculture publishes data on farm sales separately from the Bureau of Commerce's agricultural census. The commodities whose commercial production is most easily measured are those which must pass through storage or processing plants for commercial or health reasons (e.g. meat to auctions or abattoirs, milk to dairies and grain to storage silos). Livestock and vegetables not subject to packaging or processing are marketed so widely that it is not easy to measure their commercial production. Companies may regard data on stocks or turnover as commercially sensitive information which they prefer not to disclose. One way round this is to use official statistics from a census of production, such as the *Business Monitor* series for the food industries published in the UK.

The end of the food chain is, of course, food consumption which is the ultimate demand to which farmers respond. Data on this are patchy. In the UK the National Food Survey (a reminder of wartime concerns over the nation's health) provides survey-based data on regionally disaggregated trends in food consumption. The Family Expenditure Survey and the General Household Survey are broader sources which allow food purchases to be set in the context of the family's income and general patterns of spending.

Agriculture is a uniquely regulated industry – there is a separate Ministry of Agriculture in nearly every country – and so studies of agricultural policy are vital for understanding the industry's evolution. Information here is likely to be textual rather than statistical. Official statements of policy are easily available in the published legislation (Acts of Parliament and Statutory Instruments in the UK) or in discussion papers put out by the government (Green Papers and White Papers in the UK). Debates in Parliament, press notices and briefing papers from pressure groups, such as food manufacturers, landowners and farmers' organizations, can also be valuable (e.g. the British National Farmers' Union's annual *Review of NFU activities*). However, the best information is likely to be obtained by personal contact with the key officials and negotiators who were present when the decisions were made. Few ground rules can be laid down for obtaining such information beyond identifying and making contacts, persistence and a very careful and sceptical interpretation of whatever the researcher is told.

Geographic information systems

A geographic information system (GIS) is not usually a source of new data, nor is it a library of information in the manner of the Economic and Social Research Council's (ESRC) Rural Areas Database. Rather, a GIS is a computer-based system of integrating spatially referenced information (both statistical and cartographic) and providing facilities for the editing, combination, processing and display of that information. The key to a GIS is the hardware and software to allow these operations to proceed swiftly. Blakemore has reviewed recent developments in this fast-moving area (Blakemore 1987). In the UK, the Chorley Report (Department of the Environment 1987) provided a stimulus to the development of GIS and the ESRC's Regional Research Laboratories will develop the ideas and techniques further. There is some element of seeking to exploit these manipulative skills in the commercial world as well as for pure research purposes.

The development of GIS has had the beneficial effect of giving further impetus to attempts to find an agreed set of spatial units for the publication of aggregated data. Whereas, currently, data may be available for several incompatible sets of spatial units (e.g. parishes, wards, enumeration districts, travel-to-work areas), two universal 'building blocks' have emerged in the UK. The first is the kilometre grid square, whose arbitrariness may be its strength. The other is the post-code sector, which is well established as a tool in market research and works well in towns. In rural areas, however, post-code sectors are at their least useful, being large and oddly shaped. However, the ability to compare and integrate different sorts of data on agricultural topics (among others) is so valuable a prize that it is likely the search for a common spatial unit will continue.

The huge variety of official sources presents the agricultural geographer with problems in finding out what is available, and so the use of the bibliographic sources described in the next section is always helpful.

Bibliographic sources

Bibliographies are lists of the works which have been published in a particular field. Usually they record the books, pamphlets, monographs and articles produced by researchers. For agricultural geographers, the most useful bibliographies are *Geo Abstracts* and *World Agricultural Economic and Rural Sociology Abstracts*, the latter being produced monthly by the Commonwealth Bureau of Agricultural Economics. This series started in 1958 and now records about 8 800 publications a year. They are grouped by topic for ease of searching and are also available from an on-line computer database via the DIALOG system. A summary of each publication is also provided so that its contents can be assessed more carefully than the title alone would permit. A similar on-line databank for North America is provided by the USDA's AGECON file within their AGRICOLA database. The EC uses a similar database (the CRONOS system) and it looks as though data, and increasingly abstracts of publications, will be available on-line (Bellamy 1984). The trend in legal publishing, to making full texts available more cheaply and quickly on-line than on paper, may soon extend to the material for the agricultural geographer. The most recent developments concern the use of viewdata services, such as Prestel in the UK, Teletel in France and Teledon in Canada, in order to make fully up-to-date information instantly available. Formerly confined to current data, Prestel can now act as a gateway into other databases throughout the world. JANET, a network for British universities, performs a similar function and links into bibliographic and statistical databases such as the 1981 population census, the ESRC's Data Archive and the National On-line Manpower Information Service (NOMIS) system of employment data at Durham University. High running costs may limit the usefulness of such systems for the present, for example, a database for British farmers (Farmlink) on market prices, disease reports and agricultural news was withdrawn in the late 1980s.

The on-line sources already described are principally useful for abstracts and statistical data. There are, however, other bibliographies on paper which are helpful to the agricultural geographer. The results of research for higher degrees are reported in *Dissertation abstracts international, Index to theses with abstracts* and *Current research in Britain,* the latter two being purely British in coverage. There are equivalent publications in most developed countries. In most countries there is a guide to official statistical sources produced by that country's central statistical organization. In the UK, the Central Statistical Office (1980) has produced a valuable *Guide to official statistics* (Ch. 6 deals with agriculture), the book by Peters (1988) is invaluable, and in the US the *Monthly catalogue of United States government publications* (USGPO, monthly) is equally useful. *Australian government publications* records both Commonwealth and state publications. *A guide to the official publications of the European Communities* (Jeffries 1981) provides a way into that agricultural material, as do the Commission's *Agricultural Statistics Yearbook* and *The agricultural situation in the Community.* European documentation centres and depository libraries throughout the Community make the EC material readily available, while Johansson's *Official publications of Western Europe* is also an invaluable starting-point. Rather more general, but still useful to the agricultural geographer, is Lilley's *Information sources in agriculture and food science* (1981); the chapters on statistics and agricultural economics are particularly valuable for their wide spatial coverage and clarity. The annual *Countryside planning yearbook* edited by Gilg (since 1987 the *International yearbook of rural planning*) is also a good guide to the most recent developments in the field.

Humanistic sources of information

One of the trends in geography has been a renewed concern for individuals: the men and women about whom one learns so little from censuses and other official sources which aggregate data. Agricultural geographers wish to know how farmers make decisions, how they perceive their physical and economic environments, how food processors reach their investment decisions, and why and how governments make decisions on agricultural policy. These concerns require that one discovers the way in which farming is experienced by the individual farmer or decision-maker. Fieldwork among contemporary farmers may be possible and this is described in the next section, but where it is not (for example, in historical studies of farming) one may need to use sources of information such as diaries, biographies, oral history, novels, and archival film and sound recordings. These provide the potential to get 'under the skin' of individual farmers who would otherwise be lost in aggregated census statistics for parishes or counties.

Such sources have much to offer but one also has to be aware of their pitfalls. Diaries, biographies and novels can illuminate the events experienced by an individual – why someone migrated, adopted a new machine or joined a co-operative. Such sources can reveal how events were perceived by a person: the farm worker's view of mechanization may be very different from the farmer's, the machinery salesman's or the farmer's wife's. However, one has to consider why a book or diary was written. Was it to be a simple chronicle of events or was it for the purpose of self-justification, interest or 'setting the record straight'? Was the novel a record of real people – in which case was the author a participant or an observer? Or was it an amalgam of various people and events? Was it designed to inform a readership, to entertain them, to shock or to argue a cause, perhaps a political one? The purpose behind a diary or novel has to be considered when it is interpreted. *The grapes of wrath* is much more than just a chronicle of some migrants from Oklahoma to California in the 1930s.

Novels, diaries and oral history also have a common problem of memory and *post hoc* rationalization. As time passes, memory naturally fades and it does so selectively: highlights and the good times are recalled particularly well. In addition, decisions become rationalized in the light of subsequent events. Reasons and attitudes which motivated behaviour in the past are forgotten or replaced by what the respondent believes he/she thought, or should have thought, at the time; decisions are justified in the light of subsequent events.

Even more beguiling are tape recordings, photographs and archival film. With these we seem to see and hear the past talking to us directly, rather than being filtered through the printed word or the dry statistics of a census. These sources are particularly useful for showing us landscape, architecture, dress and accent. The Pictorial Section of the National Library of Australia, for example, adds much to our understanding of pioneering days through its photographs of life in the outback. Caution is more necessary where events are portrayed. Did the presence of the camera or tape recorder affect how people behaved or talked? Was their behaviour typical of that sort of person or area? It may be that these people were recorded for no better reason than that they alone agreed to co-operate, or the film-maker lived nearby, or they were photogenic 'characters' who would entertain as much as inform. The intentions and attitudes of the photographer or film-maker have to be considered when viewing the scenes. Finally, there is the further problem of representativeness. Films, photographs and sound recordings have all survived sporadically. Film, particularly early nitrate film, is subject to physical decay and eventual disintegration. Photographs are more robust and abundant but are highly scattered and poorly documented (Evans and Evans 1979; Wall 1977).

Nevertheless, some agricultural geographers may find these sources useful and some large collections are particularly valuable. The Hardman collection of photographs of the Lake District, for example, shows many farming scenes, though some are slightly too picturesque to be typical. The massive photographic archive of the Farm Security Administration records American farming in the 1930s, although much of it was taken expressly to record the Depression and, by giving poverty and despair a human face, to justify New Deal policies. Film archives can be inaccessible to the general public and published information on what they contain may be hard to obtain. The British Film Institute and regional film archives (particularly for Scotland, East Anglia and north-west England) are well stocked and clearly organized for those researchers who can gain access to them. Film archives in the US tend to be both larger than those in the UK and better funded. Oral history archives are the sparsest of these media, both in terms of quantity of material and its bibliographic control. The National Sound Archive (formerly the British Institute of Recorded Sound) and the folk studies departments of some universities (e.g. Leeds, Lancaster, Belfast and Edinburgh) are the main British sources of spoken-word recordings relevant to agricultural geography.

Fieldwork

Arguably, the geographer is better trained than most social scientists to conduct fieldwork, that is, to collect original information because none of the existing sources meet the needs of the research problem. The agricultural geographer might wish to collect information on people's attitudes or the reasons they give for their actions. Alternatively, there may be a need to obtain information on individual farm businesses, or individual farm input suppliers or food processors. Also the fieldwork may be attempting to research the processes that operate within agriculture, for example the impact of forward contracts on types of farming, or the relationship between changes in agricultural practices and landscape changes.

One recurring problem which besets fieldwork is how to select who or where to study. Resources rarely permit a census to be carried out, so it is important to have a strategy which will provide a sample of data representative of the population from which it was drawn. This may require obtaining a complete and up-to-date list of all the farms or food processors in an area to form a sampling frame: this is rarely an easy task. Telephone directories (e.g. Yellow Pages) tend to under-record both the smaller businesses which do not have a telephone and those who do not wish to be pestered by salesmen. Electoral registers rarely distinguish farmers from others, while records of landownership, even if they are publicly available, are at best an imprecise guide to who controls farming when absentee landlords and renting land are prevalent. Maps rarely show as farms all the buildings which are used agriculturally, while farm amalgamation will leave maps showing farms units which have been taken over by others. Lists of the members of rural groups, such as breed societies, dairies, co-operatives or farmers' organizations, are possible sources but often incompleteness, and the list's confidentiality, limit their utility.

One way round these problems for a farm survey is to contact all the farms in an area using personal enquiry to reveal their location (Clark and Gordon 1980). The difficulty here is to find an area for these enquiries which is representative of the whole region or country. Undoubtedly the dispersed nature of farms and farming settlements makes purely random sampling methods expensive in time and money, and so the merits of a clustered sampling design are obvious. An alternative, if a personal visit to the farmer is not essential, is the postal survey. However, response rates may be unacceptably low, the non-response may bias the sample achieved and the illiterate cannot participate. A telephone survey is another option, but is only practicable where telephones are common in the countryside and the researcher can discover the numbers of all the farmers.

A common feature of many sampling frames is that large-scale businesses are overrepresented. If this happens the researcher is in a dilemma. On the one hand, statistical theory would clearly suggest that this bias was undesirable, whereas the reality of agriculture in the developed world is that most food is controlled by the larger-scale businesses. Some concentration on these businesses, even if not a formally controlled stratification, reflects the reality of food production and, for some studies, will be less harmful than might be imagined.

Once a researcher has selected who and where to study in the field, there is the question of how to conduct the enquiry. Surveys of land use and the farmed landscape tend to use straightforward skills of identification, photography and mapping. Surveys of people, however, are more problematic. The structured interview provides standardized information from all the respondents, packaged into neat analysable categories. When the enquiry is about purely factual matters, this may be the ideal methodology; response rates of over 90 per cent for personal interviews are not uncommonly achieved. If the research is seeking not facts but the respondents' attitudes, hopes, fears, opinions and their explanation for their actions, then a less structured approach, with open-ended questions, may be preferable. If people's exact words, phrasing and intonation are important for deducing their meaning, then a tape recorder may be the best way of noting the results for later analysis. However, a tape recorder may be seen as intimidating and key officials may reveal more in their less guarded moments when the tape recorder is off.

A useful strategy may be to consider a mixture of methodologies. A large-scale postal survey can provide a broad coverage of factual issues; this could be followed up by unstructured interviews with a sample of the respondents, with in-depth discussions for a handful of key decision-makers. It is never easy to predict exactly how any strategy for fieldwork will work out in practice: its success will depend on the balance between the resources it requires and the volume and quality of information it provides. A pilot survey is always a helpful tactic for improving the efficiency of fieldwork methods and raising response rates. For example, it will help to identify areas which have been oversurveyed in the past.

51

Conclusion

The kinds of information useful for studying the contemporary geography of agriculture are very varied; the recent transformation of the food supply system has posed a new set of research questions and required the range of sources of information to be widened. In addition, researchers have adopted new approaches concerned with ecology, political economy, institutions, structuralism and humanism. This variety of sources of information has some drawbacks and many strengths. The disadvantages lie in the extent of learning necessary about the sources, their origins, formats and meanings. The strengths are more important, however, since each source of information adds a new dimension to our understanding of how agriculture functions in developed market economies. There are considerable benefits in using the fullest possible range of information so as to obtain a fully rounded view of structures and processes.

Each source of information has limitations in how it can be used and interpreted. Each also guides, to an extent, the sort of questions that are asked and the kind of research that is conducted. Yet it is through what is available, and what we can find out for ourselves through fieldwork, that we come to know the world and so create and test our understanding of it. The information explosion is a misleading expression. There are indeed increasing amounts of data, although reduced government expenditure is sometimes curbing this, and yet our dependence on fieldwork for the exact kind of data we need is undiminished. The use of computers has immensely strengthened our ability to process data, yet the incompatibilities of format and definition among so many sources remain a significant barrier to research.

Finally, information should not be seen as fixed, self-evident and authoritative. Information is more than just factual statements. It is a matter of judgement as to what is relevant and this in turn depends on one's aims and philosophical stance. When these change, the body of information which is relevant to understanding the geography of agriculture alters. Information is also a matter of judgement as to what the facts mean: the extent to which any information is useful depends on how it is interpreted. Therefore, we can visualize a two-way relationship between information and the main body of work in agricultural geography which is described in the other chapters of this book. Neither can have any useful existence without the other: information animates theory and research, which in their turn stimulate the collection of information and guide the interpretation of the results. More importantly, no amount of information can substitute for clear thinking and fresh ideas in analysing the geography of agriculture.

References

Bellamy, M. (1984) The evolution of information sources and the use of information technology in agricultural economics. *Journal of Agricultural Economics* **35**: 31–8

Best, R. H. (1981) *Land use and living space.* Methuen, London

Blakemore, M. (1987) Cartography and geographic information systems. *Progress in Human Geography* **11**: 590–606

Bowler, I. R. (1980) Source materials for teaching agricultural geography. *Teaching Geography* **6**: 80–2

Budd, J. (1988) The National Countryside Monitoring Scheme. In Bunce, R. G. H. and Barr, C. J. (eds) *Rural information for forward planning.* Institute of Terrestrial Ecology, Grange-over-Sands, Cumbria

Campbell, J. B. (1983) *Mapping the land: aerial imagery for land use information.* Association of American Geographers, Washington DC

Central Statistical Office 1980 *Guide to official statistics.* HMSO, London

Chisholm, N. W. T. (1986) Spot on data. *Geographical Magazine* **58**: 438–9

Clark, G. (1982) *The Agricultural Census – United Kingdom and United States.* CATMOG 35, Geo Books, Norwich

Clark, G. (1984) The meaning of agricultural regions. *Scottish Geographical Magazine* **100**: 34–44

Clark, G. and **Gordon, D. S.** (1980) Sampling for farm studies in geography. *Geography* **65**: 101–6

Coppock, J. T. (1976a) *An agricultural atlas of England and Wales* 2nd edn. Faber and Faber, London

Coppock, J. T. (1976b) *An agricultural atlas of Scotland.* John Donald, Edinburgh

Coppock, J. T. (1978) Land use. In Maunder, W. F. (ed) *Reviews of United Kingdom statistical sources VIII.* Pergamon, Oxford

Current research in Britain (annual except for biennial volume on the humanities), British Library Board, London

Department of the Environment (1987) *Committee of inquiry into the handling of geographic information: report* (Chairman, Lord Chorley). HMSO, London

Dissertation abstracts international (annual) University Microfilms International (available on paper and via DIALOG for on-line searching), Ann Arbour, Michigan

Edwards, A. and **Rogers, A.** (1974) *Agricultural resources.* Faber and Faber, London

European Commission (Eurostat) (annual) *Agricultural price indices.* The Commission, Brussels

European Commission (annual) *The agricultural situation in the Community.* The Commission, Brussels

European Commission (Eurostat) (annual) *Agricultural Statistics Yearbook.* The Commission, Brussels

European Commission (Eurostat) (annual) *Earnings in agriculture.* The Commission, Brussels

European Commission (Eurostat) (1987) *Farm structure 1985 – main results.* The Commission, Brussels

Evans, H. and **Evans, M.** (1979) *Picture researcher's handbook.* Saturday Ventures, London

FAO (Food and Agriculture Organization) (annual) *Production yearbook.* FAO, Rome

FAO (annual) *Trade yearbook.* FAO, Rome

Gasson, R. (1983) *Gainful occupations of farm families.* School of Rural Economics, Wye College, Kent

Gasson, R. and **Hill, B.** (1984) *Farm tenure and performance.* School of Rural Economics, Wye College, Kent

Gilg, A. (ed) (annual since 1980) *Countryside planning yearbook.* Geo Books, Norwich

Gilg, A. et al. (eds) (annual since 1987) *International yearbook of rural planning.* Geo Books, Norwich

Gregor, H. F. (1982) *Industrialisation of US agriculture: an interpretive atlas.* Westview Press, Boulder, Colorado

Hill, B. (1984) Information on farmers' incomes: data from Inland Revenue sources. *Journal of Agricultural Economics* **35**: 39–50

Ilbery, B. W. and **Evans, N. J.** (1989) Estimating land loss on the urban fringe: a comparison of the agricultural census and aerial photograph/map evidence. *Geography* **74**: 214–21

Index to theses with abstracts (annual) Aslib, London

Keddie, P. D. and **Mage, J. A.** (1985) *Southern Ontario atlas of agriculture: contemporary patterns and recent changes.* Department of Geography, University of Guelph, Ontario

Jeffries, J. (1981) *A guide to the official publications of the European Communities.* Mansell, London

Johansson, E. (ed) (1984/1988) *Official publications of Western Europe* Vols 1 and 2. Mansell, London

Langdale-Brown, I. (1980) *Lowland agricultural habitats (Scotland): air photo analysis of change.* Nature Conservancy Council, Peterborough

Lilley, G. P. (ed) (1981) *Information sources in agriculture and food science.* Butterworth, London

Lintz, J. and **Simonett, D. S.** (1976) *Remote sensing of environment.* Addison-Wesley, Reading, Mass.

MAFF (1974) *Types of farm maps.* London (two series each of eight regional maps of England and Wales)

MAFF, Department of Agriculture and Fisheries for Scotland and Department of Agriculture for Northern Ireland (annual) *Farm incomes in the United Kingdom.* HMSO, London

Martin, L. R. G. (1986) Change detection in the urban fringe employing Landsat satellite imagery. *Plan Canada* **26**: 182–90

Northfield Committee (1979) *Committee of inquiry into the acquisition and occupancy of agricultural land: report* (Chairman, Lord Northfield). Cmnd 7599, HMSO, London

OECD (Organization for Economic Co-operation and Development) (1972) *Structural reform measures in agriculture.* OECD, Paris

OECD (1985) *Food consumption statistics 1973–1982.* OECD, Paris (updated frequently)

Parry, M. L. (1975) Secular climatic change and marginal agriculture. *Transactions of the Institute of British Geographers* **64**: 1–14

Peters, G. H. (1988) *Agriculture.* Reviews of United Kingdom statistical sources 23, Chapman and Hall, London

Rhind, D. and **Hudson, R.** (1980) *Land use.* Methuen, London, Chs 2–4

Tarrant, J. R. (1974) (in USA, 1980) *Agricultural geography.* David and Charles, Newton Abbot, Ch 4

Thaman, R. R. (1974) Remote sensing of agricultural resources. In Estes, J. E. and Senger, L. W. (eds) *Remote sensing: techniques for environmental analysis.* Hamilton, Santa Barbara, pp 189–223

United Nations (annual) *Statistical yearbook.* UN, New York

United Nations (annual) *Current world food situation.* UN, New York

USDA (United States Department of Agriculture) (1981) *A time to choose: summary report on the structure of agriculture.* USDA, Washington DC

USGPO (United States Government Printing Office) *Monthly catalogue of United States Government publications.* USGPO, Washington DC

Wall, J. (1977) *Directory of British photographic collections.* Heinemann, London

Wise Committee (1967) *Departmental committee of inquiry into statutory small-holdings provided by local authorities in England and Wales: final report* (Chairman, M. J. Wise). Cmnd 3303, HMSO, London

3

Factors of production in modern agriculture

Richard Munton

The theoretical discussion in Chapter 1 highlighted the changing relationships between land, labour and capital in the recent transformation of agriculture. The purpose of this chapter is to examine the factors of production in more detail so as to reveal their roles in shaping the geography of agriculture in developed market economies. To begin with, though, we should be reminded of three issues which were introduced earlier in the book. First, technological development over the last 50 years has led to a much more rapid increase in farm production than growth in domestic demand. Domestic demand is, in general, highly income inelastic while population increase in advanced economies is limited, with the constant tendency for farmers to oversupply the market, create food surpluses and force down the rate of return to those resources employed in the industry. Unfortunately, resources in agriculture, human, as well as land and capital, are only grudgingly transferred to other sectors of the economy, an inertia that governments, for perfectly proper social and political reasons, have encouraged through various forms of farm income support. This inertia is reflected in the relatively inefficient use of labour in the agricultural sector. For example, no member country of the European Community (EC) has a primary sector in which its proportion of the national workforce is less than the proportion of the gross domestic product which it supplies (Table 3.1), although the ratio varies significantly. Portugal, for example, recorded a ratio of more than twice that of Belgium, the Netherlands and the UK.

Second, technological development has consistently led to the substitution of capital for labour in the production process. Because of the cost of more powerful machinery and ever larger buildings, it has encouraged farmers to seek internal economies of scale by expanding output. In turn, the need for capital to fund business expansion has emphasized the importance of access to credit and encouraged the concentration and accumulation of capital, and proportion of total output, in the lands of a smaller number of larger producers. The distributional consequences of modern agricultural technology are only now receiving the attention they deserve.

Third, these developments have fostered a greater dependence among

Table 3.1 Primary production in the EC: proportion of GDP and working population by country, 1988

Country	% Working population in primary production (a)	% GDP† (b)	Ratio a/b
Portugal	18.9	5.1	3.63
Spain	13.0	5.1	2.54
West Germany	3.9	1.6	2.43
Italy	9.3	4.1	2.27
France	6.4	3.2	2.00
Greece	26.6	16.4	1.62
Denmark	6.0	3.8	1.58
UK	2.2	1.4	1.57
Luxemburg	3.4	2.3	1.47
Ireland	15.1	10.9	1.38
Belgium	2.8	2.2	1.27
The Netherlands	4.7	4.2	1.12
Average of EC	7.0	3.0	2.33

* Agriculture, forestry and fisheries.

† Agriculture alone.

Source: Based on data in Commission of the European Communities (1991).

farmers on external sources of finance capital, especially loans from clearing banks, and industrial capital (credit from input manufacturers, contracts with food processing and retailing companies). As Harvey (1982: 365) argues: 'if producers . . . are heavily indebted to financial institutions for both mortgage payment and credit on current operations, the nominal "owner-occupier" is probably better regarded as a manager or even a labourer who receives a kind of "piecework" share of the total surplus value produced'. Indeed, external capitals appear to have contributed to the declining relative importance of the farming stage within the food chain. As shown further in Chapter 9, the actions of the state have been crucial to these developments. Almost all agricultural programmes have encouraged, directly or indirectly, increased productivity and efficiency while seeking to protect producers from the full rigours of the market. Price support measures have acted differentially. In spite of government rhetoric to the contrary, they have usually favoured large as opposed to small farm businesses, owners rather than tenants (see Traill 1980), and high-income earners rather than low, although the small farm development schemes of Bavaria, the west of Ireland and Scandinavia represent important attempts to ameliorate the general trend. Broadly, the net effect of state policy has been to encourage the capitalization of agriculture, the loss of farm labour, and the specialization of production.

Empirical analyses rarely contradict the general thrust of these points, although they regularly demonstrate the importance of local circumstances in the explanation of geographical patterns. Farming areas continue to reflect the influence of their histories, cultures and natural environments, with all of these contributing to regional differences in farm size and systems of agricultural production, and therefore to particular combinations of resource use. This observation has been confirmed in Fitzsimmons's (1986) study of agriculture in California, and in Gilbert and Akor's (1988) comparison of milk production in Wisconsin and California. Individual farmers and their families respond in a multitude of ways to external pressures, partly out of choice and partly out of necessity. Much depends on their ability to raise capital, their attitude to land purchase, their family structure and their ability to obtain off-farm employment.

Some general considerations

Factors of production may be defined as those ingredients necessary to the production process. Traditionally, they have been recognized as land, labour and capital, although the forms each may take vary considerably. On occasion, a fourth factor – enterprise – has been separately identified in order to encapsulate the risk-bearing and management functions of the entrepreneur. Enterprise will not be examined separately here. One reason for this is the increasingly complex ways in which managerial tasks and responsibilities are distributed among family members, farm managers and even hired workers on large farms, and in a further sense may be said to be divided between farmers, advisers and even the managerial staff of the co-operatives to which many belong. Another is the inevitable mixing of manual and management activities on farms run by sole operators.

The three remaining factors of production are always employed in combination, even if the manner in which they are combined and the relative importance of each differs markedly between farming systems. Indeed, these differences contribute to the debate as to whether agriculture can be sensibly seen as a single industry. Extremes are illustrated in the livestock sector. In the case of farms specializing in intensive, housed livestock production, land is merely a condition of production (see p. 13), providing a site upon which buildings may be constructed and to which large quantities of working capital and labour are applied and feeding-stuffs may be brought. In the case of extensive grazing systems, very small amounts of labour and capital are applied per unit area, often to large areas of poor land realizing low forage yields, but upon which the livestock may be wholly dependent.

Within general limits imposed by financial viability, the actual combination of factors of production on particular farms strays widely from the optimal proportions anticipated by economists' models. Many farmers are not primarily concerned with maximizing profits, according priority to other considerations such as business continuity or risk minimization, and many are not able to optimize the use of their resources even if they wanted or knew how to do so. Certain factors of production are only available in fixed, indivisible quantities. Capital in the form of machinery, for example, comes in fixed amounts (size and operating capacity of the machine); also as the need for labour changes around the year, farmers may not be in a position to hire or dispense with labour to fit their precise seasonal requirements. Generally, farmers with small businesses have less flexibility than large producers over both the divisibility and substitutability of one factor of production for another, which in turn affects their ability to respond to technological change (Walford 1983).

Changes in technology lead to changes in methods of production, enterprise mix and a new optimum combination of factors of production. It is generally accepted that farm business turnover has had to increase over the last 50 years if full advantage were to be obtained from the economies of scale available from a more heavily capitalized industry. But with what implications for resource mix? Research into the relationship between farm size and farming efficiency in the UK, for example, suggests that most scale economies are obtained today on farms with a labour equivalent of no more than two or three full-time workers (Britton and Hill 1975), and that this has been the case for some time. What has changed is that with the increased capitalization of production, the optimum *area* of the farm upon which the two or three workers should be employed has grown several times. At the structural level, farmers can be viewed as being trapped on a 'technological treadmill', as explained on p. 14.

Two other features of technological development should also be noted. First, many aspects of modern technology are interdependent and it is quite usual for farms to find it necessary to invest in further items of fixed or working capital in order to acquire the full economic advantage obtainable from their original investment decision. For example, modern varieties of cereals require extra additions of fertilizers if their full potential is to be realized; new labour-saving machinery may not only lead to reductions in the labour force but at the same time require greater additions of herbicides to offset the loss of hand labour. Secondly, because technological development does not advance evenly across all farming sectors, the technological status of farming will vary between regions even if all producers are experiencing the same general market forces (see Gilbert and Akor 1988). Today, these differences in status more reliably reflect the opportunities and constraints on adoption than the availability of information (Jarrett 1985).

Land

Analyses of the role of land represent an excellent illustration of the different approaches to theorizing the dynamics of agriculture which were identified in Chapter 1. This is because, in practice, it is impossible to separate the issues of production and ownership. The debate surrounding the effects of different land tenure systems on land use and productivity is discussed in Chapter 4, but of wider significance are the strong and even irreconcilable political opinions held about landownership itself. Colin Clark (1973: vii), for example, puts it this way: 'Certain monosyllabic words have the emotional impact of a bullet, with the power to kill dead any rational discussion of the economics of the subject. . . . One of these words is *land*.' At the same time, this quotation reveals Clark's own position. He refers to the need for 'rational discussion of the economics' of land, although elsewhere he acknowledges the wider social and political relations attributable to landownership; but, for him, these considerations come second as their existence 'in no way absolves the economist from his duty of analysing the circumstances under which land is valuable, or less valuable, in the sense that people are or are not willing to pay high rents and prices for it' (1973: viii). Others would reverse the order, discussing first how rights to land have been established, how they are presently distributed and how that distribution corresponds with social class (e.g. Clark 1982; Massey and Catalano 1978; Barlow 1986). These fundamental differences in approach are articulated through the use of particular terminologies, some using the word 'land' to represent an input into the economic functioning of agriculture and others the term 'property' to draw attention to the socially and legally defined rights attaching to land.

In capitalist societies land consists of much more than its topography, drainage and fixed improvements. There are rights associated with its ownership. As Pearce (1980: 118) says: 'it is more correct to refer to the set of land rights in a land parcel as being the object of ownership than merely the land itself as a physical entity', because it is from the ability to exploit the land, as conferred by rights of ownership, that power, financial benefit and pleasure are obtained – and these rights are subject to continuous dispute and redefinition. Indeed, the ability to define and protect private monopoly rights over particular parcels of land provides the basis of land value in a capitalist system. Without the power to alienate the land from other users, the land would be of little value to the owner. Equally essential are the rights to use and improve the land, and freely to sell or transfer to others all or part of the rights attached to it (see Newby et al. 1978; Braden 1982). Several recent trends have increased demands upon the state to restrict certain of the owner's rights. In the agricultural context these include the substantial rise in land prices since 1950, effectively transferring wealth from non-owners to owners, and growing concern over the external public costs arising out of the actions of private owners, most notably in the fields of environmental damage and pollution.

When viewed in its narrower sense, as a factor of production, Ricardo defined land as the 'original and indestructable powers of the soil', although experience shows that soil fertility is exhaustible, land can be destroyed beyond realistic redemption through mismanagement, and poor land can be improved. Sometimes, land is regarded as a 'free good', because it has not required the application of labour and capital to produce it. This is a misunderstanding inasmuch as various forms of improvement – drainage, irrigation, buildings – have usually had to be made in order for it to be brought into cultivation. Likewise, in the short run at least, the supply of farmland may be regarded as fixed and its price determined by demand alone. More precisely, price must be at a level where those who hold the land do not wish to sell (i.e. there is a supply price), while the price of land is not such that additional land will be brought into cultivation. Other things being equal, advancing technology, reducing costs of production and thus raising demand for food, or an increase in the demand for food leading to higher product prices, will tend to expand the area under production and lead to a more intensive use of the land.

Similarly, the area under cultivation may expand or decline depending on the competitive strengths of other land users. Throughout most of the twentieth century the area of farmland has declined in the face of urban expansion. Thus, although land is geographically immobile, its use can change frequently. At one extreme farmers may take two or three crops a year, while at the other tree crops may take 10 years to mature and then have a productive life of a further 30. Flexibility of use depends in part on the technological status of the industry and the land's inherent fertility. Although the law of diminishing returns operates on land of all qualities, generally the more fertile the land the greater will be the optimum level of labour or capital input. Fertile land not only gives higher yields but is also more adaptable.

All these points indicate that land holds a particular significance for agricultural production. In most other productive processes land acts as a *condition* of production, providing the space or site for that activity; but in much of agriculture, land is also a *means* of production entering directly into the process. For most systems of farming the soil itself provides the growth medium, while acting as a store for capital inputs of varying duration, ranging from the ephemeral (chemical nutrients) to the long term (irrigation systems). As land varies in its fertility and in its relative location, these characteristics confer advantages on some parcels of land at the expense of others. The owner, whether a landlord or an occupier, is able to extract different rates of profit from them, independent of the level of capital invested; and exclusive property rights lead to the realization of 'surplus profits' or 'economic rents', the size of these 'profits' or 'rents' being determined by the costs of the marginal producer (see Fig. 3.1).

In Fig. 3.1 'economic rent' is treated as a residual amount under perfect market conditions. Its quantity is only established when the farmer's normal profit and payment made to the landlord to cover the interest charges levied on the fixed improvement he or she has made to the land are deducted. The farmer's rate of profit remains unchanged irrespective of the quality of the land. It is the level of economic rent that goes to the owner which varies. If an owner cannot realize some level of economic rent there will be no incentive to make improvements to the land or even to let it. It is this 'unearned

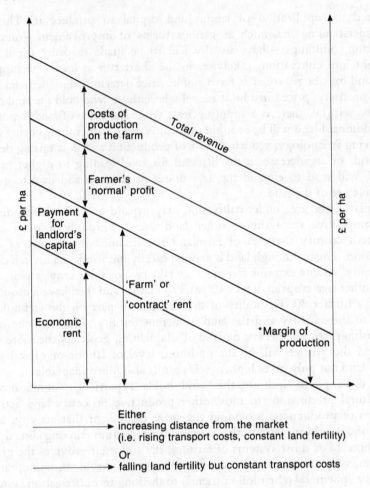

Fig. 3.1 Economic rent.
* assuming that there is an element of 'absolute' rent within economic rent then the margin of production will be some distance to the left on the horizontal line (see text)

element' (or absolute rent) in the return that is most heavily criticized by land reformers, especially in situations of land shortage where farmers may be forced to concede part of their profits in order to gain access to land. It will also be evident that a rise in farmers' revenues will, *ceteris paribus*, increase the area under production; but assuming there is sufficient demand from competent tenants to farm all the additional land that comes into production, then competitive bidding for the land will force up rents to the point that all the increase in revenue will end up in the landowners' pockets as higher economic rents, without reducing the average rate of profit to the farmer. Again, it should be noted that technological change may well change the relative advantage, and thus the economic rents, of particular land types and locations over time (for discussion see Harvey 1982: Ch. 11).

Although the actions of the state are examined in more detail in Chapter 9, no discussion of land would be complete without mention of the wide range of powers taken by the state, even in market economies, to regulate the management, ownership and loss of farmland to other uses. For historical and political reasons the nature and extent of regulation differ significantly between states, and powers are variably distributed between the national, provincial and local states. Almost all land policies seek to curtail the rights of owners and occupiers in order to meet some larger public interest. The most widely enacted policies attempt:

(a) to arbitrate between landlords and tenants over rent levels, compensation for fixed improvements, and security of occupation for tenants;
(b) to avoid the excessive concentration of land in too few hands;
(c) to limit the loss of high-quality agricultural land to other uses;
(d) to prevent the purchase of land by foreigners (Laband 1984) and by absentee landlords such as financial institutions (Munton 1985), and its occupation by those thought unqualified to farm (as in Denmark); and
(e) to protect the amenity of rural areas and to prevent environmental degradation arising from farming practices, such as soil erosion and the pollution of rivers by chemicals.

The range of instruments is equally large. They extend from advice and persuasion, to financial and fiscal penalties and incentives, to land-use zoning, to the pre-emptive right of purchase at the point of sale, and even to extensive programmes of public landownership (these matters are discussed in Jackson 1981; Mather 1986; Rose 1984; Steiner and Theilacker 1984).

Land values and land prices

In a fully developed capitalist economy all rights to land can be freely bought and sold. Market participants treat land as a pure financial asset and purchasers buy the right to obtain a stream of future incomes or rents from a particular use of the land. These future benefits are capitalized back to the present at an appropriate rate of interest to establish the land's current value, and in this way state support for agriculture can contribute significantly to the value of farmland (Traill 1980). This description does not do full justice to the operation of land markets as some of those buying and selling do not restrict their assessments to this single financial criterion, and neither are they necessarily very well informed. Not only are other financial criteria invoked, such as the value of land as an 'inflation proof' store of wealth or as a source of collateral, but many owners retain an emotional commitment to particular pieces of land. 'Pride of ownership', or an altruistic outlook on the welfare of tenants, employees and even local communities may influence decisions (see Newby et al. 1978).

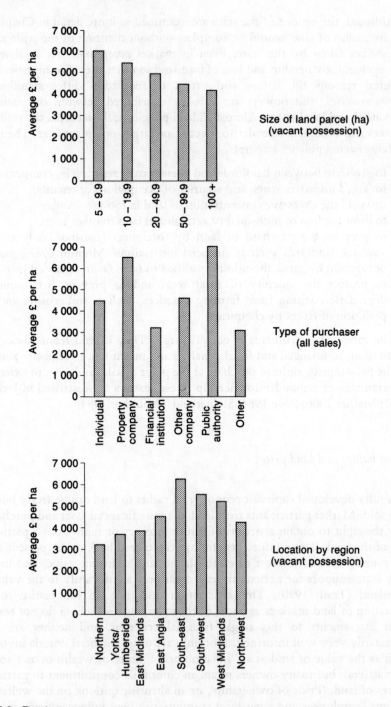

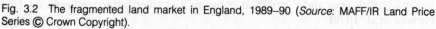

Fig. 3.2 The fragmented land market in England, 1989–90 (*Source*: MAFF/IR Land Price Series © Crown Copyright).

There is, then, no single market in land; the market is fragmented and complex. Not only do properties vary in size, quality, level of fixed investment, location and tenure, but those entering into transactions differ significantly in their motives and thus their assessments of the worth of particular parcels. Figure 3.2, for example, shows two dimensions of the fragmented land market – size of land parcel and type of purchaser – for England during 1989/90. The bid price per hectare of land tends to fall with the increasing size of the land parcel: competition is intense over smaller parcels of land where several potential buyers may feel able to finance the total value of the transaction. Public authorities and property companies are prepared to purchase higher-priced land compared with other purchasers: either because the land is needed in the urban fringe for public utilities and housing, or high-quality land is sought as an investment. Nevertheless, research continues to show that future expected farming profits remain the primary determinant of agricultural land prices (i.e. what is paid, rather than an 'objective' assessment of the land's economic worth).

Except during short periods of market instability most owners of farmland are reluctant to sell irrespective of changes in land price. As a group, farmers are occupationally immobile and wish to see their family names 'kept on the land'. Most sell to acquire additional working capital, to pay tax bills or repay debts. Others are forced into liquidation or sell because they have no successor. But for those wanting to stay in farming there is little point in selling just to realize a capital gain. This consideration may have the reverse effect on the decisions of investors; they treat land as a investment whose performance is continuously compared with other portfolio assets. In all advanced economies, however, the great majority of farmland remains in the hands of farming families, and most changes in the ownership and occupancy of land arise out of succession within the family (Fennell 1981; Harrison 1982; Marsden 1984). Over 80 per cent of farmland transacted in the UK is bought and sold by farmers.

It is often suggested that land prices are too high, by which it is meant they cannot be sustained on the basis of current levels of income or rent. Partly this is because prices are based on future expected returns, and confidence in the future of farming, while other explanations include the effects of high urban development land prices feeding back into the agricultural land market and the purchasing of land by financial institutions and hobby farmers. All these considerations emphasize the importance of seeing land as a store of wealth or capital, as well as a means of generating income, especially during periods of inflation. More particularly, owners use the value of land as collateral against which to borrow. As Shalit and Schmitz (1982: 718) argue: 'the price of farmland is determined not only by the profit it generates (agricultural income and capital gains) but also by the debt it can carry. The latter depends critically on the extent to which the banks are willing to lend farmers money to purchase farmland'; and all the evidence suggests that both bankers and farmers view land as a safe investment resulting in the low returns, by market

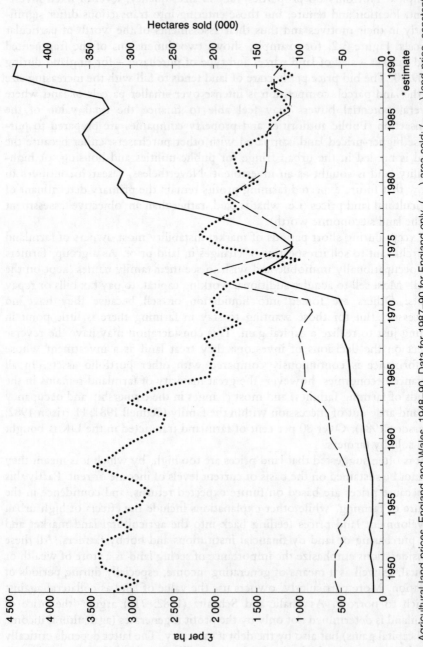

Fig. 3.3 Agricultural land prices, England and Wales, 1948–90. Data for 1987–90 for England only. (...) area sold; (– – –) land price; (——) actual price. * = estimate (*Source*: MAFF/IR Land Price Series © Crown Copyright)

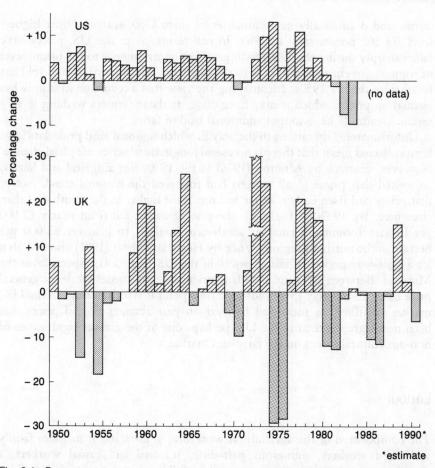

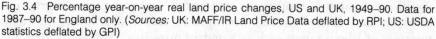

Fig. 3.4 Percentage year-on-year real land price changes, US and UK, 1949–90. Data for 1987–90 for England only. (*Sources:* UK: MAFF/IR Land Price Data deflated by RPI; US: USDA statistics deflated by GPI)

standards, which farmland yields. Throughout advanced capitalist societies, deeply engrained cultural values surrounding the social status of property ownership, held by lenders as well as owners, have ensured that land has not been reduced to a pure financial asset, and to the presumption that land 'will hold its value'. However, the sometimes damaging consequences of an extended involvement in the wider circuits of capital in the economy (see p. 78), that borrowing for land purchase often brings, are today all too evident in farming regions as diverse as the Midwest of the US (*Rural Sociology* 1986) and South Island, New Zealand (Fairweather 1989: 8).

Some evidence on movements in land prices in the US and England and Wales during the recent transformation of agriculture is contained in Fig 3.3 and 3.4. In England and Wales, land prices have risen substantially in real terms, and dramatically in nominal terms since 1950, reaching their highest level (£4 562 per hectare) in 1989. In real terms, as in the US, prices have fallen sharply during the 1980s with growing uncertainty as to the future level of support government will give to the industry. The amount of land sold has fallen since the late 1950s, encouraging the view that a reduction in supply has pushed up prices, which it may have done, as those farmers wishing to stay on the 'treadmill' have sought additional land to farm.

Unfortunately, alterations to the way in which regional land price data have been collected mean that there is no useful longitudinal series of spatial change. However, research by Munton (1976) in the 1970s, for England and Wales, suggested that prices in all regions had followed the national trend, except that prices had risen earlier, faster and remained higher in the South-east than elsewhere. By 1989/90 (Fig. 3.2), the price gradient fell from nearly £7 000 per hectare (nominal terms) in south-east England to just over £3 000 per hectare in the northern region. Work by Healy and Short (1981) also reveals a long post-war period of rising prices in the US (Fig. 3.4), especially in the Midwest. Between 1950 and 1979 land prices rose nine times while the general price index rose barely threefold. One final point is worth noting. In the UK, market volatility, as measured by year-on-year changes in real prices, has been much greater than in the US, perhaps due to the greater significance of non-agricultural forces on the farmland market.

Labour

The composition of the agricultural workforce is intricate. It includes family and hired workers, numerous part-time, seasonal and casual workers, a significant number of women, as well as full- and part-time farmers, farm managers and farm directors. Attempts to reduce this variety to single measures of labour input, such as 'standard man-days' (SMDs) (p. 106), can be misleading. They draw attention away from the economic dependencies that exist between these different groups and obscure the social context in which farm work takes place.

Throughout the 'third agricultural revolution' the numbers working in agriculture have declined substantially, whether hired workers, family members or farmers, even if the rate and timing of the decline have varied between advanced economies. This net outflow of labour is related to the inflow of industrial capital in the form of labour-saving machinery, buildings and chemicals, and is little associated with changes in farming prosperity. Outflow is much more a function of the level of job opportunities elsewhere within the economy.

During the 1970s the rate of decline in labour slowed with the onset of world recession. In the US the number of full-time hired workers actually remained stable, and as a proportion of full-time workers rose from 12.3 to

16.0 per cent between 1956 and 1979 (Perry 1982). In Western Europe, on the other hand, there has been no stabilization in the numbers of full-time hired workers. The annual rate of drift from the land in the EC was 4.5 per cent in the 1960s, falling to 3.4 per cent in the later 1970s and 3.1 per cent between 1980 and 1989. The comparable figures for the UK, with its sluggish economy, are 3.5, 1.8 and 1.4 per cent respectively. Detailed information on change in the composition of the farm workforce in England and Wales is given in Table 3.2. Total farm labour fell by 15.2 per cent between 1977 and 1989. The decline among farmers is significantly less than among hired workers, and greatest among whole-time, female family workers. One reason for this pattern is the growth in number of recorded partners and directors, with the increasingly complex business structures of family farms helping to explain the greater relative decline among family workers. The numbers of part-time farmers rose by 13 per cent, while part-time and seasonal workers declined by between 12.5 and 15.3 per cent (Table 3.2). The increased relative importance of part-time and casual labour in the workforce, partly because of the seasonality of requirements (see below), is an important trend (Ball 1987). Labour *input* fell by even more than is implied by the statistics, a further 11 per cent in the case of full-time hired workers as a result of a shortening of the working week, although Errington (1988) argues that in areas of high unemployment there may be disguised under-employment among family members who might otherwise work off-farm.

Figures such as these also do not do justice to the full effects of the industrialization of agriculture. For example, by reducing the importance of the farm stage in food production, it has transferred agricultural employment off the farm and beyond conventional agricultural statistics. Put bluntly, it has in effect made farm workers out of people who refine gasoline, make pesticides, or build machinery (Errington and Harrison 1990). Furthermore,

Table 3.2   Percentage changes in the composition of the farm workforce in the UK, 1977–89*

Farmers, partners and directors*		Family workers		Hired workers	
Whole-time farmers	−10.8	Whole-time male	−24.3	Whole-time male	−41.5
Part-time farmers	+13.1	Whole-time female	−47.4	Whole-time female	−9.1
Spouses (as directors)	−3.8	Part-time male	−22.1	Part-time male	−14.7
Farm managers	0	Part-time female	−26.3	Part-time female	−15.3
				Seasonal/casual	−12.5
Overall change	− 4.0		−26.7		−26.8

* The total workforce fell by 15.2%, from 755 000 to 640 000, over the period.
† Farmers, partners and directors made up 51% of the workforce in 1977, increasing to 58% in 1989; the proportion of family workers declined from 9 to 8%, and hired workers from 40 to 34% of the total.

Source: Calculated from agricultural census data.

Lund et al. (1982) estimate that perhaps 43 000 agricultural contract workers (or 7 per cent of the farm labour force) are not returned in the UK agricultural statistics. An even higher rate of omission exists for temporary, migrant labour, especially where illegal and unregistered immigrants are concerned as in southern California. More generally, modern farming technology has simplified many tasks making them more repetitive – a trend reinforced by the increased specialization of production on farms – counteracting the effect of reduced numbers of workers on all but the largest farms, a development that requires the remaining workers to acquire an increased range of skills. On family farms there are few opportunities for specialization and the farmer has had to *extend* the range of his skills in order to cope with both more complex accounting methods and with more sophisticated machinery. In so far as these skills cannot be learned by members of the family, additional opportunities have arisen for specialist, contracting firms, bringing some of the advantages of new technology to smaller businesses but also taking some of the decision-making and supervisory responsibility out of the hands of the farmer. In the case of the large agribusiness, specialized management, scientific and professional skills are now required of staff, helping to explain the existence of a growing managerial class.

Farm work, rewards and labour mobility

Farm work varies largely according to system of production. Some systems demand specific skills, such as milking, require tasks to be carried out in trying outdoor conditions, or include boring, repetitive operations; and technical change continues to reduce labour requirements (see Table 3.3).

Table 3.3 Labour requirements in British agriculture, 1971–91

Enterprise (hrs per ha per yr)	1971		1983		1991	
	Average	Premium*	Average	Premium	Average	Premium
Winter wheat†	18.5	12.4	13.5	9.1	13.1	9.1
Spring barley†	17.8	11.6	12.4	8.5	12.7	8.5
Maincrop potatoes‡	90.9	69.5	72.5	50.5	71.2	50.1
Sugar-beet	143.3	82.5	67.3	29.7	49.5	29.7
Dairying§ (hrs per cow per yr)	54.0	—	40.0	—	36.0	—

* Premium figures refer to the top quartile of farms where labour input is least. Farms with more than 250 ha of arable land in 1983/91 with big equipment would reduce the premium figures by a further 20%.
† Excludes bailing and carting of straw.
‡ Field operations only.
§ Figures for 60-cow herd. A 100-cow herd would lower the 1983/91 figures by 25%. Figures do not include fieldwork, such as hay and silage making.

Source: Based on figures in Nix (various dates).

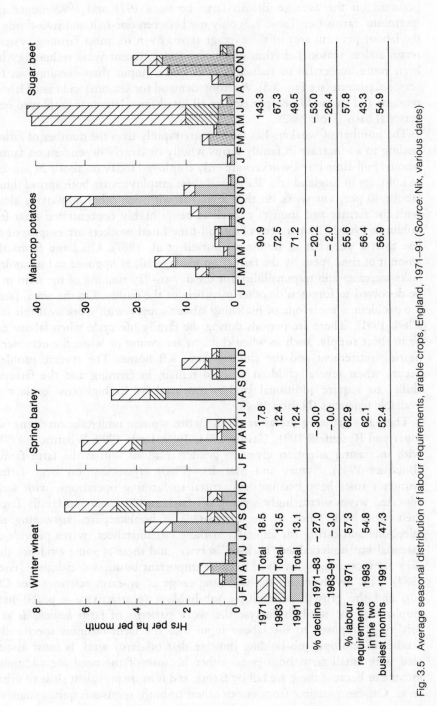

Fig. 3.5 Average seasonal distribution of labour requirements, arable crops, England, 1971–91 (*Source:* Nix, various dates)

Labour requirements for the main arable enterprise fell by between 20 and 50 per cent on the average British farm between 1971 and 1983 alone, and 'premium' farms (see Table 3.3) only use between one-half and two-thirds of the labour per unit area of the average farm. Even so, most farming systems retain a clear seasonal rhythm of activity and in recent years technology has been more successful in reducing total labour input than evening out the seasonal pattern (see Fig. 3.5). A strong demand for seasonal and casual labour remains and in the US 85 per cent of all hired farm labour is employed on a seasonal basis (Holt 1982).

The number of workers has fallen more rapidly than the number of farms, leading to an increase in family farms wholly or largely dependent on family labour. Full-time hired workers are only employed today on about 26 per cent of holdings in England and Wales, and these employees are both spread thinly (about 40 per cent work on their own, or with one other employee, along with the farmer and spouse) as well as being highly concentrated on a few holdings. Nearly 25 per cent of all full-time hired workers are employed on less than 1 per cent of holdings (Burrell et al. 1987). On large farms the amount of time spent by the farmer on managerial, as opposed to husbandry, tasks increases and responsibility for the day-to-day running of the farm may be devolved to foremen or other members of the family. On the small farm, the problem is more one of matching labour supply with work available (see Clark 1991). There are periods during the family life cycle when labour can be in short supply, such as when children are young or when the occupier is nearing retirement and the children have left home. The reverse problem occurs when several children wish to remain in farming and the farmer's ability to acquire additional land is restricted by its high cost or lack of available tenancies (Nalson 1968).

There is a growing literature on the work women undertake on farms (see Jones and Rosenfeld 1981; Gasson 1984, 1989; Little 1990; Whatmore 1991), with increasing attention given to gender relations within the farm family (Bouquet 1985). Symes and Marsden (1983) argue that on large farms, women's roles have become less central to farming operations, with well-educated wives increasingly seeking off-farm employment; on family farms their work still tends to be dominated by bookkeeping, answering the telephone and caring for animals. In these circumstances, wives provide an essential but undervalued 'back-up' service, and there is some evidence that they are increasingly involved in more important business decisions (Gasson 1984). This involvement, the widening range of on-farm activities (see Ch. 10), and the growth of multiple job-holding (pluriactivity – p. 23) have emphasized the need to analyse the work patterns of farm *households* as a totality as opposed to the labour input into the farm business specifically. Studies of multiple job-holding indicate that off-farm work is most associated with small farm businesses, either because of the need for additional income or because these are hobby farms, and is more prevalent close to urban areas. Off-farm income from one or other, or both, spouses is quite extensive, and if the farmer has to hold down an off-farm job, as well as doing farm work, the spouse often contributes substantially to the running of the farm. Gasson's (1986) nationwide survey of England and Wales suggests that off-farm work is important to 30 per cent of farm families, while Buttel and Gillespie (1984), working in the urbanized environment of New York State, established that in over half of farm families either the husband or wife had an off-farm job, and in 14.4 per cent of cases both worked off the farm (see also Buttel 1982). All these sample studies indicate the growing complexity of work patterns within farm households, although Gasson's longitudinal surveys suggest that most farmers remain either in full or part-time farming throughout their working lives (see also Barlett 1986).

As small businesses, farms are often assumed to represent intimate workplaces where tensions between employer and employee are limited to what Newby (1977: 309) describes as 'We have our ups and downs, but . . .', followed by a number of analogies drawn with marriage. Except on the largest of farms, the work situation is still unlike that in factories, with close working relations between employer and employee. At one level, the hired worker and farmer may express similar advantages and disadvantages from working in the farming industry (compare Newby et al. 1978: 153–6, with Newby 1977: 290–3), reflecting a common pride in their work; but at another, their interests are quite different. Farmers are the employers and suppliers of capital who talk of their independence as entrepeneurs and the fact that they are risk takers. They expect, and normally obtain, a greater financial reward for their efforts than the hired worker, and they are more likely to cling to the industry.

> Farmers are the least mobile group in agriculture: they have the greatest security in the industry and therefore the greatest commitment to it. On average, they are also appreciably older than the other employment status groups, which adds to their immobility both in terms of their own attitudes to non-farm work and their ability to train for another non-farm occupation (Fennell 1981: 22).

The position of family workers is more ambiguous, depending on their age, experience and prospects.

Farm work has a number of other unusual characteristics, not least that many farmers and workers live on the job and in tied housing. The small size of the business can make the farmer dependent on the skills and commitment of a particular employee, although this dependence rarely extends to migrant or seasonal workers who often fall outside labour laws designed to protect their conditions of employment. Unionization among farm employees has always been limited and no more so than among casual workers. Geographically, the distribution of casual workers mirrors the pattern of labour-intensive farming enterprises, especially fruit and vegetable production. Very largely, however, it is the hired worker rather than the farmer who is economically dependent, and Newby (1977) argues that it is this dependence, rather than a simple love of the job, which has been the traditional basis of the worker's quiescence in the face of low wage levels by industrial standards (usually in the range 50–80 per cent of industrial wages).

Low wages and the availability of alternative employment opportunities are the main determinants of the spatial pattern of job decline in agriculture. Farm wages are higher and labour loss is greater close to urban areas, and even in the rural areas of metropolitan regions those employed in agriculture make up a small and declining minority of those in work. Research by Newby (1977), among others, indicates that not only are older farm workers repelled by the idea of factory work but many also feel trapped in their present jobs by a lack of transferable skills, age or family considerations. The same arguments also apply to many middle-aged occupiers of small farms, their reluctance to leave agriculture being further buttressed by feelings of independence and a strong commitment to continuity of the family concern (Gasson 1969). These attitudes are not always reflected in the aspirations of their children. Children are often more perceptive and realistic as to the long-term economic viability of the farm, at least as a source of full-time employment, and a reluctance on their part to succeed helps to explain the high average age of farmers on small units.

Capital

Strictly speaking, capital represents those man-made inputs deliberately created to aid production. As such they range from seeds and fertilizers to machinery and buildings, to farm roads and drainage systems. It also includes 'stocks' of raw materials such as hay, wheat in store and breeding animals. Conventionally, it is divided into fixed items of capital, such as buildings, or working capital such as machinery and pesticides. Under a landlord–tenant system of tenure, the landlord is normally responsible for providing the land and fixed investment and the tenant working capital, although today, in a search for flexibility, this hard and fast distinction is less regularly adhered to. The amount of working capital deployed depends on the technological status of the industry and will be influenced by government incentives (capital grants or tax reliefs) to invest. Even at a national scale there are major differences in the amount of working capital used in the farming industries of the EC (Table 3.4). Too much should not be made of the precise figures as there are difficulties in assessing the real worth of capital assets and conventions differ. The value of the existing stock constantly depreciates and rules, which are always open to dispute, have to be agreed for the rate at which particular items of capital should be written down. None the less, balance sheets provide orders of magnitude and one starting-point for analysing national, regional and farm-level differences in the resources structure of farming.

Because of the way land enters into the production process, it is difficult to divorce it from an assessment of capital in farming. At a minimum landowners

Table 3.4 Amount of working capital on farms in the EC, 1981

Country	Working capital (ECU per ha)	Index
The Netherlands	4 500	100
West Germany	3 400	76
Italy	3 100	69
Belgium	2 700	60
Denmark	2 450	54
Luxemburg	2 300	51
France	2 010	45
UK	1 850	41
Greece	1 750	39
Ireland	1 010	32

Source: Based on data in Commission of the European Communities (1985) (comparable data not available for later years).

regard land as a capital asset against which they can borrow; Harvey (1982: 347) argues that with the full development of capitalist relations of production, to which land in particular is resistant (see Ch. 1), 'land becomes a form of fictitious capital, and the land market functions simply as a particular branch . . . of the circulation of interest-bearing capital'. At an empirical level, it is also impossible to ignore land as it holds a pre-eminent position within the balance sheets. In the case of the UK, land and buildings constitute approximately 70 per cent of the industry's assets by value (Johnson 1990), although only 30 years ago the proportion was closer to 50 per cent. Clearly, major differences exist between farms, depending on whether the occupier owns the land and, in the case of working capital, its quantity and form according to system of production.

Measurement of changes to the amount of capital invested per annum is usually expressed as gross capital formation, part of which covers depreciation of the existing stock and part (which may be a negative figure) net gains or losses to that stock. Because of the difficulties of measurement, it would be of considerable benefit to have regular inventories of capital items on farms, such as pieces of machinery or plant. Even then problems would remain. Information would be needed on the age and condition of machines and buildings, and some estimate of their working capacity. For example, over the last two decades there has been a reduction in the number of pieces of cultivating equipment on British farms but an increase in their total work capacity.

In spite of these measurement difficulties, some observations can be made on investment in working capital. First, rates of return differ considerably between systems of production, depending very largely on their relative profitability, between one year and the next due to the effects of weather, and between the efficient and less efficient farmer. Returns to working capital are not so much low as highly variable, but on the whole small businesses achieve lower rates of return than large ones because of the small farmer's inability to obtain the same economies of scale (Hill and Ray 1987: Ch. 12). Second, the amount of capital invested per unit area depends to some degree on land tenure. Owner-occupiers invest more heavily in buildings than tenants, indicative of their greater security of occupation. One result is that they tend to plan over longer time periods and have a greater capacity to borrow. On the other hand, tenants often invest more in working capital, so that the productivity of rented and owner-occupied land is not always that different. There is also some evidence that the occupiers of mixed-tenure farms are among the most dynamic of entrepreneurs, making a high use of working capital (Hill and Gasson 1985).

Capital investment is subject to the effects of inflation, changes in the rate of interest on borrowed money, and is encouraged or discouraged by specific agricultural policy measures and general changes in taxation. For example, capital deployment is influenced by the availability of capital grants and fiscal policies relating to capital taxation, such as inheritance taxes, and income tax. Some of these measures may be changed to meet wider national economic objectives and may be introduced without much appreciation of how they will affect the farming industry. In short, the agricultural consequences of these numerous, and sometimes conflicting policy measures, are largely specific to particular countries which means international comparisions need to be made with care. More generally, it is often argued that government policy favours the large producer with high marginal rates of taxation and that grants and capital allowances 'cheapen' the cost of capital relative to labour. Likewise, most economists believe that tax concessions are largely capitalized back into higher land prices, while rising farm incomes encourage investment in working capital, although this investment may be undertaken for reasons of fiscal gain or tax avoidance rather than economic efficiency. Nevertheless, in all advanced economies, capital investment has resulted in startling increases in output per hectare and per worker. In the case of the UK, the value of gross product per whole-time equivalent has risen by more than 5 per cent per annum at constant prices since 1971, while assets per worker have risen at a rate in excess of 13 per cent per annum.

Credit and debt

The demand for capital by technologically advanced farming systems has turned attention to sources of credit (see Fig. 1.1) and, more recently, as farming incomes have declined, to the level of debt among farmers (see e.g. Marsden et al. 1990; Mooney 1986). Traditionally, farmers as opposed to landlords, were reluctant to borrow and did so whenever possible from other members of their families. They tried to find sufficient capital for reinvestment from retained profits, or from extended credit from trade suppliers. In 1953, 74 per cent of liabilities of UK farmers were owed to trade sources or were private mortgages, only 3 per cent were owed to long-term institutional lenders. The remaining 23 per cent had been borrowed from the clearing banks (Burrell et al. 1987: 64). Even as late as 1969, Harrison (1975) discovered that 35 per cent of UK farm businesses carried no liabilities beyond unsecured advances from trade suppliers. Liabilities amounted to 16.9 per cent of assets on tenanted farms and about 10 per cent on other farms. The difference in percentage was accounted for by the asset value of the land and not by differences in attitude towards borrowing. Liabilities declined as farmers got older, in an equity cycle that paralleled the family cycle, and were heavily concentrated among those who had recently purchased land (see Reid 1981; Moran 1988).

Increasingly, it has proved impossible for farmers to meet their borrowing needs in this way, leading to a greater dependence on sources of finance capital. Lending from trade and family sources has not fallen in amount, but by 1986 it represented only 23 per cent of all farm borrowing in the UK; institutional sources were owed 14 per cent and the clearing banks a massive 63 per cent (see Marsden et al. 1990: 42). As a result of falling land values, liabilities have increased from 9 to 60 per cent of total assets and have increased substantially in real terms (Table 3.5). Much more significant, therefore, is their distribution and, with declining farm incomes, the ability of farmers to service the loans. Global figures disguise major differences in the security of individual businesses. Again, in spite of falling land prices, owner-occupiers retain an advantage over tenants in that they usually have some property assets they can sell to reduce the drain of interest repayments on business cash flow and incomes. In all situations, however, the focus of risk management is no longer on controlling or managing variability in income, but protection from failure or termination of the firm and on the cash flow and liquidity characteristics of a firm and its asset base (Boehlje and Eidman 1983: 938).

Table 3.5   Farm income and debt indicators in real terms (UK)*

Year	Farming income†	Cash flow‡	Total assets	Total liabilities
1975	186	180	79	86
1977	175	120	97	88
1979	130	109	120	103
1981	116	115	91	102
1983	107	89	94	116
1985	73	86	74	124
1987§	88	103	71¶	114¶

* 1980 = 100, deflated by the Retail Price Index.
† The return to farmers and spouses for their labour, management skills and own capital invested after providing for depreciation.
‡ Spending on material inputs and services and on capital items.
§ Forecast.
¶ 1986.

Source: Based on data in Marsden et al. (1990: 39, 41).

Borrowing rose during the 1970s in spite of high interest rates, partly because interest repayments could be set off against taxable income and partly because inflation reduced the impact of repayments on farming incomes – while the latter continued to rise. The position of the late 1970s and 1980s was very different. Some of the salient features of the British situation are described in detail in Fig. 3.6. For example, on income gearing, while interest payments absorbed only 9 per cent of farm business income in 1976, the proportion had risen to 31 per cent by 1985. These broad trends were replicated in the agricultural economies of many advanced capitalist countries. Falling farm incomes and the return of positive rates of interest reduced the cash flows of many businesses and increased the number of farm bankruptcies. According to Shephard and Collins (1982), by 1978, bankruptcies in the US had risen to 26.3 per 100 000 farms, compared with 3.8 per 100 000 farms in 1949. Liabilities had also risen to 17.5 per cent of assets in 1978 compared to 11 per cent in 1946, in spite of rising land values. Failure faced large farmers nearly as often as it did small farmers, because many had borrowed much more heavily during the 1970s. By early 1985, 5 per cent of Midwest farms were going bankrupt each year, contributing to the collapse of the International Harvester Company, and the figure could have reached 15 per cent but for the rescheduling of loans. Moreover, with its fragmented banking system, some rural banks in the US are highly dependent upon agricultural business and they too have experienced closure where they had loaned money too freely against rising land prices. Between 1983 and 1985, 107 banks specializing in agricultural business had collapsed, 64 in the Corn Belt and Northern Plains alone (Gajewski 1986).

To varying degrees this situation is replicated within the EC, although the more concentrated and diversified banking system is not threatened by the misfortunes of farmers to anything like the same degree as in the US. By way of example, Crédit Agricole, traditionally the farmers' bank in France, now makes less than one-third of its loans to the farming sector. However, farm incomes are falling and interest payments as a proportion of income are rising as borrowing continues to grow. In the UK interest repayments were the equivalent of 25 per cent of farming income by the early 1980s, compared with less than 10 per cent in the early 1970s (see Fig. 3.6), while in the Netherlands the figure for 1981 was 19 per cent and in Denmark a massive 58 per cent (Commission of the European Communities 1985). Interest repayments also vary widely, although they have increased everywhere, such that by 1985 they represented a staggering 47 and 33 per cent of total farm outgoings on farms in Denmark and the Netherlands respectively, but only 5 per cent in Ireland and 2 per cent in Italy.

The key points to draw out of this discussion are the growing dependence of agriculture on the movement and cost of capital within the economy as a whole; the growing penetration of the industry by finance capital; and the problems that are being created, not least for some industrial companies in the food chain and some rural banks, by the increasing cost of working capital at

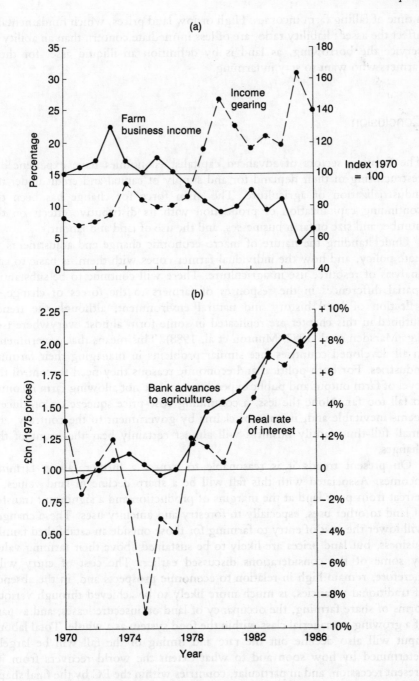

Fig. 3.6   Trends in UK agricultural incomes and borrowing, 1970–86. (a) Index of real farm business income (1970–100) and income gearing (%) ; (b) bank advances to agriculture and real rate of interest. (Changes in definition of data after 1986.)

a time of falling farm incomes. High or low land prices, which fundamentally affect the asset : liability ratio, are of less immediate concern than an ability to service the borrowing, as land is by definition an illiquid asset for those farmers who want to stay in farming.

Conclusion

The farming sectors of advanced capitalist economies have experienced a restructuring of their demand for and supply of capital and credit under the industrialization of agriculture. The main force for change has been the continuing capitalization of production with its distributive effects on the number and size of farm businesses, and the use of land and labour.

Understanding the nature of macro-economic change and adjustments to state policy, and how the individual farmer copes with them, is basic to any analysis of resource use in agriculture. There will continue to be substantial spatial differences in the responses of farmers to the forces of change, a reflection of local history and natural environment, although the trends outlined in this chapter are replicated in some form almost everywhere (see e.g. Marsden et al. 1987; Munton et al. 1988). This means that governments in all developed countries face similar problems in managing their farming industries. For both political and economic reasons they need to control the level of farm output and public expenditure, while not allowing farm incomes to fall too far. None the less, a continuing cost–price squeeze on producers seems inevitable and, in spite of claims by government to the contrary, the small full-time family business will almost certainly bear the brunt of the changes.

On present trends it is reasonable to expect a further fall in farming incomes. Associated with this fall will be a sharp decline in land values, a retreat from poor land at the margins of production and a significant transfer of land to other uses, especially to forestry and amenity uses. These changes will lower the cost of entry to farming for those outside an established family business, but land prices are likely to be sustained above their farming value by some of the considerations discussed earlier. The cost of entry will, therefore, remain high in relation to economic prospects and, in the absence of traditional tenancies, is much more likely to be achieved through various forms of share farming, the occupancy of land on insecure leases, and as part of a growing managerial class within the food system as a whole. Total labour input will also decline but the rate and timing of the fall will be largely determined by how soon and to what extent the world recovers from its present recession, and in particular, countries within the EC by the final shape of the MacSharry reforms for the CAP. In any event, if the North American experience is replicated in Europe, part-time and seasonal farm work for families interested in living on the land will grow. For those operating full-time businesses, a continuing cost–price squeeze will not reduce the need to borrow. Borrowing will reflect the trajectory of the 'technological treadmill', favouring the large producer with access to credit and the necessary technological expertise and marketing skills as the industry becomes more exposed to competitive market forces.

One of the most significant by-products of the rising financial problems of the agricultural sector has been the increased power of banking capital. Input supply firms, for example, have become engaged with financial institutions in order to provide credit packages designed to maintain market share. The main banks, faced by a rising level of farm bankruptcies, have become far more selective in their loan arrangements: they now favour farm businesses which are seen as 'productive' and 'efficient'. This narrow definition of efficiency will promote those farm businesses following the industrial model of agricultural production and so strengthen the divergent paths of development within the farm sector.

In the long run, the most important trend is the declining ability of the farming industry itself to determine its own future. While not wishing to underestimate the residual political influence of the farm lobby, its power is waning and agricultural production is becoming increasingly dependent on the financial and industrial sectors of the urban economy, both within and beyond the food chain; in the countryside, growing indebtedness, concern for the environment, competition for labour from new industries, and increased production on contract provide the entrepreneurial farmer with new economic opportunities, while at the same time reducing the degree to which farmers and their families are on their own able to decide the allocation and management of farm resources.

References

Ball, R. M. (1987) Intermittent labour forms in UK agriculture: some implications of rural areas. *Journal of Rural Studies* **3**: 133–50

Barlett, P. F. (1986) Part-time farming: saving the farm or saving the lifestyle? *Rural Sociology* **51**: 289–313

Barlow, J. (1986) Landowners, property ownership and the rural locality. *International Journal of Urban and Regional Research* **10**: 309–29

Boehlje, M. and **Eidman, V.** (1983) Financial stress in agriculture: implications for producers. *American Journal of Agricultural Economics* **65**: 937–44

Bouquet, M. (1985) *Servants and visitors: the farm household in nineteenth and twentieth century Devon.* Geo Books, Norwich

Braden, J. B. (1982) Some emerging rights in agricultural land. *American Journal of Agricultural Economics* **64**: 19–27

Britton, D. K. and **Hill, B.** (1975) *Size and efficiency in farming.* Saxon House, Farnborough

Burrell, A. M., Hill, B. and **Medland, J.** (1987) *A statistical handbook of UK agriculture*. Macmillan, London

Buttel, F. H. (1982) The political economy of part-time farming. *GeoJournal* **6**: 293–300

Buttel, F. H. and **Gillespie, G. W.** (1984) The sexual division of farm household labour. *Rural Sociology* **49**: 183–209

Clark, C. (1973) *The value of agricultural land*. Pergamon, Oxford

Clark G. (1991) People working in farming: the changing nature of farm work. In Watkins C. and Champion A. (eds) *People in the countryside: Social change in rural Britain*. Paul Chapman Publishing, London, pp 67–83

Clark, G. L. (1982) Rights, property and the community. *Economic Geography* **58**: 120–38

Commission of the European Communities (1985) *The agricultural situation in the Community: 1985 report*. The Commission, Luxemburg

Errington, A. (1988) Disguised unemployment in British agriculture. *Journal of Rural Studies* **4**: 1–7

Errington, A. and **Harrison, L.** (1990) Employment in the food system. In Marsden, T. and Little, J. (eds) *Political, social and economic perspectives on the international food system*. Avebury, Aldershot, pp 115–40

Fairweather, J. R. (1989) *Some recent changes in rural society in New Zealand*. Discussion Paper 124. Agribusiness and Economics Research Unit, Lincoln College, Canterbury (NZ)

Fennell, R. (1981) Farm succession in the European Community. *Sociologia Ruralis* **21**: 19–42

Fitzsimmons, M. (1986) The new industrial agriculture: the regional integration of speciality crop production. *Economic Geography* **62**: 334–53

Gajewski, G. (1986) Rural bank failures not a big problem – so far. *Rural Development Perspectives* June: 1–8

Gasson, R. (1969) Occupational immobility of small farmers. *Journal of Agricultural Economics* **20**: 179–88

Gasson, R. (1984) Farm women in Europe: their need for off-farm employment. *Sociologia Ruralis* **24**: 216–28

Gasson, R. (1986) Part-time farming: strategy for survival? *Sociologia Ruralis* **26**: 364–76

Gasson, R. (1989) *Farmwork by farmers' wives*. Farm Business Unit, Department of Agricultural Economics, Wye College, Kent

Gilbert, J. and **Akor, R.** (1988) Increasing structural divergence in US dairying: California and Wisconsin since 1950. *Rural Sociology* **53**: 56–72

Harrison, A. (1975) *Farmers and farm businesses in England*. Misc. Study 62, Department of Agricultural Economics, University of Reading, Reading

Harrison, A. (1982) Factors influencing ownership, tenancy, mobility and use of farmland in the member states of the European Community. *Information on Agriculture* **86**, The Commission: Luxemburg

Harvey, D. W. (1982) *The limits to capital*. Blackwell, Oxford

Healy, R. W. and **Short, J. L.** (1981) *The market for rural land: trends, issues, policies.* Conservation Foundation, Washington DC

Hill, B. and **Gasson, R.** (1985) Farm tenure and farming practice. *Journal of Agricultural Economics* **36**: 187–200

Hill, B. and **Ray, D.** (1987) *Economics for agriculture: food, farming and the rural economy.* Macmillan, London

Holt, J. S. (1982) Labour market policies and institutions in an industrializing agriculture. *American Journal of Agricultural Economics* **64**: 999–1006

Jackson, R. H. (1981) *Land use in America.* Edward Arnold, London

Jarrett, F. G. (1985) Sources and models of agricultural innovation in developed and developing countries. *Agricultural Administration* **18**: 217–34

Johnson, C. (1990) Farmland as a business asset. *Journal of Agricultural Economics* **41**: 135–48

Jones, C. and **Rosenfeld R. A.** (1981) *American farm women: findings from a national survey.* National Opinion Research Centre, University of Chicago

Laband, D. N. (1984) *Foreign ownership of US farmland.* Lexington Books, Lexington, Mass

Little, J. (1990) Women's employment in the food system. In Marsden, T. and Little, J. (eds) *Political, social and economic perspectives on the international food system.* Avebury, Aldershot, pp 141–56

Lund, P. J., Morris, T. H., Temple, J. D. and **Watson, J. M.** (1982) *Wages and employment in agriculture: England and Wales: 1960–1980.* MAFF, London

Marsden, T. (1984) Capitalist farming and the farm family: a case study. *Sociology* **18**: 205–24

Marsden, T., Whatmore, S. and **Munton, R.** (1987) Uneven development and the restructuring process in British agriculture: a preliminary exploration. *Journal of Rural Studies* **3**: 297–308

Marsden, T., Whatmore, S. and **Munton, R.** (1990) The role of banking capital and credit relations in British food production. In Marsden, T. and Little, J. (eds) *Political, social and economic perspectives on the international food system.* Avebury, Aldershot, pp 36–56

Massey, D. W. and **Catalano, A.** (1978) *Capital and land: landownership by capital in Great Britain.* Edward Arnold, London

Mather, A. S. (1986) *Land use.* Longman, London

Mooney, P. H. (1986) The political economy of credit in American agriculture. *Rural Sociology* **51**: 449–70

Moran, W. (1988) The farm equity cycle and enterprise choice. *Geographical Analysis* **20**: 84–91

Munton, R. (1976) An analysis of price trends in the agricultural land market of England and Wales. *Tijdschrift voor Economische en Sociale Geografie* **67**: 202–12

Munton, R. (1985) Investment in British agriculture by the financial institutions. *Sociologia Ruralis* **25**: 155–73

Munton, R., Whatmore S. and **Marsden, T.** (1988) Reconsidering urban-fringe agriculture: a longitudinal analysis of capital restructuring on farms in the Metropolitan Green Belt. *Transactions of the Institute of British Geographers* **13**: 324–36

Nalson, J. (1968) *The mobility of farm families.* Manchester University Press, Manchester

Newby, H. (1977) *The deferential worker: a study of farm workers in East Anglia.* Allen Lane, London

Newby, H., Bell, C., Rose, D. and **Saunders, P.** (1978) *Property, paternalism and power: class control in rural England.* Hutchinson, London

Nix, J. S. (various dates) *Farm management pocket book.* Wye College, Kent

Pearce, B. J. (1980) Instruments for land policy: a classification. *Urban Law and Policy* **3**: 115–56

Perry, C. S. (1982) The rationalisation of U.S. farm labor: trends between 1956 and 1979. *Rural Sociology* **47**: 670–91

Reid, I. G. (1981) Farm finance and farm indebtedness in the EEC. *Journal of Agricultural Economics* **32**: 265–74

Rose, J. G. (1984) Farmland preservation: policy and progress. *Natural Resources Journal* **24**: 591–640

Rural Sociology (1986) The farm crisis issue. *Rural Sociology* **51**: 391–512

Shalit, H. and **Schmitz, A.** (1982) Farmland accumulation and prices. *American Journal of Agricultural Economics* **64**: 710–19

Shephard, L. E. and **Collins, R. A.** (1982) Why do farms fail? Farm bankruptcies, 1910–1978. *American Journal of Agricultural Economics* **64**: 609–15

Steiner, F. R. and **Theilacker, J. E.** (eds) (1984) *Protecting farmlands.* AVI Pub Co, Westport, Connecticut

Symes, D. E. and **Marsden, T. K.** (1983) Complementary roles and asymmetrical lives. Farmers' wives in a large farm environment. *Sociologia Ruralis* **23**: 229–41

Traill, B. (1980) An empirical model of the UK land market and the impact of price policy on land values and rents. *European Review of Agricultural Economics* **6**: 209–32

Walford, N. (1983) The future size of farms: modelling the effect of change in labour and machinery. *Journal of Agricultural Economics* **34**: 407–16

Whatmore, S. (1991) *Farming women: gender, work and family enterprise.* Macmillan, London

4

The agricultural significance of farm size and land tenure

Ian Bowler

Recent research on the transformation (industrialization) of agriculture has laid emphasis on 'variations and unevenness' (Marsden 1988) in the sources and mediation of the capital penetration process. This perspective directs attention to the constraints which land-based production places on industrial capital, the competitive movement of 'fractions' of capital within the food chain, and the process of adjustment within individual farm businesses. In this chapter attention is directed to two of the mediating factors in the transformation of individual farm businesses – namely farm size and land tenure (also termed 'land occupancy'). The following argument shows that (a) size and tenure vary by farm and over space, (b) the two factors mediate in the processes of transformation and, consequently (c) they help shape the spatially uneven industrialization of agriculture as the farm sector has become integrated into the wider food supply system (see Fig. 1.1).

Spatial variation in farm-size structure

Conventionally farm size is measured by the land area (acres or hectares), but as agriculture has been transformed by capital investment in new technology, so financial definitions of 'business size' have been introduced. Gregor (1982a), for example, has measured regional variations in the average business size of farms in the US using the amount of capital invested in land, buildings, labour, machinery, equipment, crops and livestock. Using this measure (see pp. 128–30 for further discussion), the greatest concentrations of large farm businesses in the US are found in California (the counties of Imperial, Kern, Monterey, Yolo and Kings), Arizona (Pinal, Yuma, Maricopa), Florida (Hendry, Palm Beach, Collier, Martin) and Texas (Reeves). Broadly similar features were recorded by Smith (1980) when using Dun and Bradstreet Inc. business directories to identify the richest farms and ranches (agribusinesses) in the US, and by Windhorst (1989) employing values of farm sales (Fig. 4.1).

Unfortunately direct financial measures are not recorded by the majority of

85

agricultural censuses, and other surrogate measures of farm business size have had to be developed. Among these, weighting the output of the farm by the amount of labour required (standard man-days – SMDs) or by gross margins (European size units – ESUs) are widely used (these terms are described in the Appendix). In the UK for example, a full-time farm is defined as creating more than 250 SMDs of labour requirement per year, while in the EC a full-time farm is assumed to produce 4 or more ESUs per year (1 ESU = 1 200 European currency units). With the replacement of land and labour by capital, surrogate financial measures of farm size have been accepted more widely; nevertheless, problems are encountered in valuing inputs to and outputs from farming using national currencies when international comparisons are being made. Consequently the area size of farm remains the most internationally accepted index of farm size, despite its limitations in measuring the business size of a production unit.

There are considerable spatial differences in average farm size as measured by census units such as parishes, counties and countries. In the UK, for example, average farm size by county decreases progressively from east and south to west, while the national average exceeds that of most other countries in Western Europe (Table 4.1). Statistical averages, of course, disguise detailed variations and the proportions of small, medium and large farms in a census area give a more meaningful description of 'farm-size structure'. Table 4.1, for example, confirms the greater predominance of large farms in the farm-

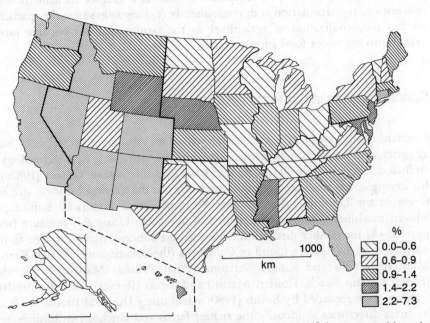

Fig. 4.1 Large-scale farms as a percentage of all farms in the US (as measured by sales > \$500 000) (*Source:* Windhorst 1989 *Tyschrift voor Economische en Sociale Geografie,* **80**: 270–82)

%
- 0.0–0.6
- 0.6–0.9
- 0.9–1.4
- 1.4–2.2
- 2.2–7.3

size structure of the UK compared with other member states of the EC, but also shows the high proportions of small farms in Portugal, Italy and Greece. Usually the larger farms in a census area represent a small proportion of all farm units but account for a disproportionately high share ('concentration') of the total area of farmland and agricultural production.

Table 4.1   National farm-size structures in the EC, 1987

Country	Number of holdings % by farm size (ha)					Average* size
	1–5	5–10	10–20	20–50	>50	
UK	14	12	15	25	33	68.9
Luxemburg	19	10	12	33	26	33.2
Denmark	2	16	25	39	17	32.5
France	18	12	19	33	18	30.7
Ireland	16	15	29	31	9	22.7
West Germany	29	18	22	25	6	17.6
Belgium	28	18	25	24	6	17.3
The Netherlands	25	18	25	27	4	17.2
Spain	53	19	12	9	6	16.0
Portugal	73	15	7	3	2	8.3
Italy	68	17	9	5	2	7.7
Greece	69	20	8	3	0	5.3
EC	49	17	14	14	7	—

* 1 ha or more.

Source: Based on data in Commission of the EC (1991: 122–5).

Theories on the spatial variation in farm size

A number of theories have been developed to account for spatial variations in farm-size structure. Following Huang (1973), those based on farm population density, stage of economic development and level of agricultural capitalization are most persuasive, although Grigg (1966) argues that the productivity of the resource base, inheritance laws and the role of the state cannot be ignored. On *farm population density*, as the population : land ratio increases, the available farmland is divided and subdivided into ever smaller farm units. Thus the large-farm structure of 'land rich' countries (historically low farm population densities), such as Australia and Canada, can be contrasted with the small-farm structure of 'land scarce' countries (historically high farm population densities), such as Norway, Italy and Greece. Through inertia in the farm-size structure, the historic relationship betwen the density of the farm population and average farm size can still be observed at the regional, as well as international, scale of analysis in industrialized countries. But as *the level of economic development* increases, particularly with the penetration of capitalist market relations, so the factors of production are transferred into the industrial from the farm sector, including the process of farm amalgamation and enlargement. It is often suggested, for example, that the UK's early industrialization under capitalism was largely responsible for the development of the country's large-farm structure. With the further *capitalization of agriculture*, and the loss of farm workers and farm families, so even larger farm businesses are created under advanced capitalism.

Other factors appear to be significant in particular national or regional contexts. For example, *the productivity of the resource base* (the combination of soil, climate and topography) can be associated with variations in farm size. Using marginal analysis, it can be shown that the farm inputs per hectare on less productive land have to be proportionally greater than on more productive land to achieve comparable outputs. On 'poor' land, therefore, the economic viability of farming is maintained by operating over a large land area with relatively low inputs and outputs per hectare. Productive land, on the other hand, tends to be apportioned between smaller farms with higher levels of inputs and outputs per hectare. This process can be reinforced where land settlement has been controlled and an optimum farm-size structure imposed by a state or regional authority. This occurred, for example, in the settlement of Australia and North America; larger farm sizes were created as the frontier of settlement was pushed into more arid environments. However, errors were made where the quality of the resource base was misjudged and farms were created which subsequently proved to be too small for the agricultural resource base. A similar mismatch between farm size and the resource base can occur in long-settled areas when there is a surge of population increase. Within the UK, for example, only marginal farmland was available for illegal squatting to absorb the rural population increases of the nineteenth century. Consequently many small farms dating from this era are to be found on economically marginal hill land in Wales and Scotland where they still pose social problems associated with low incomes.

Proximity to urban areas also influences the farm-size structure at the local level. Because of locational advantage in the costs of marketing farm produce and acquiring inputs, small farms near to urban areas can be operated with equal profit compared with larger farms at greater distances. In addition, farms near to urban areas suffer fragmentation because of urban infrastructure such as roads, canals and railways. However, the 'urban proximity' argument is complicated by the empirical observation that land around major cities is often the most productive in a region. Consequently the small average farm size may be a function of the resource base rather than urban proximity *per se*. A *behavioural* explanation can also be advanced for the relationship between land quality and farm size. Where the objective of the farm population is to maximize profits, farms will tend to be larger rather than smaller on better land. This appears to be the case in the Pacific south-west of the US (Gregor 1979). On the other hand, where the objective is to obtain only a 'sufficient' (suboptimal) level of profit, however defined, the better land can be farmed in smaller units compared with poorer land. Of these two relationships, the latter appears to be dominant in industrializ[...] regional exceptions.

Finally, *the role of the state* can be influentia[...] size structure at the national level. For exam[...] basis for land inheritance. Thus the 'primog[...] – usually male – heir) system (land equally [...] West European countries, and used to expla[...] former. The state also sponsors programm[...] breaking-up of large landed estates, for insta[...] or the creation of large state or collective farm[...]

Table 4.2 Farm-size change in the US, 1978–82

Farm size (acres)	Farm size (%)		% Change
	1978	1982	
1–9	6.7	8.4	+24
10–49	17.3	20.0	+15
50–179	33.6	31.8	−6
180–499	25.9	23.5	−10
500–999	9.9	9.1	−4
1 000–1 999	4.3	4.3	−0.4
Over 2 000	2.8	2.9	+2
Total number	2 257 775	2 241 124	—

Source: Derived from data in census of agriculture (USDA).

The farm-size structure in transition

Each national and regional farm-size structure is in a constant state of transition as 'inefficient' producers are marginalized by the competitive forces of market capitalism. With the real price of market inputs rising at a faster rate than real product prices, the farm sector is in a continuous 'cost–price squeeze' which smaller farm businesses are least able to withstand. Histori-cally, therefore, the smallest farm units have disappeared in greater number, with the land market facilitating the purchase (amalgamation) of small by larger farm businesses (pp. 59–63). This process of farm enlargement has enabled land and agricultural output to become 'concentrated' into fewer but larger production units, a process intensified under the contemporary trans-formation of agriculture. But the process of transformation is not unilinear in the sense that larger farms inevitably and uniformly displace smaller farms. For example, attention has already been drawn to the survival of small, part-time farms with members of the farm family taking off-farm employment (pp. 22–3). Moreover, there is evidence of farms being subdivided for sale as 'hobby farms' to urban migrants/commuters (Layton 1979), thereby

han decreasing the number and proportion of small farms.
US, drawing upon the type of data presented in Table 4.2,
been given to the 'disappearing middle' of medium-sized family
produce a dualistic (bimodal) farm–size structure dominated by small
ge farms. Munton and Marsden (1991), however, contest this 'dualistic'
ry of farm–size development and present farm-level data for the UK to
now a diversity of response in farm-size change (Table 4.3).

Table 4.3   Farm-level changes in farm size, 1970–85/88

Locality	Frequency of change (%)			Change in mean farm size (%)
	Increase in size	Decrease in size	No change	
Bedfordshire	51	15	34	+24.0
Cumbria	63	6	31	+16.7
Staffordshire	43	3	54	+12.7
Metropolitan green belt	41	42	17	+10.0
Dorset	58	14	28	+5.8

Source: Derived from data in Munton and Marsden (1991).

Table 4.3 also indicates that the transformation of the farm–size structure is not uniform by locality. This aspect of farm–size change is demonstrated further in Table 4.4 using data for three sample counties in the UK between 1978 and 1988. The trend towards farm enlargement is evident in all counties, both on farm numbers and the agricultural area, and especially in Norfolk. But Bedfordshire and Cumbria show increases in the proportions of small farms, while Bedfordshire shows a slight tendency towards a 'disappearing middle' in both farm numbers and area. In addition, the development of part-time farming, as shown in more detail in Chapter 10, is particularly character-istic of the urban fringe. The urban area provides alternative employment opportunities and increases the 'opportunity cost' in adjacent areas for those remaining in full-time agricultural employment. The urban fringe is also favoured by new entrants to farming who buy smallholdings as 'hobby farms', although scenically attractive areas within commuting distance of urban centres also feature in this type of farming. In both the Lake District and Peak District National Parks of England, for example, there is evidence of middle-sized farms being subdivided for sale as 'hobby farms'.

The formation of very large farming units, for example by investment companies, also operates in a spatially selective way. Higher-grade farmland, and farming types yielding better than average profits per hectare, are favoured. In the UK, the farm-size structure of eastern and south–central England is affected by this process, although there is no consistent evidence that farm type is influential in the overall spatial restructuring of farm size (Bowler and Ilbery 1989). However, Smith (1980), for the US, has shown the attraction of metropolitan regions for the development of large agribusinesses (especially Californian valleys and peninsula Florida); while research in

Ontario (Mage 1985) reveals how foreign (absentee) landowners have assembled 'super farms' on both the prime (counties of Lambton and Halton) and marginal (Muskoka, Parry Sound and Haliburton counties) farmlands of the province. Indeed marginal farming areas, whether at high altitudes and latitudes, or on poorly drained land, with their high proportions of economically marginal small farms, can be the locations with the most active farm enlargement. But also influential are the age structures of farmers in particular localities, the regional rural development programmes pursued by national or provincial governments, and the pre-existing farm-size structure (Todd 1979).

Table 4.4   Trends in farm sizes for sample counties in the UK, 1978–88

County	Farm size (total ha)				Total
	<20*	20–99.9	100–299.9	>300	
Norfolk (cereals)					
1988 number of holdings (%)	46.5	31.9	15.7	5.9	(100)
Change on 1978 (%†)	−0.9	−1.7	−0.7	+3.3	(−11.1)
1988 agricultural area (%)	3.5	20.0	35.5	41.0	(100)
Change on 1978 (%†)	−2.2	−0.1	+1.2	+1.1	(−0.4)
Bedfordshire (cropping)					
1988 number of holdings (%)	47.3	33.2	15.2	4.3	(100)
Change on 1978 (%†)	+0.3	−0.5	−0.3	+0.5	(−5.0)
1988 agricultural area (%)	4.1	23.7	38.1	34.1	(100)
Change on 1978 (%†)	−0.8	−1.0	−2.2	+4.0	(−0.1)
Cumbria (sheep and beef)					
1988 number of holdings (%)	27.9	53.0	16.7	2.4	(100)
Change on 1978 (%†)	+1.1	−2.5	+1.3	+0.1	(−3.2)
1988 agricultural area (%)	3.4	39.8	36.3	20.5	(100)
Change on 1978 (%†)	+0.1	−2.2	+2.1	0	(−0.4)

* Excluding 'agriculturally insignificant' holdings.
†Percentage points.

Source: Calculated by author from agricultural census data.

Whether or not these regionally varying processes can significantly alter variations in the spatial pattern of farm sizes is open to debate. The process of farm enlargement in the transformation of agriculture has certainly been more active in some regions compared with others; Gilbert and Akor (1988), for example, identified an increasing structural divergence in dairy farming between the family-farm structure of Wisconsin and the capitalist farm structure of California. Grigg (1987) and Hart (1987) for the UK and US respectively, however, have concluded that interregional differences in farm-size structure appear to persist. Much depends on the presence or absence of the dynamic corporate or large-farm sector (agribusiness) in a regional economy.

Farm size and the efficiency of production

A number of agricultural trends are associated with variations in farm size. Grigg (1966), for example, has drawn attention to the relationships between farm size and changing agricultural productivity (efficiency), population trends in rural areas, and social/political structures. Looking first at *agricultural productivity*, a well-known relationship exists between an increase in farm size and the increased efficiency with which capital and labour are used in agriculture (Bowler 1983). Spreading fixed costs over a larger land area is one source of economies of scale, although average efficiency appears to increase only up to the level of a four- or six-person farm (full-time labour). Thereafter, an increase in farm size seems to have more to do with the accumulation and concentration of wealth than the search for scale economies (Lund and Hill 1979). A literature also exists to show that the consolidation of farms at a theoretically optimum size varies by farm type (Found 1971: 35–49). In a Canadian (Manitoba) context, for example, Todd (1979) shows why large farms are the optimum size for cattle rearing, medium-sized farms for grain and mixed farming, and small farms for units supplying nearby urban markets.

The relationship between farm size and type is also established when the farm size is fixed and the occupier must select an appropriate type of farming. Small farms, for instance, require a high net financial profit per hectare to remain economically viable; in temperate farming this effectively limits small farms either to intensive livestock farming based on pigs, poultry and dairy cows, or to intensive horticultural and fruit production, including growing under glass or plastic (Table 4.5). On larger farms, by contrast, more extensive farming enterprises can be economically viable, including, in temperate farming, cereals, field vegetables, beef cattle and sheep. Moreover, while the

Table 4.5   Farm size: farm-type relationships in the EC, 1988/89

Type of farming	Average size (ha)	Average labour (AWU)*	Total output (1 000 ECU)
Cereals	40.5	1.13	33.4
Cropping	21.6	1.63	30.4
Horticulture	3.9	2.75	89.7
Vineyards	8.4	1.56	33.7
Fruit	7.7	1.48	16.7
Dairy	28.7	1.66	67.0
Livestock	10.4	1.70	163.6
Mixed	26.3	1.69	49.3
(EC12)	23.0	1.60	40.5

* AWU: annual work unit (agricultural work done by one full-time worker in one year – part-time and seasonal work are fractions of an AWU).

Source: Compiled from statistics in Commission of the EC (1991: 48–52).

occupiers of small farm units have to specialize in the use of their resources on just one or two enterprises, those on larger farms can usually operate more diversified farm businesses. Economies of scale also influence the use of farm machinery. The efficient operation of modern farm machinery usually requires a minimum area of land, and farmers who have less than this cannot afford to buy and operate the machinery because its fixed costs are too high.

The superior economic efficiency in the use of resources by larger over small farms is the main justification offered for farm enlargement in contemporary agriculture. But farm size is also a main mediating factor in the transfer of new farming technology and practices to the production sector. Research across a range of disciplines – sociology, geography, economics – consistently shows that the occupiers of larger farms are more likely to adopt new technologies and practices ahead of the occupiers of smaller farms. This variation in innovative behaviour, and the technical and economic superiority which it brings, can be observed in features such as the adoption of new crops and livestock, farm plant and machinery, membership of marketing co-operatives, involvement in production for food processors under forward contracts, the use of state-aided investment grants and subsidies, and farming practices such as irrigation and the use of agrichemicals (Clark 1986). The mechanisms involved in the diffusion of new farm technology are discussed in more detail in Chapter 6 (pp. 143–9). Here it is sufficient to point up the hierarchical pattern of diffusion from larger to smaller farms, and from regions with a large-farm structure to those dominated by smaller farms. However, the spatial distribution of farming types also provides a context for the diffusion of each innovation, since new technology tends to be farm-type specific. Taken together, therefore, the regional distributor of farm size and farm type provide the main contexts for the spatially varied impact of agricultural innovations on the transformation and efficiency of farming.

Farm size and farm income

As shown by Hill (1990), farm income cannot easily be defined or measured and the following discussion is inevitably simplified. At its most basic level, net farm income can be calculated as the difference between the total costs of production and the total value of agricultural output. It is the sum available for remunerating the farm family for the physical work on the farm and for a return on the capital represented by the farm and its assets (Hill and Ray 1987: 296). However, relevant data can be presented for individual farms, for averages over aggregates of farm by size or type, or for all farms in a region or country (the 'national' farm). Moreover, net income can be presented per hectare of farmland, per labour input, per £100 of inputs, or per farm business. Each calculation results in a slightly different outcome. On the first criterion, for example, smaller farms tend to outperform larger farms because of the

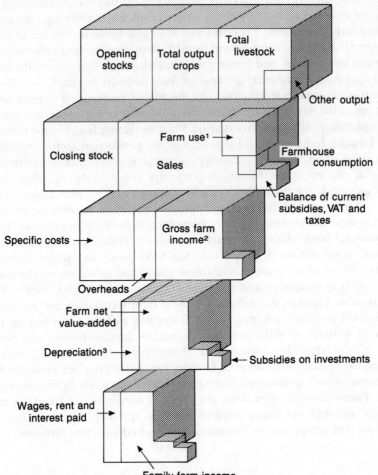

Fig. 4.2 The calculation of indicators of farm income in the EC. *Notes:* (1) Output used as inputs to other production on the farm; (2) farm gross value-added; (3) on the basis of replacement cost (*Source:* Commission of the EC 1991)

intensiveness of their production systems; but on the other criteria, larger farms tend to be superior because of their economies of scale, both internal and external. Internal economies, for instance, can be gained on technical, managerial and marketing costs (Ilbery 1985: 111); they enable unit costs of production to be reduced. External economies of scale (regional concentrations of large farms or farms of a certain type) can be obtained on the costs of farm labour (maintaining a pool of skilled labour), the acquisition of services and supplies, the provision of specialist farm advisers and credit facilities, and the creation of specialized marketing channels (co-operatives, forward contracts with food processors, packing and grading stations).

The calculation of net farm income, either through sample farms or aggregate data for the farm sector, requires a number of assumptions about the pricing and allocation of costs; consequently gross rather than net values are commonly employed. For example, the cost of farm family labour has to be imputed, since formal wage or salary transactions, and the labour input of farm wives, are not systematically recorded; the rising real value of farmland is strictly a form of 'income' and should be part of any calculation; new investment can be discounted over a varying number of years. Simplifying, under gross income calculations, a division is made between fixed and variable costs. Gross income is measured as total output less variable costs (for example, fertilizers, sprays, feed), with fixed costs ignored (for example, rent, labour, machinery, buildings). The output (gross) figure measures goods sold, but can also include any increase in the value of the farm and the livestock, together with government grants and subsidies, but less costs of wear, tear and depreciation. A variety of indicators of agricultural income, as used in the EC, are shown in Fig. 4.2.

Even gross figures (gross margin, gross output, gross income) do not capture the totality of farm income, since personal income includes earnings from non-farm employment and unearned incomes (mainly pensions and investments). Contributions from these sources can add significantly to the income from farming, both on small and large farms, and from all members of a farm family. For example, Nix (1990) has reported Inland Revenue figures for the UK (1987) showing nearly 44 per cent of all income for farmers coming from outside food production on their own farms, with non-farm occupations (20 per cent), investments (20 per cent) and pensions (5 per cent) as other sources of income. The owners of very large farms, for example, often employ a manager to run the farm while they develop off-farm businesses. Clark (1991) found this feature to be most prevalent in the UK in the south-eastern counties of England (Fig. 4.3). Moreover, Gasson's (1989) research suggests that 29 per cent of farm wives have off-farm employment, a figure estimated at 45 per cent for the US by Deseran et al. (1984).

With trends towards pluriactivity, just what constitutes 'farm income' is increasingly problematic. It also throws into question one of the bases (low farm incomes) for state involvement in, and financial support of, the farm sector. At present, strictly 'farm income' is used when setting product prices supports; as shown by both Nix (1990) and Hill (1990), there is an argument that 'farm family income' rather than 'farm profitability' should be the basis for calculating support measures.

There can be little doubt that the transformation of agriculture over the last four decades has increased rather than decreased the income differentials within the farm sector. The continued loss of farms, farm families and farm workers is a measure of the low incomes earned in some sections of farming, especially on the smallest holdings (see Table 3.2). At the same time, the profits declared by the huge agribusinesses, together with the generous net farm incomes reported for larger farms, demonstrate the real gains made

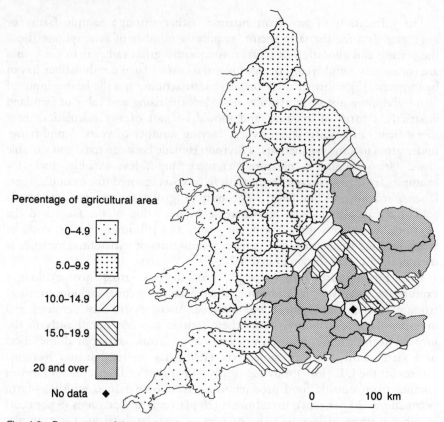

Percentage of agricultural area

0–4.9

5.0–9.9

10.0–14.9

15.0–19.9

20 and over

No data ◆

0   100 km

Fig. 4.3 Percentage of the total agricultural area farmed by salaried managers, England and Wales, 1985 (*Source:* Clark 1991: 79)

elsewhere. Even if the returns to capital investment on most farms still remain below those obtainable elsewhere in the economy, the rising asset value of farmland has been a considerable compensation. In addition, manufacturing firms supplying farm inputs, or processing food outputs, have enjoyed a financially rewarding trade in the last four decades. More recently, though, the world-wide recession of the 1980s, and its accompanying farm crisis has had a variable impact on economic returns in agriculture. At an aggregate level, real farm incomes have fallen in the UK since 1980, while varying from year to year; in Italy and France, however, they have remained at a stable level; while in countries such as the Netherlands, Ireland and Spain real incomes have risen. At the individual farm level, businesses attempting to service high levels of debt have suffered the most depressed net financial returns in recent years.

Table 4.6 shows income disparities by farm size and type for the EC, and is interrelated with Table 4.5: vineyards, for example, are characterized by small farms (see Table 4.5). Table 4.6 enables some appreciation to be gained of the disparities in farm incomes between groups of farms. The largest farms appear

Table 4.6   Financial returns to farming in the EC, 1988/89

Farm size* or type	Financial indicator (average/holding in 1 000 ECU)		
	Farm net value-added (FNVA)	FNVA per annual work unit (FNVA/AWU)†	Family farm income per unit of unpaid labour (i.e. not hired)
Farm size			
Below 4	2.3	1.5	0.9
4–8	9.2	5.8	5.1
8–12	15.3	9.8	8.6
18–24	28.5	17.0	15.0
Over 24	64.0	36.2	35.4
Farm type			
Cereals	11.7	10.3	7.4
Cropping	12.6	7.7	6.1
Horticulture	38.6	14.0	13.1
Vineyards	18.4	11.8	10.5
Fruit	10.2	6.9	6.7
Dairy	26.7	16.1	13.6
Livestock	8.8	17.0	12.8
Mixed	15.0	8.9	6.8

* Farm net value-added per annual work unit in 1 000 ECU.
†See Table 4.5 for definition.
Source: Based on data in Commission of the EC (1991: 48–53).

to have 30 times the income of the smallest farms on most indicators, while livestock farms outperform fruit farms by a factor of nearly 3, at least as measured by net value-added per work unit. The available evidence points towards increasing rather than decreasing income disparities, despite the intervention of the Common Agricultural Policy (CAP), while, with farm type and size having varied spatial distributions, income disparities by region within the EC are evident (Bowler 1985). Figure 4.4, for example, shows the financial returns to farm labour as three to four times greater in farming regions within Denmark, northern Germany, the Netherlands, Belgium north-east France and eastern England, compared with regions throughout Portugal, Spain, Ireland, Italy and Greece. The former areas have the advantage of larger business sizes as well as higher levels of price support for their farm products under the CAP. The consequences of these income differentials for state intervention in the food supply system are considered in more detail in Chapter 9.

Farm size and rural social trends

The transformation of farm size has thrown a number of social trends in rural areas into relief. Perhaps the most significant has been the intensification, in

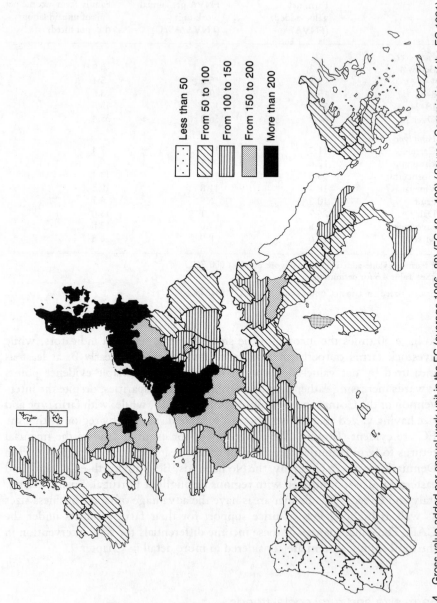

Fig. 4.4   Gross value added per annual work unit in the EC (average 1986–88) (EC 12 = 100) (*Source:* Commission of the EC 1991)

some rural regions, of the long-term trend in the outward migration of farm families and farm workers. Some writers, such as Brierley and Todd (1990) for the Canadian prairies, maintain that the further decline in the farm population has so undermined rural services in the small central places serving agricultural areas as to threaten their sustainability. The services include public transport, hospitals, schools and retail outlets. Hart (1991), however, is more sceptical, pointing out that the employment base of many small rural areas has changed from farm services to manufacturing. But this is not the consensus view: the economic and social sustainability of many rural communities remains in doubt, especially at the economic margins of commercial agriculture (Fuller et al. 1990).

The changing size structure of farming has also had implications for social relations within rural communities. Gregor (1982b), for example, has examined the clash between technological change in agriculture and the cultural ideals of western societies as regards equal opportunity and the right to till the soil in family-sized units. With the economics of modern farming working to the disadvantage of small, full-time family farms, and with significant proportions of farmland being concentrated in the hands of large landowners, the cultural ideals are threatened and the interests of large and small family farms within the rural community lie on a collision course. Moreover, financial obstacles (high land prices) are placed on the recruitment of new farmers into agriculture; farm inheritance and college training for farm management have become socially selective mechanisms determining the characteristics of the next generation of farmers.

Rural sociologists in the US have also been concerned for many decades with the quality of life in rural communities faced by the process of farm enlargement. Using indices such as social and physical amenities, participation in social and religious institutions, and the degree of local control over the political process, the 'Goldschmidt hypothesis' (Goldschmidt 1978; Green 1985) contends that communities within a family-farm structure enjoy a superior quality of life compared with those located within a structure of large, capitalist (corporate) farms. Dependent wage workers, on large-scale farms, suffering low earnings and social alienation, are considered to act as the intervening variable between farm size and quality of life, although Heffernan and Lasley (1978) also detected less involvement in the community by the managers of large, corporate farms compared with the family farmers they displaced. While the latter are interpreted as forming the old middle class, in which wealth is a means to social goals, the former are a transition class to a new middle class motivated mainly by economic goals, in which the acquisition of wealth is a goal in itself. There is some doubt as to the broad applicability of these features: not only has subsequent research questioned the methodological basis of Goldschmidt's work, but relatively few rural communities, especially in Western Europe, are still dominated by the farming population.

Changing gender roles in the farm family appear to have accompanied the transformation of agriculture, but in ways that vary between farms of different sizes. This theme has already been introduced in Chapter 3 (pp. 72–3): it is sufficient here to emphasize the increased dependence on the labour of the farm wife on small, pluriactive farms where the husband has an off-farm job (Pfeffer 1989); the joint role of husband and wife in the development of new, often non-farming enterprises, such as farm accommodation and farm shops; and the displacement of farm wives from active participation in the farm business on the larger farms (Symes 1991). The latter trend is associated with the specialized technological skills required in modern farming for which farm wives have little training. Whether or not these developments represent a major shift in the patriarchal farm family structure is open to debate (Whatmore 1991); certainly Gasson (1989) found a majority of farm wives still restricted to managing farm accounts, rendering assistance in the farm office, and provided back-up manual labour at times of peak demand or business crisis.

Land tenure (occupancy) under the transformation of agriculture

In developed market economies, a broad division can be drawn between land owned and farmed by the same person (owner-occupied) and land rented for farming from the owner (cash tenancy). Different national and regional histories, social conventions and legal institutions have produced considerable spatial variations in the percentage of land in owner-occupation: in the EC, for example, the proportions vary from over 90 per cent in Ireland, through 70–90 per cent in Denmark, (West) Germany and Italy, to under 30 per cent in Belgium (Harrison 1985). Over the last four decades, an increasing proportion of farmland has passed into owner-occupation with the demise of the landlord–tenant system. In England and Wales, for instance, the proportion of farms rented fell from 82 per cent in 1891, to 49 per cent in 1950 and to 19 per cent by 1985 (Table 4.7). Two processes appear to be at work. Firstly, most countries impose taxes on the intergenerational transfer of wealth, including land, and many large landed estates have been subdivided

Table 4.7 Changing land tenure in England and Wales, 1971 and 1985

Year	Farm numbers (%)			Agricultural area (%)		
	Owner-occupied	Mixed	Rented	Owner-occupied	Mixed	Rented
1971	46	24	30	35	33	32
1985	59	22	19	42	35	23

Source: Derived from agricultural census data.

and sold in part or in whole to pay death duties. Secondly, tenant farmers have received increasing security of tenure, including limits on the extent to which agricultural rents can be increased, thereby reducing the financial incentive for letting farmland. Consequently, as tenancies fall vacant, many landlords take their land 'in hand' to farm under their own management; as a result the rate of farm enlargement can be greater among former tenanted as compared with owner-occupied farms.

Under the recent transformation of agriculture, a far more complex structure of land tenure (occupancy) has begun to emerge as a survival strategy for family farm businesses. Most importantly, the proportion of farms comprised by owned and rented land has increased significantly in most countries, a feature often disguised in agricultural censuses when farmland has to be recorded as either 'mainly' owned or tenanted. The category of mixed tenure (or part-ownership), for example, now accounts for 35 per cent of farmland in England and Wales (wholly owned 42 per cent – Table 4.7), and nearly 60 per cent in the four Corn Belt states (Iowa, Illinois, Indiana and Ohio) studied by Hart (1991). A clear trend is for owner-occupied family farms to be enlarged by renting rather than purchasing more land; in general terms, the larger farms tend to become part-owned. Given the historically high real value of farmland in recent years, together with high rates of interest on capital needed to purchase land, renting rather than purchasing has been the preferred option in most countries.

However, not all farming regions are characterized by this new structure of land occupancy. Hart (1991), for the Corn Belt of the US, for example, found a spatially varying pattern based on farm type (cash-grain farming areas) and urban fringe location (land awaiting development). Part-ownership was found not to be so prevalent where fixed capital investment is needed on rented land (improvement of pastures for cattle farming) or where land is relatively cheap and can be purchased (dissected uplands). However, competition for let land is intense, and the resulting pattern of occupied land for an individual farm can be highly fragmented. This process of part-ownership appears to be intensifying the existing fragmentation of owned land, as reported for example by Smith (1975), together with the diseconomies that accompany farming widely dispersed parcels of land. In addition, the falling proportion of land for rent, but its relatively high value in relation to farm incomes in the 1980s, has resulted in most land being offered on 'insecure' rather than 'secure' tenancies (i.e. rents are renegotiated at short intervals and heirs to the farm have no rights to continue as tenants). Munton and Marsden (1991) found that the proportion of let land held on insecure leases had reached 13 per cent within the Metropolitan (London) Green Belt, 14 per cent in Staffordshire and 17 per cent in Dorset.

The legal category of owner-occupation is also taking on a variety of forms. For example, so as to spread the taxation liability, property has become more widely distributed among members of the farm family. Partnerships (between spouses, siblings and other members of the extended family), trusts, share farming and fiscal tenancies now account for more land than sole occupation, but to a greater extent in lowland as compared with upland farming areas in the context of the UK (Munton and Marsden 1991). Share farming (cropping) involves the landowner retaining the property rights, with another farmer working the land. While profits are shared, the main risks are borne by the person managing the land. In addition, share farming has been interpreted for the southern states of the US as a form of control mechanism by capital (landowners) over wager labourers (Reiss 1985). Fiscal tenancy is a legal arrangement for tax avoidance whereby the 'tenant' also owns the land. These complex legal arrangements increasingly characterize the accumulation and concentration of land in family farm businesses. Parents and their children can operate a multiple-unit family business by occupying separate landholdings but farm the land together as one farm business. Increasingly complex legal arrangements, often for tax avoidance, bind the farm family together in partnerships, shared ownership, share farming, trusts and companies (Gasson et al. 1988).

Land tenure has a broad relationship with farm size; rented holdings in the UK, for example, account for almost half the total holdings over 200 ha in size. Tenure also influences the complexity of decision-making for the farm business (sole occupier as compared with partners or share owners), and has consequences for capital investment on the farm and the financial returns obtained. On the last two points, conventionally the landlord is responsible for fixed capital on a tenanted farm (such as the land itself, land drainage and farm buildings), with the tenant providing the variable or working capital. Sometimes the terms of a tenancy agreement can prescribe the type of farming and farming practices (e.g. crop rotation) that must be followed, or exclude certain enterprises or non-farm uses of the land, for example under farm diversification (see p. 156). Also the rent for the farm must be renegotiated periodically (commonly at 3–5-year intervals) so as to reflect changing land values and the profitability of farming. Generally the pattern of farmland rental values follows the price of farmland itself. Owner-occupiers, by comparison, have exclusive rights over the use of their land and buildings, enjoy the capital gains from any increase in land values and retain all farm profits. The last two aspects can be important for gaining access to credit facilities and capital for investment.

Research on the relative profitability of owner-occupied as compared with tenanted farms is inconclusive. As Hill and Ray (1987: 148) observe,

> Tenants are not clearly more efficient than owner-occupiers except in certain size groups (e.g. small specialist dairy farms) . . . large scale arable and mixed farming seem more efficient under owner-occupation . . . no strong support is available on grounds of efficiency for the superiority of one form of land tenure or the other . . . neither is there clear evidence that the landlord and tenant system is superior to owner-occupation in its provision of capital . . . [nor] is there conclusive evidence on the greater flexibility of tenancy in responding to structural (i.e. farm size) adjustment.

Rather part-ownership farms have been found to be most financially success-ful, especially cropping, capital-intensive and medium-sized farms (Gasson and Hill 1984). Even here, though, it is not clear whether the form of land tenure (and the associated structuring of capital) is the factor responsible, or whether the superior economic returns are a function of the more dynamic, and economically successful, farmers involved in enlarging their holdings through mixed tenure.

Conclusion

This chapter has explored the agricultural significance of farm size and land tenure. Not only do these features vary between individual farm businesses, but in aggregate they vary over space. In addition farm size and land tenure operate as mediators in the industrialization of agriculture, contributing to the uneven transformation of the farm sector between farm businesses and farming regions. Reinterpreting the work of Marsden et al. (1987), they mediate in the external relations between the farm business and external capital as regards:

(a) technology (manufactured inputs, advice, technical assistance);
(b) finance (mortgage and credit facilities); and
(c) marketing (linkages of products to market outlets).

Larger farms, for example, have been shown to be more innovative with regard to new technology, with consequences for the dynamism and transfor-mation of regions with a large-farm structure. The discussion has also examined further the relations between farm businesses and finance capital, as introduced in Chapter 3. Capital structure, for example, has been shown to vary with land tenure with consequences for the economic performance of farm businesses under different types of land tenure. However, the linkages between farm businesses and marketing structures are explored later in Chapter 7.

Farm size and land tenure also have a differential influence on the internal relations of the farm business. Four dimensions to such relations can be identified: the ownership of farm capital, the ownership of land rights, control over business management and the structure of farm labour. Marsden et al. (1987) show how the degree of subsumption (pp. 24–5), either formally by external agencies or as real subsumption within the farm business (i.e. the diffusion of control away from a single person), can be placed on an ordinal scale. Table 4.8, for example, shows the scaling of the subsumption in the internal organization of the farm business. As discussed earlier in this chapter, land rights (land tenure) vary from sole occupation, through corporate family ownership, to complex family tenancy and corporate tenancy; business management varies from single-unit farms under single management, through

Table 4.8   An ordinal scale of subsumption in the internal organization of farm businesses

| Score | Farm capital | Internal relations | | Labour relations |
		Land rights	Business management structure	
1	Farm family, individual head of household	Simple owner-occupation, sole owner-operator	Single unit farm managed by head of household	Family labour farm, 1 full-time family worker
2	Farm family, shared and nominal corporate family ownership	Family owner-occupation	Single unit farm, joint family management	Family labour farm, more than 1 full-time family worker
3	Farm family, corporate ownership of single farm unit	Corporate family ownership (nominally renting)	Single unit farm, family owned but employing manager	Family labour farm, + casual hired labour only
4	Farm family, corporate ownership of multi-farm unit	Mixed family ownership and tenancy (where both are significant)	Multi-unit/farm, family owned and managed	Mixed labour farm, 1–3 full or part-time hired worker (hired < family)
5	More than family corporate, involving non-family capital	Simple family tenancy, (renting from 1 non-family owner)	Multi-unit/ enterprise business, family owned employing manager	Mixed labour farm, over 3 full or part-time hired labour (hired < family)
6	Family and landowner corporate business	Complex family tenancy (renting from 2 + non-family owners)	Multi-unit enterprise business, non-family owned and managed	Mixed labour farm employing over 50% contract hired labour
7	Non-family, single corporate owner	Non-family corporate (with manager)	Multi-unit/ enterprise business, non-family owned, employing manager	Wage labour farm, all hired labour
8	Non-family, multi-corporate owner	Non-family corporate tenancy	Multi-unit/ enterprise business, non-family owned, contract co-managed	Wage labour farm, all contract labour

Source: Marsden *et al.* (1987).

multi-unit but family-owned farms, to managed (corporate) farm businesses. With each increase in the scale value, there is an increase in the internal control exercised over the farm business by agents other than the head of the household. Also there is an implicit but general relationship between the scale values and farm size: larger farms have scores in the range from 4 to 8. Small farms, for instance, tend to be financed by capital from the farm family, while

Table 4.9 A typology of the internal and external relations of the farm business (percentage in each category)

External relations (formal or indirect subsumption) Score range of 0–12	Score 10–12	4–9	10–16	17–25	26–32 Score
	10–12	0	2	3	2 IV
	7–9	2	10	8 III	1
	4–6	12	26 II	4	1
	0–3	16 I	12	1	0

Internal relations (real or direct subsumption) Score range 4–32

  I : Marginal closed farm unit
 II : Transitional dependent farm unit
III : Integrated farm unit
IV : Subsumed farm unit

Data from a sample of 264 farms in west Dorset, east Bedfordshire and the metropolitan (London) green belt.

Source: Derived from Whatmore et al. (1987).

larger farms progressively draw upon non-family capital. On labour relations, small farms depend mainly on family workers, whereas larger farms have an increased dependence on hired and contracted labour.

When placed together in a matrix, measures of the subsumption of the external and internal relations of the farm business produce a typology of farms. In Table 4.9 Whatmore et al. (1987) identify four types as having particular significance: (I) marginal closed; (II) transitional dependent; (III) integrated, and (IV) subsumed farm units. They represent increasing levels of subsumption, although in their farm sample, 66 per cent of farms were found to have relatively low levels of subsumption on both dimensions. The empirical data also support the contention that the subsumption of farm businesses also varies regionally. While 64 per cent of farms in the study regions experienced no change in the degree of subsumption between 1970 and 1985, 28 per cent showed an increased level (8 per cent decreased). Increased subsumption was less evident on farms in west Dorset compared with east Bedfordshire and the metropolitan (London) green belt, a feature explained by the small size of dairy farms found in that study area. In the next two chapters a more detailed examination is made of the role of type of farm as a mediating factor in the transformation of agriculture.

Appendix

A *standard man-day* (SMD) is defined as the average amount of labour needed each year to raise 1 ha of a crop or one head of livestock using the average current technology available. Thus an intensive crop, such as raspberries, may need 32 SMDs per hectare, whereas a more extensive crop like wheat may need only 0.8 SMD per hectare. The total SMD size of a farm is obtained by calculating the SMD requirement for each enterprise and then summing for all the farm's enterprises. The greater the total number of SMDs, the bigger the farm's business size is deemed to be. Thus, a horticultural farm might be small in area but large in terms of labour input, whereas a hill farm might be physically extensive but small in size of business when assessed in terms of SMDs, because of its low stocking rate on poor moorland grazings.

A *standard gross margin* (SGM) for an enterprise is the average per hectare (for crops) or per head (for livestock) gross margin calculated for a number of sample farms; the gross margin is the difference between the value of output and the variable costs of production such as seed, fertilizer and feed, but before deducting fixed costs such as rent or regular labour. The 'standard' can be applied to any farm.

The *European size unit* (ESU) for a farm is calculated by multiplying its crop areas and livestock numbers by the relevant SGMs, with financial values expressed in European currency units. The values are summed and divided by 1 200, so that 1 ESU = 1 200 ECU (European currency units). The ESU needs to be recalculated from time to time to take account of changing input costs and output prices. In the UK, a parellel procedure is adopted in calculating the business size unit (BSU). The total ESUs on a farm define its business size.

References

Bowler, I. R. (1983) Structural change in agriculture. In Pacione M. (ed) *Progress in rural geography*. Croom Helm, London, pp 46–73

Bowler, I. R. (1985) *Agriculture under the Common Agricultural Policy: a geography*. Manchester University Press, Manchester

Bowler, I. R. and **Ilbery, B. W.** (1989) The spatial restructuring of farm types in the English counties, 1976–85. *Tijdschrift voor Economische en Sociale Geografie* **80**: 203–10

Brierley, J. S. and **Todd, D.** (1990) *Prairie small-town survival: the challenge of Agro-Manitoba*. Edwin Mellen Press, Lampeter

Clark, G. (1986) Diffusion of agricultural innovations. In Pacione M. (ed) *Progress in agricultural geography*. Croom Helm, London, pp. 70–92

Clark, G. (1991) People working in farming: the changing nature of farmwork. In Champion, T. and Watkins C. (eds) *People in the countryside.* Paul Chapman, London, pp 67–83

Commission of the European Communities (1991) *The agricultural situation in the Community. 1990 Report.* The Commission, Brussels

Deseran, F. A., Falk, W. W. and **Jenkins, P.** (1984) Determinants of earnings of farm families in the US. *Rural Sociology* **49**: 210–29

Found, W. C. (1971) *A theoretical approach to rural land-use patterns.* Edward Arnold, London

Fuller, A. M., Ehrensaft, P. and **Gertler, M.** (1990) Sustainable rural communities in Canada: issues and prospects. In Gertler, M. and Baker, H. R. (eds) *Sustainable rural communities in Canada.* The Canadian Agriculture and Rural Restructuring Group, Saskatoon, pp. 1–41

Gasson, R. (1989) *Farm work by farmers' wives.* Farm Business Unit, Wye College

Gasson, R., Crow, G., Errington, A., Hutson, J., Marsden, T. and **Winter, M.** (1988) The farm is a family business: a review. *Journal of Agricultural Economics* **39**: 1–42

Gasson, R. and **Hill, B.** (1984) *Farm tenure and performance.* School of Rural Economics, Wye College

Gilbert, J. and **Akor, R.** (1988) Increasing structural divergence in US dairying: California and Wisconsin since 1950. *Rural Sociology* **53**: 56–72

Goldschmidt, W. (1978) Large-scale farming and rural social structure. *Rural Sociology* **43**: 362–66

Green, G. P. (1985) Large-scale farming and quality of life in rural communities: further specification of the Goldschmidt hypothesis. *Rural Sociology* **50**: 262–74

Gregor, H. F. (1979) The large farm as a stereotype: a look at the Pacific Southwest. *Economic Geography* **55**: 71–87

Gregor, H. F. (1982a) *Industrialisation of US agriculture. An interpretive atlas.* Westview Press, Colorado

Gregor, H. F. (1982b) Large-scale farming as a cultural dilemma in US rural development. *Geoforum* **13**: 1–10

Grigg, D. B. (1966) The geography of farm size: a preliminary survey. *Economic Geography* **42**: 205–35

Grigg, D. B. (1987) Farm size in England and Wales from early Victorian times to the present. *Agricultural History Review* **35**: 179–89

Harrison, A. (1985) Farm tenancy arrangements in Western Europe and the USA. *Agricultural Administration* **19**: 179–87

Hart, J. F. (1987) The persistence of family farming areas. *Journal of Geography* **86**: 198–203

Hart, J. F. (1991) Part-ownership and farm enlargement in the Midwest. *Annals of the Association of American Geographers* **81**: 66–79

Heffernan, W. D. and **Lasley, P.** (1978) Agricultural structure and interaction in the local community: a case study. *Rural Sociology* **43**: 348–61

Hill, B. (1990) *Farm incomes, wealth and agricultural policy.* Avebury, Aldershot

Hill, B. and **Ray, D.** (1987) *Economics for agriculture. Food, farming and the rural community*. Macmillan, London

Huang, Y. (1973) On some determinants of farm size across countries. *American Journal of Farm Economics* **55**: 89–92

Ilbery, B. W. (1985) *Agricultural geography: a social and economic analysis.* Oxford University Press, Oxford

King, R. L. (1977) *Land reform: a world survey*. Bell, London

Layton, R. L. (1979) Hobby farming. *Geography* **65**: 220–3

Lund, P. J. and **Hill, P. G.** (1979) Farm size, efficiency and economics of size. *Journal of Agricultural Economics* **30**: 145–58

Mage, J. A. (1985) Absentee ownership of Ontario's farmland. In Fuller, A. M. (ed) *Farming and the rural community in Ontario: an introduction.* Foundation for Rural Living, Toronto, pp 197–217

Marsden, T. (1988) Exploring political economy approaches in agriculture. *Area* **20**: 315–22

Marsden, T., Whatmore S. and **Munton, R.** (1987) Uneven development and the restructuring process in British agriculture: a preliminary exploration. *Journal of Rural Studies* **3**: 297–308

Munton, R. and **Marsden, T.** (1991) Dualism or diversity in family farming: Patterns of change in British agriculture. *Geoforum* **22**: 105–17

Nix, J. S. (1990) Aspects of farm profitability: an outmoded concept? *Journal of Agricultural Economics* **41**: 265–91

Pfeffer, M. J. (1989) The feminization of production on part-time farms in the Federal Republic of Germany. *Rural Sociology* **54**: 60–73

Reiss, F. J. (1985) Farm tenancy arrangements in the USA. *Agricultural Administration* **19**: 245–62

Smith, E. G. (1975) Fragmented farms in the United States. *Annals of the Association of American Geographers* **65**: 58–70

Smith, E. G. (1980) America's richest farms and ranches. *Annals of the Association of American Geographers* **70**: 528–41

Symes, D. (1991) Changing gender roles in productionist and post-productionist capitalist agriculture. *Journal of Rural Studies* **7**: 85–90

Todd, D. (1979) Regional and structural factors in farm-size variation: a Manitoba elucidation. *Environment and Planning A* **11**: 237: 58

Whatmore, S. (1991) Life cycle or patriarchy? Gender divisions in family farming. *Journal of Rural Studies* **7**: 71–6

Whatmore, S., Munton, R., Marsden, T. and **Little, J.** (1987) Interpreting a relational typology of farm business in Southern England. *Sociologia Ruralis* **27**: 103–22

Windhorst, H. W. (1989) Industrialization of agricultural production and the role of large-scale farms in US agriculture. *Tijschrift voor Economische en Sociale Geografie* **80**: 270–82

5

Farm types and agricultural regions

John Aitchison

This and the following chapter examine the type-of-farm context for the contemporary transformation of agriculture. In this chapter attention is turned to the problems of classifying farm types and defining agricultural regions. Chapter 6 goes on to examine the changing enterprise structure of farm businesses.

Like agricultural economists, who were among the first to develop farm classification (largely as a tool on which to base farm business advice), agricultural geographers have classified to satisfy a variety of objectives (Grigg 1969). Although the precise reasons for which individual classificatory studies have been carried out can be difficult to disentangle, often because they have not been explicity articulated, four overarching objectives can be distinguished.

To present ordered information

Farms and farming regions form complex, dynamic systems. To appreciate their character, to track trends and changes, and to facilitate comparative analysis, such systems need to be classified. Classification serves a fundamental descriptive purpose; it filters information with a view to isolating meaningful patterns of ordered variation. Needless to say, for many agricultural geographers it is patterns of spatial or regional variation in types of farming that are of major interest. While they might be criticized for being too heavily descriptive, there is no doubting the fact that most taxonomic studies are at least factually informative. This is especially true of investigations carried out in less well-documented regions of the world, and in regions where previously uncharted developments in farming practices have taken place. This knowledge-generating role of classificatory inquiry should not be underrated.

To seek explanations and to generate hypotheses

Patterns not only invite description, more importantly they invite explanation. In delimiting types-of-farming regions, for instance, agricultural geographers will normally endeavour to account for the spatial structures that emerge by citing the possible influence of particular processes or forces. Such factors include: soil conditions and climate; land, labour and capital availability; the size, demands and distance of relevant markets, including food processors; historical and cultural traditions; ideological constraints; systems of land tenure and the behavioural characteristics of farmers and growers.

To test hypotheses

Not all classifications have pattern descriptions as their dominant focus; many are effected with a view to a more formal testing of specific hypotheses. For instance, in testing the effect which distance to markets and the movement of factors of production have on farming patterns, geographers and others have been obliged to categorize, ordinate and regionalize types and intensities of agricultural activities (Chisholm 1979). Another example is provided by studies which have sought to test Boserup's (1965) model relating the process of agricultural change in regions of subsistence agriculture to that of population change. Here, farm systems need once again to be differentiated according to the type and intensity of agricultural practice. It can be argued that if classifications are to contribute towards a more profound understanding of agricultural space then they will need, as in the above instances, to be allied with particular models of explanation; merely processing data in the hope that meaningful (interpretable) patterns will emerge is not a strategy that will ensure the establishment of sound theory in agricultural geography. This viewpoint has much to commend it, as long as it is recognized that inductive classificatory explorations of data sets still have a role to play.

To appraise alternative conceptual and methodological approaches

Although the majority of taxonomic studies in agricultural geography have substantive aims, a considerable proportion also focus on issues of a conceptual and methodological nature. Problems relating to the definition and measurement of the variables (attributes) to be used in a classification, for instance, have excited much debate in the literature, and while such matters may appear to be rather dry and esoteric, they are in fact of crucial theoretical significance. How to weight crop and livestock enterprises on common scales, how to measure farm size, and how to derive composite scales for attributes that are complex and multivariate in character, are just some of the important conceptual questions to which agricultural geographers have addressed themselves in seeking to classify farms and farming regions. Perhaps even more attention has been directed towards the problem of actually creating typologies. Agricultural geographers have developed and experimented with a wide range of taxonomic procedures – from qualitative and loosely structured categorizations to sophisticated statistical treatments of large, multivariate data sets.

Over the years numerous typologies and regionalizations of agricultural systems have been developed, and interest in such matters continues to figure prominently in research. Despite this, it has to be said that a number of agricultural geographers have questioned the utility of certain types of classificatory exercise. Chisholm (1964) and Morgan and Munton (1971), for instance, have been especially critical of those investigations which offer simple descriptions of regional patterns and fail to probe, with any profundity, into the processes that help account for structural and functional variations in farming systems and their spatial expression. Such exercises, it is argued, add little to the cumulative store of theoretical understanding, and are essentially 'one-off' empirical contributions of limited value. Furthermore, because of the way in which they have been conceptualized and their dependence upon highly subjective decisions, studies of this type do not provide a sufficiently robust foundation for further research. Other critics have also claimed that far too much emphasis has been placed on narrow, static conceptions of farming systems, with insufficient attention being directed towards their dynamic, evolving character within varying cultural contexts. Advocating a more applied emphasis, Aitchison (1986) has suggested that too few classificatory studies have addressed themselves to practical problems or served policy-oriented needs.

While there is clearly much scope for development and improvement on these and other counts, it cannot be denied that classificatory endeavours have contributed much to our understanding of agricultural landscapes. Agricultural geographers will always find themselves in situations where it is appropriate or necessary to identify types of farms and to distinguish types-of-farming regions. Since the conceptual and theoretical significance of classification to geographical inquiry in general has been subjected to considerable critical scrutiny (e.g. Harvey 1969), the intention here is to focus more specifically on the practical difficulties that arise in seeking to build typologies and to structure regionalizations of farming systems – the questions that need to be asked at various stages in the classificatory process and the crucial, but often highly subjective, decisions that have to be taken. In so doing illustrative reference will also be made to a range of empirical studies.

The process of classification

The quality of research in agricultural geography is greatly influenced by the
way in which the features being investigated are defined, measured and
categorized. These issues are essentially classificatory (taxonomic) in nature
and warrant very careful consideration. The key problem facing would-be
taxonomists is that of choosing between alternative courses of action. Classification is a demanding and often vexatious decision-making process, and all
the more so because in the end it is never possible to say that the derived
solution is a 'right' or 'true' one. Classifications can only be appraised and

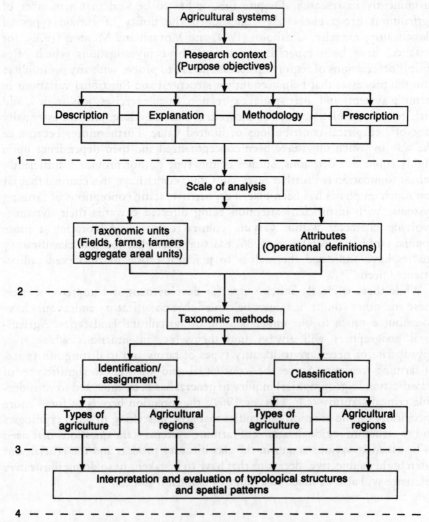

Fig. 5.1   Stages in the process of classification (*Source:* Aitchison 1986)

defended in the light of the *purposes* for which they have been structured. It is important, therefore, that these purposes be specified explicitly and without ambiguity. Then, and only then, is it possible to move on to the formulation of an appropriate taxonomic strategy, a strategy which requires taxonomists to make difficult and key decisions concerning the units to be classified, the attributes to be used in the classification, and the methods by means of which the classifications themselves are to be effected (Fig. 5.1).

Study areas and taxonomic units

Empirical studies of farming systems relate to particular regional *context*. The choice of context – its size and character – is clearly of some relevance for it must provide a suitable setting in which to explore the issues being examined. At a more practical level it must also take into account the availability of resources, and the constraints that they might impose (e.g. the time and cost involved in data collection). Classifications of farms and farming-type regions have been effected at almost every conceivable scale – from typological analyses of small farming communities, at one extreme, to regionalizations of world agriculture at the other. Perhaps not surprisingly, the size of study areas is often determined by the nature of available data and the spatial format in which it is collated. It is for this reason that so many typologies and regionalizations relate to spatial domains, the boundaries of which follow administrative limits recognized by data-collecting agencies (e.g. states, provinces, counties, parishes or communes). Over recent years the vast amount of data gathered by such agencies, and the relative ease with which it can now be processed, have encouraged agricultural geographers to undertake highly detailed statistical analyses of very extensive areas. Studies of farming patterns in Europe (Kostrowicki 1984), the US (Gregor 1982) and Canada (Troughton 1982) are cases in point.

A critical stage in the process of classification, and one which is not unrelated to the dimensions of the study area, concerns the choice and definition of taxonomic units – those elements or features within the agricultural landscape that are the essential focus of interest. While numerous studies have classified systems of land use using data at 'field' level (e.g. in testing the applicability of the von Thünen model – Chisholm 1979), it is commonly claimed that in typological and regional analysis the 'farm' constitutes the most meaningful unit of taxonomic reference (Birch 1954). This derives from the fact that agricultural patterns actually emerge as a result of decisions taken by individual farmers concerning the organization of their production systems – the crops and grasses to be grown, the livestock to be husbanded and the markets to be served. The assumption is that by setting the analysis at farm level, the prospects for more penetrating explanations of the resultant patterns are that much greater.

113

There is much truth in this viewpoint, but certain caveats need to be advanced when it comes to considering whether or not the 'farm' is always the most appropriate analytic unit. If the goals of an inquiry seek to account for typological or regional patterns by highlighting decision-making processes, then it has to be recognized that increased vertical and horizontal co-ordination within the agricultural industry has contributed to a substantial shifting of the locus of decision-making beyond the individual farmer (see p. 163). The growth of joint farming systems, where the resources of individual farms are pooled and managed in an integrated manner following group decisions (e.g. *groupements agricoles en commun* in France), and of contract farming, where production practices at farm level can be dictated by food-processing agencies and agribusinesses, are but two developments which underline the significance of this trend. The role of the state in agriculture and its often powerful control on decision-making complicates the picture even further. Finally, it also has to be appreciated that in many regions of the world the notion of singular units of landholding does not apply (e.g. customary systems of land tenure in Africa).

While these caveats are worth emphasizing, it is evident that for many studies a farm-based approach will remain attractive. A practical problem which arises here, however, is that census returns for individual holdings are seldom readily available (see p. 40). This means that agricultural geographers are frequently obliged to gather farm information through questionnaire surveys, and often for only small samples of holdings. Even this can prove difficult since lists of populations of holdings and their locations can themselves be difficult to come by. An associated dilemma is that without information on farms within a study area it is impossible either to make a decision as to what constitutes an agricultural unit (e.g. the distinction between hobby, part-time and full-time farms), or to ensure that an adequately representative sample is drawn (e.g. stratifications of farms in terms of such diagnostic attributes as farm size). In many parts of the world data-collecting agencies actually classify agricultural holdings into types, but this information is frequently only available in aggregate form (i.e. total numbers of farms of specified types within sets of administrative areas). In the EC, for instance, an elaborate typological system has been established. Here, using standard gross margin (SGM) coefficients (see p. 106), agricultural holdings are assigned to types according to the proportionate significance of selected enterprises (see below).

In analysing aggregate data relating to total numbers of farms it is important to appreciate that the administrative areas to which they refer are being treated as 'collective' rather than 'singular' entities. To say that in a particular county, x per cent of holdings are engaged in cash cropping (in which case the county is a 'collective' unit), is not the same as saying that x per cent of the farm-land in that same county is under cash crops (in which case the county is a 'singular' unit). As Harvey (1969: 352) has noted, these two forms of statistical aggregation are logically quite distinct and, depending upon

114

empirical circumstances, could yield different patterns requiring differing interpretations.

Given that the majority of classificatory studies in agricultural geography are based on aggregate data for census divisions, it needs to be stressed that such divisions generally segment agricultural landscapes in an arbitrary and highly unsatisfactory manner. Summary statistics for census divisions that straddle sharply contrasting physical environments, for instance, can miss meaningful internal variations in farming practice and at the same time imply the existence of composite systems of production which have no basis in reality. Birch (1954), in an analysis of farming patterns in the Isle of Man, has neatly illustrated the distortions and inferential problems that can arise as a consequence. Also, as discussed in Chapter 2, agricultural geographers have to assume that the crops, grassland and livestock attributed to particular census divisions are actually to be found there. In reality, as Coppock (1960) has shown for parishes in England and Wales, this may not be the case; indeed, because farm and administrative boundaries are seldom coincidental there can be gross discrepancies. Combining census divisions, with a view to creating a database that is more homogeneous in terms of the size and shape of the constituent areas can help, but the feeling that one may be mapping phenomena that are not really there still lingers. These issues, coupled with those alluded to above, serve to underline the fact that in classificatory studies the selection of taxonomic units has to be exercised judiciously, with an eye not only on the aims of the investigation but also on the possible effect that the selection might have on the interpretation of results. The same also applies when it comes to choosing the attributes to be used in actually identifying types and regions of agriculture.

Taxonomic attributes

Classifications of farms and farming-type regions are attribute-specific. The task of selecting attributes is accordingly of crucial significance and must once again be approached with the aims of the investigation clearly in mind. Also of importance is the way in which the chosen attributes are scaled and the level at which they are measured (e.g. categoric, ordinal or numeric). Since such matters can have analytic implications, it is usual to associate decisions on attribute measurement with decisions on the procedures to be adopted in the classification itself. Geographers have used a wide variety of attributes in structuring typologies and regionalizations of agricultural systems. These include physical and monetary measures of various factors of production (e.g. land capability, the scale and intensity of operations in terms of inputs of land, labour and capital), the 'behavioural' characteristics of farmers and growers (e.g. motives, attitudes and values), and the level of specialization and commercialization of production systems. Of all the attributes that have been adopted over the years, however, the most frequently encountered are undoubtedly those relating to the enterprise structures of farms and farming regions.

In many classificatory studies statistics for individual crops and livestock are aggregated into more general enterprise categories (e.g. Coppock 1964) and their significance rated using standard conversion coefficients such as standard man-days (SMDs), livestock feed requirements, gross margins, gross outputs and grain-equivalent units. A problem here, at least as far as prospects for cross-study comparisons are concerned, is that the conversion coefficients themselves are seldom uniform. Officially recognized standards can vary markedly from country to country and, as in the case of the gross margin statistics adopted by the EC, can even vary within individual countries. The use of conversion coefficients is widespread and only occasionally are agricultural geographers in a position to base their classifications on actual input or output figures for individual enterprises. Whatever the scales on which enterprise attributes are finally measured, they are normally subjected to further re-expression in the form of proportions, densities or statistical coefficients (e.g. location quotients, measures of diversification in enterprise profiles). This too can have a strong influence on the classificatory structures that eventually emerge, and is yet another decision that has to be carefully considered. Although some writers working with data for census divisions have deliberately elected to base their analyses on raw total statistics, such a strategy is generally avoided since it involves a differential, and therefore biased, treatment of large and small areas. In these circumstances there is a danger that resultant variations in types of farming may be more a reflection of the size of taxonomic units than they are of any inherent agricultural differences.

A revealing insight into the difficulties that can arise in seeking to isolate meaningful sets of taxonomic attributes, together with associated operational definitions, is to be gained from a reading of conference papers prepared for the International Geographical Union Commission on Agricultural Typology between 1965 and 1979 (e.g. Vanzetti 1972, and 1975; Reeds 1973). The remit of this Commission was to structure a methodological framework for the typological analysis of world agriculture. Given the range of system types to be embraced, it is evident that such a framework would have to be flexible, not too demanding in terms of the levels at which attributes are to be measured and, in regard to the choice of attributes themselves, highly selective. Skilfully guided by the chairman of the Commission – Jerzy Kostrowicki – a working typological model has been successfully established. The model itself (see below) is based upon a balanced set of 28 attributes, organized into 4 subgroups – social, operational, production-oriented and structural. Without entering into detail, it can be noted that all of the attributes (Table 5.1) are expressed as ratios – proportions and intensities per unit area or per unit of labour – and that some involve the use of standard conversion coefficients. Thus, gross agricultural production is measured using grain-equivalent units, while livestock and draught animals are respectively rated in

Table 5.1   Typology of world agriculture: attributes

Social
  1. Percentage of total agricultural land held in common
  2. Percentage of total agricultural land in labour and share tenancy
  3. Percentage of total agricultural land in private ownership
  4. Percentage of total agricultural land operated under collective or state management
  5. Number of active workers per agricultural holding
  6. Area of agricultural land per holding (hectares)
  7. Gross agricultural production per agricultural holding

Operational
  8. Number of active agricultural workers per 100 ha of agricultural land
  9. Number of draught animals per 100 ha of cultivated land
 10. Number of tractors, harvesters, etc. in terms of total horsepower per 100 ha of cultivated land
 11. Chemical fertilizers: NPK per hectare of cultivated land
 12. Irrigated land as a percentage of total cultivated land
 13. Harvested land as a percentage of all arable land (including fallow)
 14. Livestock units per 100 ha of agricultural land

Production
 15. Gross agricultural production per hectare of agricultural land
 16. Gross agricultural production per hectare of cultivated land
 17. Gross agricultural production per active agricultural worker
 18. Gross commercial production per active agricultural worker
 19. Commercial production as a percentage of gross agricultural production
 20. Commercial production per hectare of agricultural land
 21. Degree of specialization in commercial production

Structural
 22. Perennial and semi-perennial crops as a percentage of total agricultural land
 23. Grassland (permanent and temporary) as a percentage of total agricultural land
 24. Food crops as a percentage of total agricultural land
 25. Livestock production as a percentage of gross agricultural production
 26. Commercial livestock production as a percentage of gross commercial production
 27. Gross production of industrial crops as a percentage of total agricultural production
 28. Herbivorous livestock as a percentage of total livestock

---

terms of weight equivalent units and horsepower units. The process of classification can be conceived of as a search for patterns of ordered variation within a matrix of attribute scores. There are innumerable ways in which such a search can be carried out, and once again the would-be taxonomist must be prepared to make, and justify, what are often highly subjective decisions concerning which approach (taxonomic procedure) to adopt and how many types, classes or regions to recognize.

Taxonomic procedures

In the field of numerical taxonomy a distinction is commonly drawn between methods of assignment (or identification) and methods of classification. In the

117

first of these, *individual* taxonomic units are categorized by assigning them to pre-defined classes; in the second, populations of taxonomic units are assessed *collectively* with a view to isolating meaningful clusters of individuals. This distinction is useful, if not always clear-cut, and provides a convenient framework within which to consider alternative taxonomic procedures.

Methods of assignment

Methods of assignment have been widely used by agricultural geographers, agricultural economists and agricultural data-gathering agencies. One of the main reasons for this is that once an acceptable typological grid has been established (i.e. the set of predetermined classes), the task of assigning individual taxonomic units to particular categories can normally be accomplished with speed and facility. It is the specification of the typological grid, and the definition of constituent cells (e.g. type-of-farming classes), that can be problematic. In many parts of the world agricultural census agencies have drawn up typological grids which simply distinguish type-of-farming classes by detailing critical values for selected attributes. These values are structured so as to ensure that all farms can be allocated unambiguously to single cells. In taxonomic parlance this means that the cells are 'mutually exclusive' and 'jointly exhaustive'. An example of such a typological system is that adopted by the EC (Commission of the European Communities 1978). Too complex to detail here, this typology is hierarchically structured and recognizes 17 'principal' and 58 'particular' types of farming. For purposes of assignment each of these types is characterized by a limit, or set of limits, which detail the proportion of total SGMs associated with specified enterprises. Thus, for a farm to be identified as a 'specialized dairying' unit, its herd of dairy cows would have to account for over two-thirds of the farm's total SGMs. The same threshold is used to distinguish other specialized systems of production.

The typology also differentiates 'partially dominant' types, where single enterprises (field crops, horticulture, permanent crops, grazing livestock or pigs and poultry) account for between one- and two-thirds of total gross margins, and 'bipolar' types where two enterprises satisfy this condition. While these class limits and the way in which the classes themselves have been defined might be criticized for being rather arbitrary, it must be appreciated that the typology has to cater for the categorization of large numbers of holdings and at the same time has to embrace a very wide range of system types. To propose a more data-dependent typology, with the structure and limits of classes possibly varying from year to year, would also be inappropriate since one of the main aims of the whole exercise is to establish a standard framework within which to monitor patterns of agricultural change.

An alternative approach to the adoption of seemingly arbitrary typological thresholds is to base class limits on recognized statistical parameters. In compiling distribution maps and in identifying types-of-farming systems, agricultural geographers make frequent use of central tendency statistics and

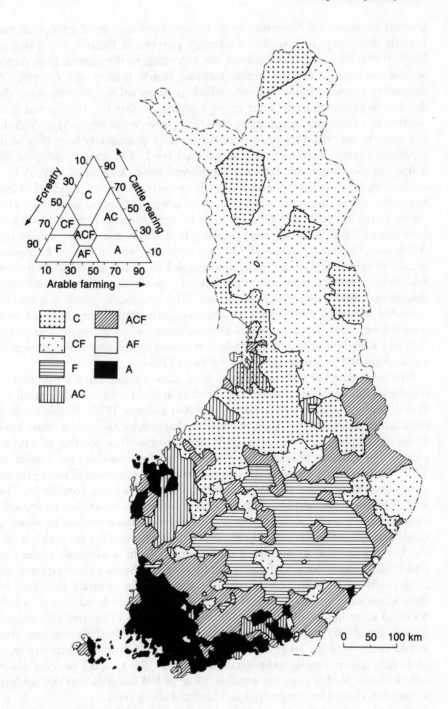

Fig. 5.2   Types of farming in Finland (*Source:* Varjo 1984)

selected measures of dispersion (e.g. means, medians, standard deviations, quartile divisions). In a study of farming patterns in Finland, for instance, Varjo (1984) has categorized communes according to the relative importance of three enterprise types – arable farming, cattle rearing and forestry. A distinctive feature of this analysis, which is based on the contributions the three enterprises make to total gross margins, is that the typological grid consists of a triangular graph partitioned into seven main sectors (Fig. 5.2). In this case the sector limits (class boundaries) have deliberately been chosen to represent average conditions at the national level. The strategy adopted by Varjo can be contrasted with that employed by Jackson et al. (1963) in a similar study of cropping and livestock systems in eastern England. Here triangular graphs are partitioned in a more arbitrary fashion, with the sector limits being set at 20 per cent for each of the cropping (grains, roots and horticulture) and livestock (dairy cattle, beef cattle and sheep) categories. Aitchison (1981) has suggested another approach to the use of triangular graphs in which taxonomic units are assigned to classes by calculating their Euclidean distances from specified sets of coordinates. These coordinates denote the centres of pre-defined classes. The procedure, which can also be extended to include four attributes (the triangular graph being replaced by a tetrahedron), involves a geometrical transformation of the basic percentage data and a method of assignment which is essentially an alternative version of a combinatorial technique devised by Weaver (1954).

First formulated to distinguish crop combination regions in the Midwest of America, Weaver's 'least squares' procedure has since been widely used by agricultural geographers (e.g. Coppock 1964; Gillmor 1977). In this method the attributes which have been selected to characterize taxonomic units have to be measured on a common scale and expressed as profiles of ranked proportions. These 'observed' profiles are then matched against a suite of 'ideal' profiles using a simple deviation statistic. The number of ideal profiles in any investigation is equal to the number of attributes being considered. The first of these profiles depicts a situation of complete concentration or specialization in the pattern of proportions (i.e. a one-attribute system in which a single attribute accounts for 100 per cent of the total value for the scale), while the last depicts one of complete uniformity (i.e. an m-attribute system in which all m attributes are of equal significance). Between these extremes are ideal profiles for two to (m-1) attribute systems. To ascertain to which of these ideals an observed profile of ranked proportions is closest, and to which it should accordingly be assigned, deviations (i.e. sums of squared differences) are calculated. By way of illustration, Table 5.2 shows the deviations that would be derived for an agricultural holding in which four enterprises (e.g. beef cattle, sheep, cereals and pigs) account for 54, 28, 13 and 5 per cent of net farm income. In this case, the smallest value of 694 suggests that the holding in question should be categorized as a beef and sheep farm.

A problem that can arise at this stage, and one which exposes an inherent weakness in the least squares procedure itself, is that in practice the number of

Table 5.2   Weaver's least squares method of assignment

| | Enterprises (in rank order) | | | |
	Beef	Sheep	Cereals	Pigs
% Net farm income	54.0	28.0	13.0	5.0
% Single attribute ideal	100.0	0.0	0.0	0.0
Deviations	46.0	−28.0	−13.0	−5.0
Sum of squared deviations = 3 094				
% Two-attribute ideal	50.0	50.0	0.0	0.0
Deviations	−4.0	22.0	−13.0	−5.0
Sum of squared deviations = 694				
% Three-attribute ideal	33.3	33.3	33.3	0.0
Deviations	−20.7	5.3	20.3	−5.0
Sum of squared deviations = 894				
% Four-attribute ideal	25.0	25.0	25.0	25.0
Deviations	−29.0	−3.0	12.0	20.0
Sum of squared deviations = 1 394				

combinations and permutations of attributes can turn out to be rather large. As a consequence the taxonomist often has to be prepared to sacrifice detail for the sake of clarity, and resort to some further aggregation of the results. In seeking to distinguish patterns of regional variation in farming systems, for example, agricultural geographers have found themselves obliged to compile maps which emphasize only first-ranking attributes. Permutations of attributes may be recorded on such maps, but only as strings of symbols located within each of the constituent areas. One example of the agricultural regions identified by this method is discussed in Chapter 6, where reference is made to a classification of agricultural enterprises in Northern Ireland.

All of the assignment strategies that have been considered thus far are limited to situations in which the attributes can be measured on common scales and expressed as percentages. Such constraints clearly narrow the compass of taxonomic inquiry. It is possible, however, to structure an assignment procedure that allows for a wide range of attributes to be considered simultaneously, even though they may be measured on completely different scales. A notable example of such a procedure is that developed by the Commission on Agricultural Typology (see above). It will be recalled that the typological framework proposed by this Commission is based on a mixed set of 28 attributes (Table 5.1), and that scores on these attributes are initially in the form of ratios (percentages and intensities). To facilitate analysis, measurement scales for each of the attributes are standardized, with world ranges being subdivided into five classes. This means that profiles of attribute scores for individual taxonomic units are accordingly reduced to strings of 28 integers. It is these strings of numbers (ranging from one for low scores to five for high scores) that are used in assigning the units to particular classes.

The assignment process itself is stylistically similar to that adopted in Weaver's 'least squares' procedure and involves the calculation of a simple deviation statistic. Whereas the classes in the 'least squares' model are theoretical ideals, in this case they have been established empirically following extensive literature surveys and field investigations. The system of classes is hierarchical in structure and allows for three levels of categorization. At the highest level of generalization, six basic types of farming are distinguished. These first-order types have been labelled as follows:

E – traditional extensive agriculture;
T – traditional intensive agriculture;
M – market-oriented agriculture;
S – socialized agriculture;
A – highly specialized livestock breeding;
L – latifundia.

To date, over 20 second-order and more than 100 third-order types of farming have been identified. Reference profiles (i.e. class scores for the set of 28 attributes) have been detailed for each and every one of these types and it is against them that observed profiles of scores are matched. Model codes for the six first-order types are displayed in Table 5.3. The matching process is based on a simple summation of the difference between observed and expected profiles of scores. Unlike the 'least squares' procedure, however, the system allows individual taxonomic units to be associated with more than one type of farming. This arises because all deviations below certain specified thresholds are deemed to be significant. These thresholds have been set at 34, 23 and 12 for the first-, second- and third-order types, respectively. Thus, if deviation statistics for a particular farming region yield values below 12 on two third-order types then the region is deemed to be 'transitional' in character and is recognized as such. A further distinctive feature of the typology is that it is open-ended and allows for new types to be introduced, especially at the third-order level. The need for new types is signalled when the matching process fails to isolate any deviations below the specified thresholds.

Table 5.3 Typology of world agriculture: first-order types and model codes

First-order types	Attribute categories			
	Social	Operational	Production	Structural
E – Extensive traditional agriculture	5321211	1111121	1221111	1333314
T – Traditional intensive agriculture	1241222	4412242	4422221	1142213
M – Market-oriented agriculture	1151233	2154343	4455544	1223313
S – Socialized agriculture	1115555	3243242	3333433	1242214
A – Highly specialized livestock	1133445	3000003*	3045535*	1515513
L – Latifundia	1551354	2211132	2334422	1323424

* Zeros allocated where attributes do not apply.

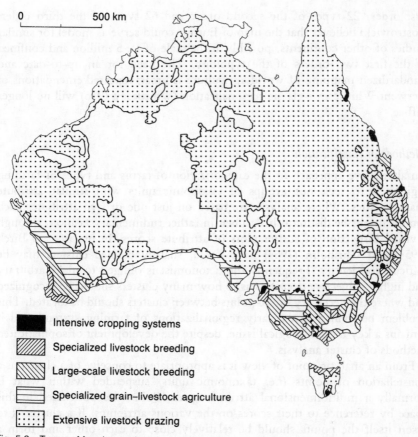

Fig. 5.3   Types of farming regions in Australia (*Source:* after Scott 1983)

Legend:
- ■ Intensive cropping systems
- ▨ Specialized livestock breeding
- ▩ Large-scale livestock breeding
- ▤ Specialized grain–livestock agriculture
- ⣿ Extensive livestock grazing

It is necessary to appreciate that the typological model described here has matured into its present form following 20 years of debate and experimentation. Agricultural geographers from many parts of the world have tested the model as it has developed. Interesting applications include studies of farming systems in France (Bonnamour and Gillette 1980), Canada (Troughton 1982), Malaysia (Hill 1982), Australia (Scott 1983), India (Singh 1979) and Europe (Kostrowicki 1984). Thus far the typology has mainly been used for purposes of regionalization. Figure 5.3 illustrates the results of an application of the model for Australian agriculture; here five cropping and livestock zones are identified (Scott 1983). A major project associated with the development of the typology has been the compilation of a farming-type regions map of Europe. This map, edited by Kostrowicki, but requiring the collaboration of agricultural geographers from many parts of Europe, was published in 1985 at the scale of 1 : 2.5 million. Using a combination of colours and shading schemes, this highly impressive map identifies all 6 types of agriculture of the first order, 22 types of the second order and 62 types of the third order. Kostrowicki believes that the map of Europe could serve as model for similar studies of other continents, possibly at the scale of 1 : 5 million and confined to the first two orders of the typology. The need for an up-to-date and standardized mapping of world agriculture is great; repeated emendations of Derwent Whittlesey's classic regionalization (Whittlesey 1936) will no longer suffice.

Methods of classification

An alternative approach to the categorization of farms and types-of-farming regions is to seek out groups of taxonomic units with similar attribute characteristics. Where analyses are based on just one or two attributes such groups might be distinguished, albeit in rather rudimentary fashion, through a visual inspection of scatter plots of attribute scores. In practice, as Birch (1954) and Chisholm (1964) have noted, sharply defined clusters may be difficult to identify. Frequently the taxonomist is obliged to make arbitrary and highly subjective decisions as to how many clusters should be recognized and where the necessary break-points between clusters should be located. This problem bedevilled many early regionalizations of farming systems and it remains a key methodological issue, despite the development of sophisticated methods of cluster analysis.

From an analytic point of view it is appropriate to conceive of a cluster as a constellation of points (i.e. taxonomic units) suspended within what is normally a multidimensional attribute space. Points are positioned in this space by reference to their scores on the various attributes. If a cluster is to assert itself the points should be relatively close to each other and form a compact swarm – the more compact the swarm, the more distinct the cluster. One way of measuring the compactness of a cluster is to calculate average inter-point distances and their variances. Not all cluster-analytic methods use these measures in isolating clusters, but those that do (e.g. Ward's error sum of squares procedure) have proved to be particularly popular among agricultural geographers. Such 'minimum variance' solutions, as they are often termed, have been widely employed in building agricultural typologies and for purposes of regionalization (e.g. Anderson 1975; Ilbery 1981).

While the selection of an appropriate clustering routine, from the large number currently available, is important, and can have implications in regard to the patterns that eventually emerge (Byfuglien and Nordgard 1973; Aitchison 1975), it is by no means the only critical decision that has to be taken in carrying out a taxonomic analysis. Perhaps even more important is the approach adopted by the taxonomist in ensuring that a reliable classification has been achieved. The increased availability of powerful computing systems and the necessary software may have been clustering a very straightforward analytic exercise, but the taxonomist is still obliged to make vital decisions at various stages in the manipulation of the data. The interactive nature of the classificatory process and the significance of the issues involved can best be demonstrated through a practical application.

The method of cluster analysis is best explained through a worked example. The central objective of the exercise presented here is simply to effect a 'minimum variance' regionalization of farming systems in England and Wales using aggregate statistics for 54 counties (excluding the Isles of Scilly). The statistics record the numbers of agricultural holdings within each of the counties, classified according to the typology adopted by the EC (see p. 118). For purposes of analysis, six basic types of holdings are distinguished – dairying, cattle and sheep, cropping, horticulture, pigs/poultry and unclassified (i.e. holdings without any of the enterprises recognized by the EC typology). Rather than base the clustering on raw aggregate statistics or on percentages, because of statistical problems that can arise when working with closed percentage sets, the decision has been taken to use intensity ratios which record the number of farms of particular types per 1 000 ha of agricultural land. Given the way in which the data are subsequently to be processed, it is of no importance that, because of variations in the areal size of production units, the range of intensity measures is likely to be different for each of the six type-of-holding categories.

Having defined the set of attributes, the next stage in the investigation is to consider the geometry of the taxonomic space to which they relate. In so doing it has to be appreciated that 'minimum variance' clustering procedures operate on dissimilarity matrices the coefficients of which are inter-point Euclidean distance measures. If these distances are derived using untransformed attribute scores then it has to be assumed that the attributes themselves define a Euclidean space. For this to apply the attributes must all be statistically independent of each other. Only if the dimension (attributes) of the taxonomic space are orthogonal (i.e. uncorrelated) can distances be meaningfully calculated using a Pythagorean equation. Whether or not the attributes being employed in a particular inquiry are indeed orthogonal can be determined by ascertaining the degree to which they covary. The matrix of correlation coefficients presented in Tables 5.4 shows that there are in fact quite strong levels of inter-association between the six attributes being considered here. To circumvent this problem taxonomists normally 'orthonormalize' the original data set by subjecting it to a transformation procedure. For this, and other

Table 5.4 Product moment correlations for farm types in England and Wales

Farm types	1	2	3	4	5	6
1. Dairying	1.00					
2. Cattle and sheep	0.59	1.00				
3. Cropping	− 0.53	− 0.71	1.00			
4. Pigs and poultry	0.20	0.18	0.26	1.00		
5. Horticulture	− 0.14	− 0.11	0.35	0.58	1.00	
6. Unclassified	0.48	0.61	− 0.24	0.75	0.50	1.00

reasons (see below), agricultural geographers have made frequent use of various factor-analytic models. In this study the six farm-type attributes have been orthonomalized by means of a principal components analysis. The analysis suggests that 90 per cent of the information in the initial data matrix is captured by three, orthogonal, principal components. Since they define a Euclidean space within which the 54 counties can be readily located, it is appropriate to consider using scores on these three components as the basic data for the clustering phase of the inquiry.

The majority of the classificatory routines that have been employed by geographers form clusters in hierarchical and agglomerative fashion. In contrast to 'divisive' procedures, which proceed by sequentially breaking down total populations of taxonomic units into smaller and smaller clusters, those of the 'agglomerative' variety work in the other direction and fuse individual units and groups of units into broader and broader clusters. In applying both of these strategies the main task facing the taxonomist is to decide at which level in a hierarchy of clusters the most meaningful set of clusters is to be found. Various 'stopping rules' have been proposed, but in the end the decision has to be based upon the overall aims of the inquiry (e.g. the detail required in a regionalization) and on the interpretability of the results.

Apart from the problem of how many clusters to identify, taxonomists also have to recognize that computer algorithms frequently contain built-in options which affect the way in which clusters are actually formed. Decisions on such options can have quite a significant influence on the resultant configuration of clusters. To illustrate the point, Fig. 5.4 presents a series of regionalizations of farming systems in England and Wales based on the orthonormalized set of county data described above. The six clusters displayed in Fig. 5.4(a) have been derived from a straightforward application of Ward's error sum of squares procedure. In this case the sequence of fusions (normally charted as a dendrogram) has simply been halted at the six-cluster stage. This spatial pattern is disturbed, albeit only slightly, if the counties that have been allocated to the various clusters are then 'relocated' with a view to minimizing within-cluster variances (Fig. 5.4(b)). A further alternative, and one which yields markedly different results as far as this particular analysis is concerned (Fig. 5.4(c)), is to begin by allocating the counties to six clusters in random fashion and then to proceed iteratively with the relocation procedure. In certain circumstances it might be appropriate to consider imposing a contiguity constraint when forming clusters. If, for example, it is necessary to define agricultural regions for purposes of administration or planning, it might be desirable to distinguish clusters that define integrated areal units. As Fig. 5.4(d) demonstrates, such a constraint can yield patterns which are less than illuminating from a geographical point of view.

As far as the four regionalizations are concerned, the most meaningful is undoubtedly that derived from the combination of a part-optimum classification and iterative relocation (Fig. 5.4(b)). For this particular classification the

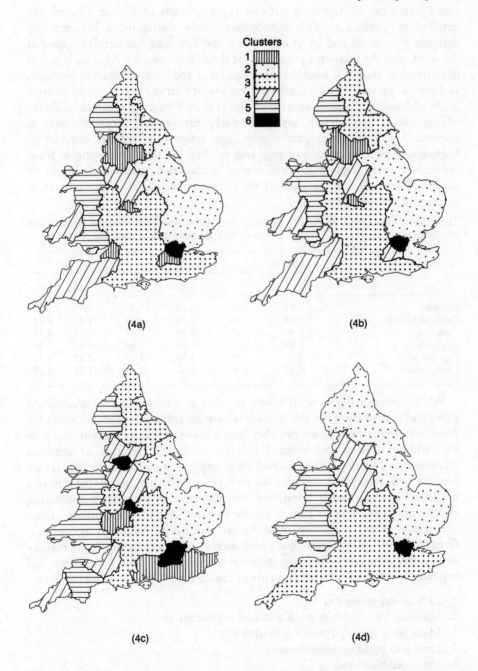

Fig. 5.4 Types of farming regions in England and Wales, 1982: (a) Ward's error sum of squares: six-cluster stage; (b) Ward's error sum of squares: six-cluster stage after relocation; (c) initial random classification with iterative relocation; (d) Ward's error sum of squares with contiguity constraint

127

mean intensities of farms of various types in each of the six clusters are detailed in Table 5.5. The distribution neatly distinguishes between the dairying (cluster 4) and livestock rearing and fattening (cluster 5) regions of the west, and the cash-cropping areas of the east (cluster 2). Also highlighted is a mixed, transitional zone, occupying a large and compact tract in southern and central parts of England, where a variety of farming types prevail (cluster 3). In addition, the classification identifies regions where more intensive forms of agricultural production are particularly prominent. Clusters 1 and 6 (Greater London) return particularly high ratios for holdings devoted to horticulture and the rearing of pigs and poultry. Equally interesting in these 'urbanized' areas is the high incidence of 'unclassified' holdings. It might be suggested that these figures reflect the strong concentration in such areas of essentially 'hobby' farms.

Table 5.5 Farming-type regions of England and Wales (minimum variance clusters with relocation at six-cluster stage: mean number of holdings per 1 000 ha – see Fig. 5.4b)

Farm types	Clusters					
	1	2	3	4	5	6
Dairying	0.81	1.03	2.60	7.56	3.47	1.53
Cattle and sheep	9.94	1.98	4.79	8.96	11.19	6.42
Cropping	3.90	7.40	3.59	2.17	0.63	4.95
Horticulture	3.81	1.56	1.00	1.38	0.66	4.40
Pigs/poultry	3.17	2.17	1.30	1.07	0.32	9.41
Unclassified	4.37	1.18	1.53	2.81	1.95	5.62

While factor-analytic models can be used as simple data transformation procedures, providing the taxonomist with standardized orthogonal scales for purposes of clustering, they can also fulfil a more profound heuristic purpose by yielding illuminating insights into the statistical structure of attribute relationships (Tarrant 1974). Indeed most applications of such procedures do actually seek to explain and label the factors (components) generated within a particular analysis. In so doing, the aim may be either to isolate underlying regularities in the data set being examined or to test specific hypotheses. In an exemplary study of the industrialization process in American agriculture, Gregor (1982) has used principal components analysis to ordinate and classify over 3 000 counties. This investigation is based on eight investment and expenditure attributes, rated in terms of capital inputs per farm and per acre:

1. Labour expenditures;
2. Machine hire, custom work and fuel expenditures;
3. Machinery and equipment investments;
4. Land and building investments;
5. Crop expenditures;
6. Crop investments;
7. Livestock–poultry expenditures;
8. Livestock–poultry investments.

Separate analyses of the two data sets (i.e. inputs per farm and per acre) indicate high degrees of correlation between the attributes, suggesting that principal components analysis might be rewardingly employed in deriving composite scales along which the counties might be arrayed. The upper and lower triangular halves of the matrix, presented in Table 5.6, show correlations between the attributes that make up the two data sets, while Table 5.7 lists loadings for their first components. The component extracted from the 'capital inputs per farm' data captures 52 per cent of the variation in the correlation matrix and serves to identify a 'scale structure of general agricultural industrialization'. For the 'capital inputs per acre' data, the first

Table 5.6   Scale and intensity of agricultural industrialization: capital inputs per acre and per farm*

Attributes	Product moment correlation coefficients							
	1	2	3	4	5	6	7	8
1	1.00	0.89	0.90	0.84	0.83	0.78	0.64	0.59
2	0.66	1.00	0.91	0.85	0.90	0.85	0.59	0.47
3	0.59	0.81	1.00	0.82	0.85	0.79	0.56	0.53
4	0.65	0.73	0.59	1.00	0.83	0.78	0.57	0.47
5	0.46	0.59	0.59	0.37	1.00	0.87	0.48	0.36
6	0.55	0.66	0.68	0.44	0.76	1.00	0.34	0.21
7	0.22	0.24	0.09	0.27	−0.01	−0.11	1.00	0.80
8	0.20	0.28	0.20	0.39	−0.20	−0.18	0.66	1.00

* Upper triangular half – inputs per acre (intensity). Lower triangular half – inputs per farm (scale).
Attributes: (1) labour expenditures per acre/farm; (2) machine hire, etc. per acre/farm; (3) machinery and equipment investments per acre/farm; (4) land and building investments per acre/farm; (5) crop expenditures per acre/farm; (6) crop investments per acre/farm; (7) livestock expenditures per acre/farm; (8) livestock investments per acre/farm.

Sources: Gregor (1982: 160, 175).

Table 5.7   Scale and intensity of agricultural industrialization: component loadings

Attributes	Component loadings	
	Scale component	Intensity component
1	0.7993	0.9492
2	0.9219	0.9572
3	0.8685	0.9408
4	0.7923	0.9110
5	0.7210	0.9113
6	0.7849	0.8455
7	0.2444	0.6946
8	0.2475	0.6084
% Variance explained	52	74

Attributes listed on table 5.6.

Source: Gregor (1982: 162, 176).

component explains 74 per cent of the total variation and is labelled as an 'intensity structure of general agricultural industrialization'. Gregor probes more deeply into the nature of these components (using Varimax rotations to isolate substructures), but does use these basic dimensions as scales for the classification of individual counties. This is achieved by calculating component scores. The spatial distributions of scores on each of the scales are examined, and in a final synthesis, in which the scores on the two scales are considered together, the counties are ordinated and a typology of agricultural industrialization determined. With counties being partitioned according to whether or not their scores are above 0.5, below −0.5 or between these limits, a set of nine types is distinguished. Figure 5.5 shows the distribution of counties yielding scores above 0.5 on either or both of the components.

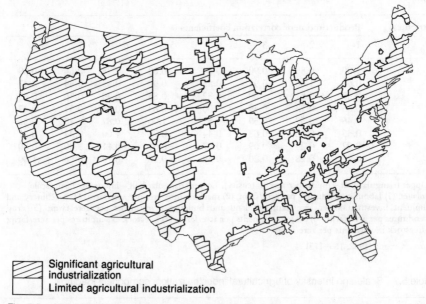

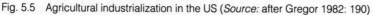

Significant agricultural industrialization
Limited agricultural industrialization

Fig. 5.5   Agricultural industrialization in the US (*Source:* after Gregor 1982: 190)

As noted in Chapter 3, the most highly industrialized agricultural regions (i.e. where scores for both the 'scale' and 'intensity' components are above 0.5) are the areas of Illinois and Iowa, North-east coastal regions, the Pacific South-west and parts of Florida and the Gulf Coast. In the Great Plains and the Intermountain West, with their areally large livestock rearing farms, counties generally achieve scores of over 0.5 only on the 'scale' component. Gregor categorizes counties that fail to achieve scores in excess of 0.5 on either of the two components as 'types of limited industrialized agriculture'. The weakest of all the counties in terms of levels of investment and expenditure are seen to be in areas where environmental conditions impose major constraints and where market opportunities are far from favourable (e.g. Alaska, the Ozark–Quachita Highlands, much of central and eastern Texas, the Colorado Plateaux, parts of the Great Plains, the Great Lakes Cutover and Appalachia).

Although cluster and factor-analytic procedures have found favour among agricultural geographers over recent years, facilitating classifications of large and complex sets of data, it has to be stressed that these procedures do need to be selected and used with considerable circumspection. As has already been noted, the main problem is that so many methods and analytic options are available. Since different methods can yield different solutions there is a strong case for the recommendation of a more experimental and interactive approach in taxonomic inquiry, with individual sets of data being subjected to a variety of tests and procedures. As to which of the resultant solutions should eventually be adopted, this is a matter that can only be decided upon by the taxonomist after having considered the statistical evidence and the meaningfulness of the findings as far as the aims of the particular investigation are concerned. It has to be accepted that this subjective stage in the taxonomic process is unavoidable and often requires a high level of intuitive evaluation (Aitchison 1986).

Conclusion

Regionalizations and typologies of agriculture have contributed greatly to our understanding of farming systems throughout the world, and have contributed significantly to the development of agricultural geography as a systematic science. However, if a single criticism were to be levelled at classificatory research in agricultural geography it would be that the vast majority of studies seek to answer essentially academic questions. Admittedly, these studies frequently contain information of practical consequence to planners and policy-makers, and to the farming communities concerned, but this seldom constitutes a central focus for discussion and debate. More often than not revelations of this type appear as incidental by-products of 'pure' research investigations. Given the socio-economic and environmental problems that currently beset so many agricultural regions of the world, this lack of a strong applied stance in taxonomic studies is regrettable and should be rectified. The adoption of such a stance would enrich the classificatory tradition in agricultural geography and give it a much-needed sense of social purpose.

References

Aitchison, J. W. (1975) Cluster analysis, regionalization and the agricultural enterprises of Wales. In Vanzetti, C. (ed) *Agricultural typology and land utilization*. Centre of Agricultural Geography, Verona, pp. 17–33

Aitchison, J. W. (1981) Triangles, tetrahedra and taxonomy. *Area* **13**: 137–43

Aitchison, J. W. (1986) Classification of agricultural systems: types and regions. In Pacione, M. (ed) *Progress in agricultural geography*. Croom Helm, London, pp. 38–69

Anderson, K. E. (1975) An agricultural classification of England and Wales. *Tijdschrift voor Economische en Sociale Geografie* **66**: 148–57

Birch, J. W. (1954) Observations on the delimitation of farming-type regions, with special reference to the Isle of Man. *Transactions of the Institute of British Geographers* **20**: 141–58

Bonnamour, J. and **Gillette, C.** (1980) *Les types d'agriculture en France 1970: essai méthodologique.* CNRS, Paris

Boserup, E. (1965) *The conditions of agricultural growth: the economics of population change under population pressure.* Allen and Unwin, London

Byfuglien, J. and **Nordgard, A.** (1973) Region-building: a comparison of methods. *Norsk Geografisk Tidsskrift* **27**: 127–51

Chisholm, M. (1964) Problems in the classification and use of farming-type regions. *Transactions of the Institute of British Geographers* **35**: 91–103

Chisholm, M. (1979) *Rural settlement and land use. An essay in location.* Hutchinson, London

Commission of the European Communities (1978) Commission decision of 7 April 1978 establishing a Community typology for agricultural holdings. *Official Journal of the European Communities* **L148**: 1–34

Coppock, J. T. (1960) The parish as a geographical–statistical unit. *Tijdschrift voor Economische en Sociale Geografie* **51**: 317–26

Coppock, J. T. (1964) Crop, livestock and enterprise combinations in England and Wales. *Economic Geography* **32**: 65–81

Gillmor, D. A. (1977) *Agriculture in the republic of Ireland.* Akademiai Kiado, Budapest

Gregor, H. F. (1982) *Industrialization of United States agriculture.* Westview, Colorado

Grigg, D. B. (1969) The agricultural regions of the world: review and reflections. *Economic Geography* **45**: 95–132

Harvey, D. (1969) *Explanation in geography.* Edward Arnold, London

Hill, R. D. (1982) *Agriculture in the Malaysia region.* Akademiai Kiado, Budapest

Ilbery, B. W. (1981) Dorset agriculture: regional types. *Transactions of the Institute of British Geographers* **6**: 214–27

Jackson, B. G., Barnard, C. S. and **Sturrock, F. G.** (1963) *The pattern of farming in the eastern counties: a report on a classification of farms in eastern England.* School of Agriculture, Cambridge

Kostrowicki, J. (1984) Types of agriculture in Europe: a preliminary outline. *Geographia Polonica* **50**: 132–49

Morgan, W. B. and **Munton, R. J. C.** (1971) *Agricultural geography.* Methuen, London

Reeds, L. G. (ed) (1973) *Agricultural typology and land use.* McMaster University, Ontario

Scott, P. (1983) The typology of Australian agriculture. *Geographia Polonica* **46**: 7–19

Singh, V. R. (1979) Agricultural typology of India. *Geographia Polonica* **40**: 113–31

Tarrant, J. R. (1974) *Agricultural geography.* David and Charles, Newton Abbot

Troughton, M. (1982) *Canadian agriculture.* Akademial Kiado, Budapest

Vanzetti, C. (1972) *Agricultural typology and land utilization.* Centre of Agricultural Geography, Verona

Vanzetti, C. (1975) *Agricultural typology and land utilization.* Centre of Agricultural Typology, Verona

Varjo, U. (1984) Changes in farming in Finland 1969–1975. *Fennia* **162**: 103–15

Weaver, J. C. (1954) Crop combination regions in the Middle West. *Geographical Review* **44**: 175–200

Whittlesey, D. (1936) Major agricultural regions of the earth. *Annals of the Association of American Geographers* **26**: 190–240

6

Changing farm enterprises

Christopher Edwards

The previous chapter has shown the heterogeneous nature of agriculture as far as type of farming is concerned, although there are considerable problems of classification and regionalization. The diversity of agriculture provides varying opportunities for the accumulation of capital by those working in the production sector, as well as for those agro-industries supplying farm inputs and processing farm outputs. Suppliers of agrichemicals for crops, for example, are faced by different marketing opportunities in any country compared with the suppliers of concentrated feed for dairy cows. But farming regions are not fixed, they vary in their composition and location over time. This chapter looks in more detail at these features, taking the structure of the individual farm enterprise as its focus.

Enterprise choice in agriculture

One of the fundamental changes in modern agricultural geography has been the recognition that, in order to understand the patterns of agricultural production in any area, greater emphasis must be placed upon the way in which individual crop and livestock enterprises are combined together, to form distinctive production systems on individual farms, and the way in which these combinations of enterprises vary both spatially and over time. A traditional approach of agricultural geographers in attempting to describe and explain spatial diversity in agriculture has been to study the spatial distribution of individual crop or livestock types, and to attempt to explain their pattern of distribution in terms of variations in the factors of production which affect them. It is possible, for example, to study the spatial distribution of barley growing and from this to derive some explanation of the environmental, economic and social factors which affect the distribution of this enterprise. Though this approach has some value, it also has serious limitations. Individual crops or livestock types which make up the production enterprises on farms are rarely produced in isolation. In the previous chapter we saw how individual enterprises may be commonly combined to form distinctive types of farming. In such an orderly association of interrelated and interdependent enterprises, none can be modified without causing related changes to other enterprises in the farm production system. Thus the distribution of barley may well be as much a function of factors affecting other enterprises with which it is associated on the farm, as of factors which directly affect barley production.

It must be remembered that the enterprise combination shown on a type-of-farming map are the cumulative result of all the individual farm enterprise structures encountered in a census district. What is not shown is the range of diversity of individual farm structures which may exist within each region. Gillmor's (1977) study of enterprise combination regions in the Republic of Ireland illustrates this point (Fig. 6.1). In districts in the north and west of the country the combination are essentially simple; they are a result of all or most farms having broadly similar structures based on a limited range of enterprise choice. The regions surrounding Dublin on the other hand, where choice is greater, exhibit more complex structures. But as Gillmor points out, these are derived not as a result of all farms operating complex mixed farming systems, but as a cumulative result of individual farms within the regions operating a wide range of simple specialist structures geared to the nearby Dublin market. Similar patterns were observed by Scott (1981) around the major cities of eastern Australia, where again the complex structures identified in particular regions were often a net result of individual farms opting for different types of specialist enterprises, rather than all farms having complex enterprise structures.

Though a considerable body of published evidence exists which illustrates the spatial variation in enterprise structures which occur at a regional and national level, there is surprisingly little published evidence of spatial variation in enterprise structures at the individual farm level. A study by Edwards (1974), however, serves to exemplify the spatial variations and similarities which may occur between farms within a relatively small geographical area. A survey was made of the enterprise structure of individual farms in a 550 km² area of Co. Londonderry, Northern Ireland. The area consisted of the lowland valleys of the rivers Roe and Bann, where most types of livestock enterprise and limited arable cropping of potatoes and barley are possible, separated by a north–south trending upland area rising to over 300 m, where only hill sheep and cattle enterprises are viable. Results revealed, firstly, a variation between farms in the number of enterprises combined into their enterprise structure (Table 6.1) with some bias towards simple specialist or two-enterprise structures. When actual types of enterprise combined on each sample farm were considered, however, a considerable variety of different individual structures were revealed (Table 6.2). Some 46 distinctive structures were identified for a small sample of 176 farms, illustrating the considerable diversity of enterprise structures which may be encountered among farms within a relatively small area of land.

135

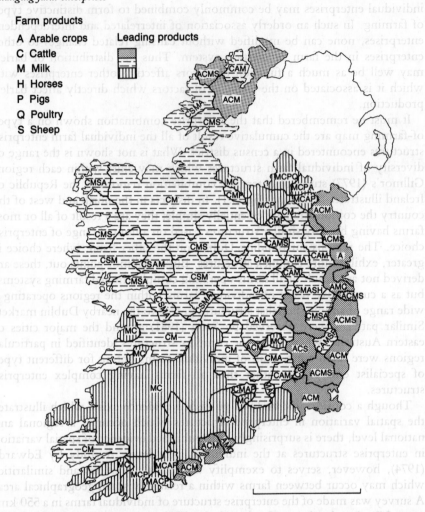

Farm products

A Arable crops   Leading products
C Cattle
M Milk
H Horses
P Pigs
Q Poultry
S Sheep

Fig. 6.1   Enterprise combinations in the Republic of Ireland. Farm products: A = arable crops; C = cattle; M = milk; P = pigs; H = horses; Q = poultry; S = sheep (*Source:* Gillmor 1977)

One way of understanding the varied combination of enterprise on any farm, and the varied response by farmers to their changing circumstances, is to adopt a 'constrained choice model' for the farm business. At any location, a farmer will be offered a varied choice of possible crops and livestock production options, from which a mix of enterprises will be selected which best fit the farmer's personal production goals at any particular time. The mix of enterprises chosen for incorporation into the farm production system may be termed the farm enterprise structure.

Such structures will vary in complexity dependent on a variety of factors. Firstly, the range of possible enterprises which the farmer may be able to incorporate successfully into the enterprise structure will be affected by the

Table 6.1   Number of enterprises in combination, Co. Londonderry, Northern Ireland

Number of enterprises	Number of farms	Percentage of sample farms
1	56	31.8
2	68	38.6
3	39	22.2
4	13	7.4
Total	176	100.0

Source: Edwards (1974: 35).

Table 6.2   Farm enterprise structure, Co. Londonderry, Northern Ireland

Type of enterprise structure	Percentage of sample	
Dairy	16.6	
Beef/sheep	8.3	
Beef	7.2	Plus 30 other distinctive
Beef/pigs	6.1	structures, each
Dairy/pigs/beef	5.0	accounting for <2.0%
Pigs	4.4	of sample,
Dairy/potatoes	3.9	including:
Dairy/pigs	3.9	3 single enterprises,
Pigs/beef/sheep	3.4	7 two-enterprise,
Beef/potatoes	2.8	12 three-enterprise and
Pigs/poultry	2.8	8 four-enterprise structures
Poultry/dairy/pigs	2.8	
Dairy/beef	2.2	
Total	69.4	

Source: Edwards (1974: 50).

opportunities offered and the constraints imposed by a wide variety of external environmental, economic, social and political forces, as they operate at a particular farm location. The farmer can therefore be seen to be acting within a constrained choice model. Factors of the physical environment, for example, may in certain cases impose such severe constraints that the choice of enterprises which can be selected may be strictly limited. In hill and mountain regions a combination of low temperatures, high rainfall, poor soils and low-quality vegetation often reduce the farmer's choice to the simple option of selecting an enterprise structure specializing entirely upon extensive hill sheep production. Similarly, in the interior of Australia and on the rangelands of the south-western US, shortage of water limits vegetation growth and again leaves little alternative to simple enterprise structures based on extensive sheep grazing or cattle ranching. Even where physical conditions are less extreme, limitations in any one of the environmental factors of production may restrict enterprise choice; in the UK, for example, heavy clay soils in much of the western lowlands make arable enterprises difficult, while the lighter soils in the east of the country favour arable farming and restrict good grass growth, consequently reducing the attractiveness of grazing livestock enterprises.

Secondly, variations in other factors of production may limit or extend the choices open to the farmer within the same constrained choice model. Restrictions of farm size, for example, preclude successful cash cereal production on small farms where mechanization cannot be used effectively, while lack of capital or shortage of farm labour may restrict choice to those enterprises with low fixed costs or minimum labour requirements. The degree of complexity of farm enterprise structures, therefore, will be to a large extent governed by limits imposed by particular factors of production.

Thirdly, a major factor which governs the degree of complexity of enterprise structures is the degree of complementarity between individual enterprises, those which interact closely together being more often combined than those which have few functional links. Thus on many dairy farms in the past, the keeping of a subsidiary pig unit formed a natural complement to dairy production, the pigs being fed on skimmed milk or whey from farmhouse butter and cheese making. The development of factory butter and cheese production, and the consequent increased difficulty of obtaining these types of pig feed, coupled with the gradual intensification of pig production into larger specialized units, has reduced this functional link; the pig unit no longer forms as natural a complement to the dairy herd, and a noticeable decline in distinctive dairy/pig enterprise structures has occurred (Edwards 1976: 170).

Finally, the goal motivation and values of farmers can be seen to play a vital part in instituting and explaining enterprise choice in agriculture. Economists have constructed a variety of sophisticated models based upon the assumption of an Economic Man who behaves in a perfectly rational way, based upon perfect knowledge, to achieve the optimum single defined goal of maximizing profit. Simon (1952), however, suggested that decision-makers simplify reality to fit their own capabilities, regard decisions in terms of either satisfactory or unsatisfactory outcomes, and can thus be better described as 'satisficers' rather than 'optimizers'. In an investigation into farming in central Sweden, Wolpert (1964) was able to demonstrate the differences between actual productivity and what might be expected from farmers reacting as optimizing Economic Man. Simon further suggested that the decision-maker rarely seeks a single profit-maximizing objective, but may in fact combine different goals. Gasson (1973), basing her findings on empirical research among farmers in Cambridgeshire, illustrates this wide variety of goals (Table 6.3). Work by Ilbery (1979) also confirms that farmers base their decisions on a variety of concurrent goals within a satisficer model, suggesting that the economic goal of security and stability of the farm business dominates farm decisions; but once this appears to be achieved, farmers are strongly influenced by personal and social considerations, choosing to work with enterprises which give them most satisfaction.

The personal preference and experience of farmers, therefore, together with

Table 6.3   Goals and values of farmers

Instrumental	Maximization of income; satisfactory income; future security of income; risk avoidance; expansion of farm business; provision of good working conditions
Social	Gaining recognition/prestige in farming and outside community; belonging to the farming community; continuing family tradition; working with family
Expressive	Pride of ownership; self-respect; use of special skills/abilities, chance to be creative; meeting a challenge or objective
Intrinsic	Variety of work; preference for outdoor life; value of hard work; independence and freedom

Source: Based on Gasson (1973).

their attitude to risk taking, may affect types of enterprise chosen and combined into enterprise structures. Given a range of equally viable enterprise options, final choice may be governed by whether a farmer has a personal preference for a particular enterprise. Similarly, experience of particular types of production may inhibit or extend the range of choice; a farmer with wide practical experience of a range of different enterprises may more easily adopt a complex farm structure, whereas one whose whole farming life has been devoted, for example, to arable crops may feel inhibited by lack of experience of stockmanship from incorporating a livestock system on to the farm. The farmer's attitude to risk can also be important, some farmers preferring to adopt a more mixed enterprise structure to spread the risk of any one enterprise becoming unprofitable, while others are more willing to accept the greater risk of devoting all their resources to a single specialist enterprise.

The degree of complexity of the final enterprise structure chosen, therefore, will be a product of a constrained choice model of opportunities offered to the farmer by a combination of factors of production at the particular farm location, the degree of functional link or complementarity between chosen enterprises, and the farmer's personal goals, preferences and experience.

Farm enterprises and agricultural land use

The choice of enterprises on a farm finds visible expression in the landscape through agricultural land use and livestock. While there is considerable field-to-field variation, not least because of the demands of crop rotations, when taken in aggregate for census units such as parishes, broad patterns of land use and stock raising can be identified. The analysis of such patterns has been a traditional concern of agricultural geographers. Recently Wrathall (1988) has continued this tradition by studying changes in the distribution of a range of arable crops in England and Wales. His analysis shows, for example, an increasing concentration of the declining barley crop in western counties, relatively unchanged locations for the production of crops such as wheat, potatoes and sugar-beet, but the spatial diffusion of a range of 'new' crops such as oilseed rape, fodder maize, linseed and evening primrose (see pp. 149–50). Figure 6.6 contains information on land-use patterns for Northern Ireland, showing the spatial effects of the decline in oats, the increased significance of barley, and the localized production of crops such as fruit and potatoes over the study period. Another tradition has been the application of agricultural location theory to land-use patterns. Von Thünen's model has been widely applied to examine the impact of distance from market on the structure of land use. Earlier textbooks, such as by Morgan and Munton (1971: 79–88), have treated this topic exhaustively and it is not developed here. Indeed many of the processes that shape contemporary land-use patterns, as examined in this book, are not amenable to analysis by von Thünen's model, and 'rings' of production with increasing distance from urban market centres are of decreasing relevance for industrialized agriculture.

Changing the farm enterprise structure

In order to maintain the farm production system at a level which is deemed satisfactory in terms both of personal goals and opportunities offered, the farmer is faced with decision-making at two levels. Firstly, continual short-term or operational decisions must be made in order simply to maintain the efficiency of the existing production system. When exactly to plough a field, sow or harvest a crop, move livestock indoors or outdoors, for example, will involve the farmer in decisions of a day-to-day, or even hour-by-hour nature, dependent on acquired knowledge of the condition of the farmland or livestock as they are affected by rapid changes in weather and soil conditions. The ability to make these operational decisions will, of course, depend to a large extent on the farmer's level of experience of a particular type of production; where levels of experience are high, the farmer may not even be aware in fact that a decision has been made, as instinctive reactions are made to changes based on a wealth of experience.

At a second level, however, farmers must also be involved in a whole series of longer-term strategic or planning decisions which are essential if they are to be able to adjust production processes and enterprise structure in response to changing external forces. Decision-making may thus be thought of as a process of mental activity brought about by either external stimuli, or by the growth of an individual's desire to change. The decision process itself may be conveniently subdivided into three main stages: 'recognition' or awareness, 'preparation', and 'acceptance' or action (Thornton 1962). Progress through the decision process has been usefully illustrated by Jones (1967) and is summarized in Fig. 6.2. In an initial 'recognition' stage, the farmer responds to a variety of different stimuli, which induce an awareness that some change from the status quo of the present production system may be necessary. The farmer may become aware, for example, of some problem existing or arising in the production system which requires a solution. Such problem-solving motivation to the decision process may be aroused, for example, in a farmer who, realizing that yields obtained from a particular crop are below those obtained by other farmers, begins to seek ways by which they might be improved.

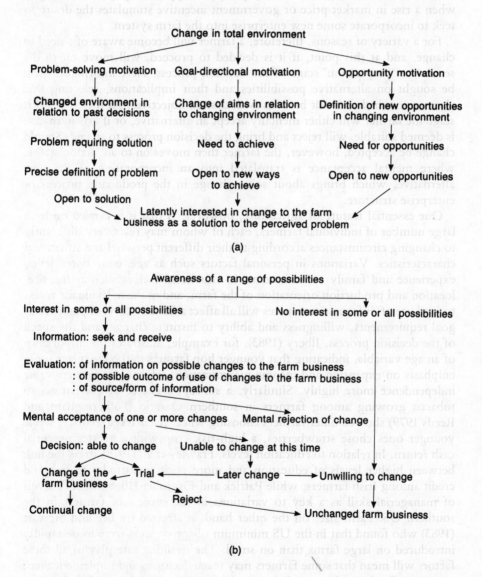

Fig. 6.2 A model of farmer decision-making: (a) recognition stage; (b) preparation action stage (*Source:* after Jones 1967)

141

Alternatively, the farmer may wish to change a system in order to achieve some new perceived goal. An example of this type of goal motivation would be the dairy farmer who decides, perhaps because of increasing age, to seek some alternative form of farming which would not involve the twice-daily routine of morning and evening milking. On the other hand, the farmer may wish to take advantage of a newly perceived opportunity as, for example, when a rise in market price or government incentive stimulates the desire to seek to incorporate some new enterprise into the farm system.

For a variety of reasons, therefore, a farmer will become aware of a need to change, and at this point, if it is decided to proceed, will move on to the second or 'preparation' stage of the decision process. Here information will be sought on alternative possibilities and their implications, selecting that alternative which it is felt best fits with desired objectives. At the end of this stage the farmer will either mentally accept an alternative, or if no alternative is deemed suitable, will reject and bring the decision process to an end. Should change be accepted, however, the farmer then moves on to an 'action' stage, where mental acceptance is translated into an incorporation of the new alternative, which brings about some change in the production process or enterprise structure.

One essential feature of agriculture, however, is that it is carried on by a large number of individual farmers, each of whom may react very differently to changing circumstances according to their different personal and situational characteristics. Variations in personal factors such as age, educational level, experience and family commitments, in situational factors such as the size, location and production orientation of the farm, and in the individual's access to, and use of, information sources will all affect a farmer's level of perception, goal requirements, willingness and ability to institute change, and the speed of the decision process. Ilbery (1983), for example, illustrates the importance of an age variable, indicating that younger hop farmers in the UK place more emphasis on expansion and income maximization, while older farmers value independence more highly. Similarly, a survey of choice of alternatives to tobacco growing among farmers in southern Ontario (Fotheringham and Reeds 1979) showed older farmers choosing wheat for market security, while younger ones chose strawberries, a high-risk crop with a greater potential cash return. In relation to education levels, Frawley et al. (1975) stress the link between higher levels of education and more positive attitudes to risk and credit among Irish farmers, while Patrick and Eisgruber (1968) identify levels of managerial skill as a key to variations in the success of farmers in the southern US. Farm size, on the other hand, is stressed by Lee and Stewart (1983) who found that in the US minimum tillage systems were more rapidly introduced on large farms than on small. The resulting interplay of all these factors will mean that some farmers may reach decisions and implement them more rapidly than others, and changes in farming will tend to take place as a gradual diffusion through the farming community.

The adoption and diffusion of agricultural innovations

If change to the farm business structure is related to the adoption of an innovation by the farmer concerned, we can turn to the literature on the adoption and diffusion of agricultural innovations to obtain further insights into the processes of enterprise change. Three models of innovation diffusion are available based on social, market and resource theories.

The social theory of innovation diffusion

The social diffusion model is based on empirical results from studies of a variety of different innovations which suggest that there is a time-lag in the spread of acceptance of an innovation through a farming community. The rate of adoption conforms to a symmetrical S-shaped logistic cumulative frequency curve. In its early stages relatively few farmers take up the innovation, but its acceptance gains momentum until half the farmers have adopted and half have not, from which point numbers adopting decline and finally slowly 'tail off' until complete adoption is achieved (Fig. 6.3a). By replotting these data against actual mean time of adoption, Rogers (1958) produced a 'bell-shaped' curve which is seen to conform closely to statistical normality (Fig. 6.3b). This being so, it is possible to subdivide this normal curve into time bands on the basis of the standard deviation of the distribution about the mean date of adoption, with a known proportion of the total population of adopters falling into each category, and from this, derive a classification of innovation adoption groups.

The usefulness of this model is that individual farmers can be apportioned to one of the categories on the basis of their relative time of adoption, and each individual subsequently studied to see to what extent certain personal and situational characteristics are common within each group. Using Rogers' (1958) terminology of five adopter categories, Jones (1967) summarized the findings of a large number of individual studies throughout the world to reveal the similarity in personal and situational characteristics of farmers within each category (a 'trait complex'). 'Innovators', for example, tend to have the largest and most specialized farm businesses, a high social status, high levels of agricultural and general education, and to be young and wealthy with relatively easy access to risk capital. They have an ability to make decisions and assess an innovation using impersonal information sources, such as scientific reports, and interact mainly with other innovators often over considerable distances. They tend to be willing to accept risks, though this latter characteristic may lower their opinion leadership in the eyes of other farmers. 'Early adopters', on the other hand, while also tending to have larger and more specialized farms, derive most of their information on a personal basis from local commercial or government advisory officers or from the mass media and farming press. They are less willing to take risks, but often

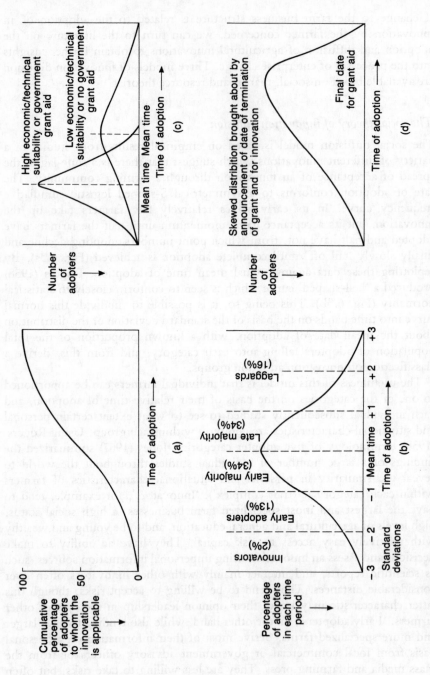

Fig. 6.3 The social theory of innovation diffusion (*Source:* Rogers and Shoemaker 1971: 182)

because of this are respected by farmers in the adopter categories below them, and have the greatest opinion leadership of all groups. In the 'early majority' group are found farmers with average social status and average sized and less specialized farm businesses; they obtain information from local advisory sources, but also from interpersonal contact with 'early adopters', and often only adopt an innovation when it has been demonstrated to work on neighbouring farms. Beyond the mean time of adoption lie the 'late majority'; they are characterized by lower social status, small-scale operations often of a less specialized nature, and need considerable pressure from others before considering adoption. Finally there are the so-called 'laggards' who tend to have the smallest farm businesses, low incomes and little system specialization, who are often older, more traditional or conservative in their attitude, least willing to take risks, and who often only interact with other farmers in the same category with similar personal and situational characteristics.

The social diffusion model, therefore, identifies two important sets of factors which govern decision-making and innovation adoption: namely, the personal and situational characteristics of potential adopters, and their varied use of information sources. The latter factors include a well-documented (Katz and Lazarsfeld 1955), two-step flow of communication in which higher-order decision-makers accept more formal information sources and transmit information informally to lower-order groups. It should be realized, however, that the model is built around the basic assumption that a particular innovation is universally applicable and available to all potential adopters. In reality, as Jones (1963) points out, interrelated economic and technical characteristics of the innovation itself may be of considerable importance in modifying the rate at which its adoption and diffusion are likely to proceed (Fig. 6.3c). Economic (or capital) characteristics include the costs of initial adoption together with the continued maintenance of the innovation; these costs are balanced not only against expected returns to the farm system, but also against the costs and returns of existing techniques. Expensive innovations, or ones with a slow rate of return on investment, may be adopted more slowly, while delay may also be caused where an existing technique is to be replaced which has involved high capital investment, and which has not yet paid for itself or become obsolete.

Hill and Ray (1987) cite a useful threefold classification, with examples, of the relationship between the capital characteristics of an innovation and its likely speed of diffusion. First, there are those innovations which save capital and are generally taken up very rapidly. One example has been the development of an artificial insemination scheme in the UK for cattle, which meant that most small dairy farmers no longer needed to buy or maintain a breeding bull. Second are those innovations which involve minimal change in capital investment, such as a new seed variety or fertilizer type, and again spread quickly. Third, there are those innovations, such as the introduction of a new type of milking parlour, which require much additional capital not only to finance the innovation itself, but also to write off existing milking machinery at a negligible scrap value. Such types are seen to proceed much more slowly.

The technical attributes of an innovation, in terms of its complexity, divisibility, conspicuousness and compatibility with existing techniques, will also affect its rate of adoption. A technique which is mentally difficult to understand, difficult to demonstrate or operationally complex will tend to diffuse more slowly, particularly among the later adopter groups. In a study of 50 innovations available to dairy farmers in Pennsylvania, for example, Fliegel and Kivlin (1962) demonstrated that the more complex a practice the slower its diffusion was likely to be. Similarly, an innovation which is divisible, in that it can be tested or tried out on a small scale before complete adoption, for example a new seed variety, will tend to diffuse more quickly than one which requires full adoption without a trial stage, such as the purchase of a new item of farm machinery. There is also evidence to suggest that the degree of conspicuousness will also affect the adoption rate. An innovation which is readily seen, such as a new item of outdoor machinery or a feed silo, may be regarded as enhancing the local prestige of the adopter and will be more noticeable to other potential adopters; though more costly, the innovation may be diffused more quickly than relatively cheap but largely unseen innovations such as changing management techniques, for example animal feed recording (Moore 1978).

The degree of technical compatability of the innovation with existing techniques will also be important. A new technique which readily fits into an existing system, where the farmers already have considerable experience, will tend to diffuse more quickly. Jones (1962), for example, demonstrated that the adoption of bulk milk collection tanks was closely associated with those dairy farms which had previously adopted pipeline milking or milking parlour systems. Innovations which may involve the acquisition of completely new skills and the abandonment of years of acquired experience, on the other hand, will tend to diffuse more slowly. A dairy farmer, for example, may more readily adopt a bulk milk tank as a natural extension of his system, than change his farm system to cereal production of which he has little or no previous experience, even though this latter change may in fact be more profitable on purely economic grounds.

Integrating social and spatial processes

A major interest to agricultural geographers, however, is the way in which innovations are diffused spatially and bring about changes in patterns of agriculture. The first geographer to study spatial diffusion was Hägerstrand (1953) who, by studying the acceptance of a number of new agricultural techniques, including farm subsidies and tuberculosis inoculation in cattle in an area of central Sweden, suggested that innovations tended to spread outwards from an initial centre where the technique was first introduced. His model of spatial diffusion proposed that an innovation is spread by one farmer adopting at a particular location, with information about the innovation spreading outwards from that site by face-to-face ('pairwise telling') communication between farmers. The distance decay function in the probability of contact between 'tellers' and recipient farmers produces a wave-like, or ripple, pattern of innovation away from the initial site of first adoption. The greater the physical distance between a potential adopter and the initial site of introduction, the later will that potential adopter become aware of, and adopt, the innovation. The basic assumption of this model is that personal contact between farmers acts as the main agent of diffusion, and that the speed and direction of diffusion are dependent upon distance between adopters, together with physical, social and political barriers to information flow. It is doubtful if this assumption has ever been valid and certainly the development of modern means of communication has done much to reduce any constraints imposed by a model based on interpersonal information flow. The mass media, for example, make it theoretically possible for all potential adopters to hear of an innovation at much the same time. More recent research, therefore, casts doubt upon the central assumptions of the Hägerstrand model.

A principal shortcoming of the simple distance decay model is that it tends to pay little attention to the possible influence of the different personal characteristics of potential adopters, and tends to ignore, or at least play down, the findings of social diffusion studies. Differences in types of information channel used by different adopter groups, for example, call into question the simple farmer-to-farmer contact assumed in the Hägerstrand model. Innovators are seen to make greater use of impersonal information sources, while it is only among the later adopters that increasing importance is attached to interpersonal contact. The effect of these differences is thought to bring about a more hierarchical pattern of spatial diffusion, the innovation spreading rapidly among innovators, often at widely separate centres, from which it expands by a process of infilling as time proceeds and different information sources are used by different adopter groups.

The market theory of innovation diffusion

Further complexity can also be introduced into spatial diffusion by the nature of the innovation itself and by market supply factors. Griliches (1960), for example, when examining the adoption of hybrid corn in the US, observed a pattern of early adoption in the heart of the Corn Belt spreading outward in time to peripheral areas like Texas and Alabama, but suggested that this could be attributed to the economic and technical attributes of the innovation and to supply factors. As special types of hybrid seed had to be developed to suit different environments, seed producers concentrated first on producing and supplying a seed type suitable for those areas where open-pollinated corn was already an established and important crop, where the technical and economic advantages of the new variety were most likely to be perceived, and where the innovation was most likely to be easily and swiftly incorporated into existing farm systems. Thus adoption occurred first in Iowa and spread

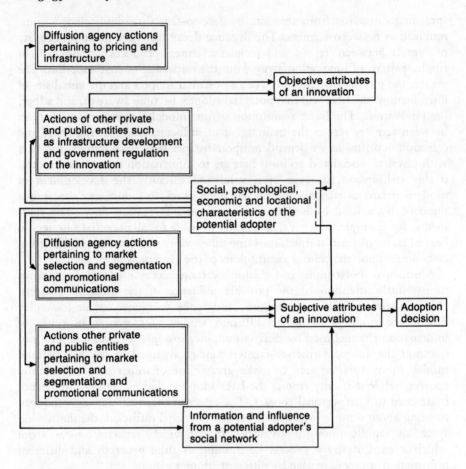

Fig. 6.4 The interface between diffusion agency actions and adoption behaviour (*Source:* after Brown 1981: 102)

quickly to environmentally similar areas to the west and east. The new seeds only spread later to the south-west, into areas like Texas where harsher environmental conditions required the development of different hybrids, and later still to the south-east where only a small proportion of farmland was already in corn, and where plant diseases and low yields reduced potential profits from adoption, both to the farmer and the seed supplier.

Further theoretical advances have been made, with Brown (1981) defining a *market and infrastructure theory* of agricultural innovation. This approach emphasizes the strategies adopted by diffusion agencies (business firms and public and private sector extension/advice agencies) in the supply of innovations. Four main activities can be identified: infrastructure provision (to enable the use of an innovation), pricing policy (level and spatial pattern), promotional communications, and market selection and segmentation. These activities help define the 'objective' and 'subjective' attributes of an innovation (Fig. 6.4), the former being constrained by infrastructure and pricing, and the latter by communications and market selection. Ilbery (1985: 90–2) has reported the empirical findings of geographical research using the market and infrastructure approach and they are not repeated here. However, it is clear that the marketing goals of diffusion agencies can determine their entry into segmented market areas in order of profitability, density of sales and volume of potential sales. Also, centralized as compared with decentralized organizational structures can influence the marketing strategy and its spatial outcome. A decentralized organization, for example based on a franchise system, can have a varied spatial impact depending on which areas produce the entrepreneurs to take up the franchises. Moreover, within the defined areas, the diffusion agencies can determine which farms and farmers first receive information about an innovation, as well as access to capital and credit to purchase the innovation. The available evidence suggests that the larger, rather than smaller farms, are the first to be approached, thereby reinforcing their technical and economic superiority in agricultural production (Brown 1981).

The resource theory of innovation diffusion

Here innovation and diffusion are interpreted as the unequal access to the means of production among the farm population, especially as regards land, capital and credit (Blaikie 1978). But differential access to public infrastructure (e.g. transport), market information, education and government grants and subsidies (Bowler 1979) can be additional constraints on the adoption of an innovation. In a resource theory approach to technological change in agriculture, therefore, non-adoption among smaller farms is interpreted as the consequence of differential access to the means of production rather than a lack of innovative attitudes by the farm occupiers. Thus access to and acceptance of an innovation depend on the class position of the adopter. In addition, the spatial pattern of farm size, farm type and land tenure can constrain the spatial diffusion of an innovation, thereby producing the spatial regularities observed by numerous empirical studies (e.g. Dalrymple 1979; Joseph and Keddie 1981; Winsberg 1970).

Case studies of enterprise change in agriculture

The discussion thus far has produced two perspectives on the processes by which individual farm enterprise structures are changed through time. One is a decision-making perspective (Fig. 6.2), while the other is an innovation diffusion perspective (Figs. 6.3 and 6.4). Looking first at enterprise change as the product of the diffusion of an innovation, the adoption of maize (corn) in the UK illustrates a number of important points. Figure 6.5 shows maize being introduced into the most suitable arable farming areas of south-east England, with a gradual spread into less suitable areas in the north and west. Thus the pattern reflects the suitability of the resource base for growing maize. But the relatively early adoption of the crop in some northern and western counties, for example Cheshire, Yorkshire and Devon, suggests a hierarchical element in the diffusion pattern derived from longer-distance contacts between innovative farmers. Tarrant (1975) confirms these impressions in a study of the early years of the diffusion process. A significant number of innovative farmers had contacts extending up to 70 km from their farms. However, the market model of diffusion was also relevant since many farmers were persuaded to grow maize after contact with seed merchants, government advisory officers and farmers' organizations such as a Maize Development Association. One group of farmers had adopted as a result of personal visits to maize-growing areas in Europe and North America.

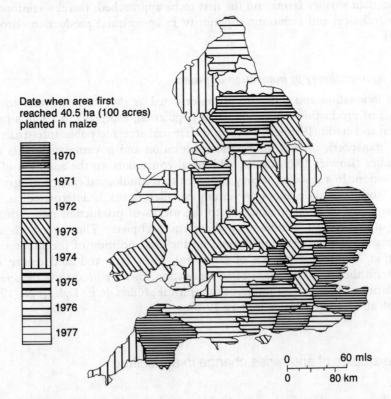

Date when area first reached 40.5 ha (100 acres) planted in maize

1970
1971
1972
1973
1974
1975
1976
1977

```
0          60 mls
0          80 km
```

Fig. 6.5 Spatial diffusion of maize in England and Wales (*Source:* MAFF Agricultural Statistics for England and Wales)

We now turn to studies of the changing enterprise structure on farms bearing in mind the decision-making process summarized in Fig. 6.2. Farm business structures need to be kept flexible to respond to temporal changes in such features as market prices, input costs, new technology and government policy. In order to achieve flexibility, individual enterprises may be expanded or contracted from time to time as circumstances change, which leads to changes in the emphasis placed upon particular enterprises within a farm's production system, with the result that all farm enterprise structures should be regarded as potentially changing phenomena. The type, direction, speed and magnitude of change will be governed by constraints imposed and opportunities offered by the factors of production present on the farm, and by the individual farmer's decision-making capabilities and perceptions.

Relatively few studies of such changes at the farm level have been published until recently, and there is little evidence of such changes being linked empirically to decision-making models. Two studies, however, illustrate the scale and nature of these structural adjustments. In the first (Edwards 1976), enterprise changes were investigated on a sample of farms in Northern Ireland over a 3-year period (1972–75). The survey was undertaken in an area of Co. Londonderry in the north of the province, a region of mixed livestock and cash-crop farming; physical conditions ranged from hill environments to high-quality lowland areas. A considerable amount of short-term change was recorded: 55.6 per cent of farms registered a change in the number of individual enterprises in their production system. Among the remaining 44.4 per cent whose total number of enterprises did not change, specific enterprises often changed their relative importance within the enterprise structure. Only 27.5 per cent of farms recorded exactly similar structures at both survey dates, the majority of these being specialist single-enterprise farms at the outset of the survey. The actual changes in enterprise structures between the two survey dates revealed a trend towards greater specialization at the expense of mixed farm structures; second, a change in emphasis in the types of enterprise (dairying, beef and sheep being favoured at the expense of pigs, poultry and arable cash crops of potatoes and barley); and third, a tendency for change to be less apparent among already specialist farms, where land, labour and capital resources had already been concentrated, and among hill farms, where the options for change were limited by physically marginal conditions.

The second study (Edwards 1980) was carried out in an area of central Somerset in the west of England. Though physical contrasts offer a range of land quality, farming in this area has been mainly orientated towards dairying from at least the beginning of the eighteenth century. Previous studies stressed the dominance of milk production on most farms with few viable alternatives, so that change in enterprise structure was expected to be minimal. Again, however, the study revealed considerable change between 1963 and 1976. Dairy farms dominated the area at the earlier period, but were rarely totally specialized, often running a number of minor but complementary subsidiary enterprises, particularly pigs and poultry. By 1976, however, a fall in the

151

overall dominance of dairying was noted, nearly 30 per cent of farms having opted for an alternative to dairying as their main enterprise. On the dairy farms which remained, the tendency was either to increase specialization by dropping subsidiary enterprises, or to change subsidiary enterprises from small pig and poultry units which were no longer commercially viable, to a more economically viable beef enterprise, fattening surplus calves from the dairy herd. These results reveal that even in a traditional 'core' area of specialist dairy production, where change might have been expected to be minimal, only 31.0 per cent of farms recorded no change in their enterprise structure over the survey period, reinforcing the dynamic nature of enterprise structures on farms.

These studies by Edwards capture the farm-level impact of the processes of specialization and concentration (see pp. 13–16) that were gathering momentum in the 1960s and 1970s. When aggregated for whole farming regions, spatial patterns of specialization and concentration are revealed. In Europe, for example, beef and sheep enterprises have not been strongly influenced by scale economies and are thus to be found on a wide variety of farms. In semi-arid areas of the US and Australia, in contrast, these same enterprises have been increasingly concentrated on to large, specialist, intensive 'feed-lots' as the most economic method of production. Such a system is illustrated by Gregor (1974) in California, where cattle are kept in groups in regular-shaped pens, not grazing but being fed from a centrally located feed point; the aim of the system is to maintain a continuous throughput of a product fattened to optimum weight, in the minimum time, and with the minimum input of costly labour. In cereal production, wheat farming has tended to become concentrated on fewer but larger farms in Europe (Britton 1969, 1977), the US (Winsberg 1980, 1982) and Australia (Scott 1981). In contrast, barley production has tended to become more dispersed as its value for stock feeding has increased, spreading production, for example in the UK, from traditional specialist cereal-growing farms in the south and east, on to livestock farms in the north and west. The best examples of enterprise concentration, however, are seen in intensive pig and poultry production. Scale economies achieved by the production of pigs and poultry in large numbers, in intensively housed 'factory-farming' conditions, has lowered unit costs, such that pig and poultry products have become relatively cheaper for the consumer and increased their market potential (Haines 1982). The effect of this has been to concentrate these enterprises, which often in the past formed small subsidiary units on a large number and wide variety of farms, on to a relatively small number of very large specialist farms located in just a few regions (for example, pig production in the Humberside region of England). The farms are but one stage in a vertically integrated industry which is dominated by national food-processing and meat-packing corporations.

Nor has specialization stopped here. Increasingly in the livestock sector, individual farms are specializing in only part of the production process. This is not necessarily a new phenomenon. In the past, for example, certain areas

Table 6.4   Types of full-time farms in Northern Ireland

Farm type	Percentage of all farms			Percentage change by type	
	1972	1976	1980*	1972/1976	1976/1980
Dairy	34.2	37.2	40.7	8.8	9.4
Dairy, pigs and poultry	8.3	3.6	2.1	−56.6	−41.7
Beef, sheep and pigs	7.9	5.2	4.3	−34.2	−17.3
Beef and sheep	25.0	32.6	31.9	30.4	−2.0
Pigs and poultry	6.8	5.4	5.9	−20.6	9.2
Cropping	5.2	5.1	6.3	−1.9	23.5
Mixed arable and livestock	12.6	10.9	8.8	−13.5	−19.3
Total	100.0	100.0	100.0		

* No update is possible because of the reallocation of farm types into nine new type categories after 1983.

Source: Based on data from the Department of Agriculture for Northern Ireland.

Table 6.5   Enterprise concentration in Northern Ireland

	Average size of enterprise					Percentage of total numbers/area on above-average size units				
	1972	1976	1980	1984	1988	1972	1976	1980	1984	1988
Dairy cows (number)	16	22	29	37	39	69.8	74.2	70.7	75.5	69.5
Beef cows (number)	10	12	11	12	13	74.5	71.9	73.4	70.9	71.0
Sheep (number)	103	111	126	161	196	72.8	76.7	74.3	73.2	71.6
Pigs (number)	59	69	94	130	185	71.2	73.5	77.7	81.5	83.3
Poultry (commercial flocks of over 50 birds)	1 607	2 736	4 346	5 634	7 377	82.3	83.6	85.6	80.4	85.7
Oats (ha)	1.9	1.7	2.0	2.2	—	67.0	64.1	67.1	52.8	—
Barley (ha)	6.0	5.6	6.3	6.6	71	70.5	70.0	71.8	65.7	71.4
Potatoes (ha)	1.3	1.3	2.0	2.5	31	80.1	77.1	77.0	78.1	79.7

Source: Based on data from the Department of Agriculture for Northern Ireland.

carried out the whole process of beef breeding, rearing and fattening on the one farm; in other instances these separate elements of beef production were carried on different farms at widely different locations. In the Republic of Ireland, for example, there has been a long tradition of calf breeding on small farms in the remote west; these calves were moved to better land on midland farms for rearing as stores, before being sold on to fattening farms around Dublin, or transhipped live across the Irish Sea to fattening areas in eastern England (Coppock and Gillmor 1967; Gillmor 1969). Similarly in the UK, the entire sheep industry is based upon a complex stratified system of production which interrelates specialist mountain, marginal and lowland sheep systems (Haines 1982). This separation of parts of the production process, or

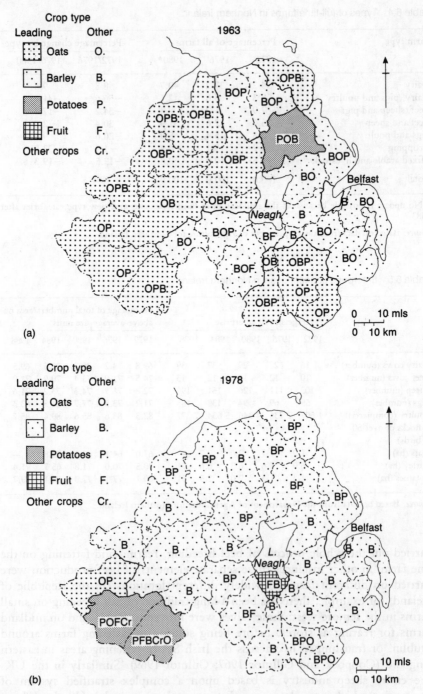

Fig. 6.6 Changing arable crop combinations in Northern Ireland: (a) 1963; (b) 1978 (*Source:* Based on data from the Department of Agriculture for Northern Ireland)

vertical integration, is perhaps best exemplified again by pig and poultry production. In the poultry sector, for example, it is now rare to find a commercial poultry farm that undertakes the whole production process, from hatching chicks to final production of poultrymeat or eggs. The supply of chicks for particular purposes is carried out by specialist hatcheries, while others specialize in rearing-on the chicks for broiler production, or to point-of-lay, while other units carry only laying fowl; the whole process is organized on a contract basis, either between the farmers concerned, or increasingly with a large agribusiness comprised of commercial livestock feed firms or food-processing and marketing companies (Haines 1982; Vogeler 1981).

The continuous changes in enterprise structures, and the trend towards specialization, are further illustrated in Table 6.4 and Fig. 6.6 for Northern Ireland. The data reveal a marked and continuing decline in the number of complex crop combinations and mixed farm types; specialist farms, or those with simple combinations of enterprises, increased in importance, typified by the slow but consistent rise in the number of specialist dairy farms. Within this general trend, however, individual enterprises and structure fluctuated in importance in response to short-term changes in circumstances and perceptions. Beef/sheep types of farm structure, for example, increased rapidly in number between 1972 and 1976 in response to expectations of new markets and higher prices on the UK's entry into the EC, but declined in popularity between 1976 and 1980 in response to low beef prices caused by overproduction in the mid-1970s. The rise in number of cropping-type farms between 1976 and 1980 reflected an increase in barley prices in the province and, in particular, increased potato production in Northern Ireland for 'export' to mainland UK following a shortfall in British production in the dry summer of 1976.

Data from Northern Ireland can also be used to illustrate the trend to increased enterprise size on farms by enterprise concentration (Table 6.5). The most dramatic increase has been in the poultry sector where, by 1980, a mere 3.7 per cent of large poultry farms accounted for 85.6 per cent of the total commercial poultry flock. Evidence of similar patterns of regional or national changes in enterprise structures in other areas have been illustrated by a number of authors. Studies by Britton and Ingersent (1964), Britton (1977), Bowler (1981) and Edwards (1987), for example, reveal the effects of the changing popularity of particular enterprise structures and the trend towards specialization and enterprise concentration at regional and national levels in the UK; while Winsberg (1980, 1982) has demonstrated the same effect in the US. Gillmor (1987) has analysed the degree of spatial change in agriculture during a period of rapid modernization in the Republic of Ireland, and Bowler (1986) has illustrated the spatial change and regional specialization that has occurred within countries of the EC as a result of intensification, concentration and specialization in agriculture.

Diversifying the farm business structure

The processes of intensification and specialization are now coming to an end, at least in many of the agricultural economies of the developed world. The oversupply of markets, and the associated problem of funding such oversupply, is leading to a search for possible production alternatives. Some farmers are responding to the problem of falling farm incomes by taking employment off the farm (pluriactivity). For them the farm has become a source of only part-time employment and income (Gasson 1988). Other farmers, however, are seeking to diversify the use of their farm resources. Here the search process for new enterprises, as discussed in relation to Fig. 6.2, and the adoption of innovative enterprises (Fig. 6.3), are being actively applied (Carruthers 1986).

Looking in more detail at farm diversification, Ilbery (1988) draws a distinction between farming and non-farming enterprises. Farming diversification implies the 'conversion' of farm resources (land, labour and capital) to non-traditional enterprises. Some of these are listed in Table. 6.6 under the headings of 'adding value to conventional farm products' and 'unconventional agricultural enterprises'. To the list should be added new types of crop which are innovative for that farming system, and the planting of land to trees for commercial timber production. These trends are already under way as discussed by Lockeretz (1988) for soya beans in the US and 'new' crops in England and Wales by Wrathall (1988). Also state farm policies are in place to subsidize the enterprises. The UK, for example, has a 1988 Farm Woodland Scheme, the EC has a 'set-aside' programme to allow farmers to idle their productive land, while attempts have been made for many years in the EC to 'despecialize' certain farming regions, notably in the wine sector (Jones 1989), by diversifying the regional agricultural economy. The emerging evidence shows that the 'conversion' of farmland is proceeding at different rates and in different directions from region to region. Ilbery's (1990) analysis of the early pattern of 'set-aside' in the UK, for example, found a concentrated pattern of such land in the south-eastern counties of England.

On non-farming diversification, the main categories are 'tourism and recreation', 'use of ancillary buildings and resources' and 'public goods' (Table 6.6). Research in the UK, mainly in relation to the response by farmers to the 1988 Farm Diversification Grant Scheme (Dalton and Wilson 1989; Ilbery and Stiell 1991), shows the main development to be in the areas of farm accommodation for tourists (Evans and Ilbery 1989), enterprises involving horses and farm shops. Sport fishing, shooting and farm museums are so far of minor importance. Equally it is clear that most farmers are intent on pursuing all possible economies in running conventional farm businesses as their priority, taking on farm (conversion) diversification as a second option, and only moving to non-farming diversification when those options appear to be exhausted (Halliday 1989). Moreover, non-farming diversification is selectively benefiting agriculture in urban fringe (Ilbery 1991) and tourist areas.

Table 6.6 Alternative enterprises for the farm business

Tourism and recreation	*Tourism*	Bed and breakfast
		Cottages/chalets
		Caravans/camping
		Activity holidays
	Recreation enterprises	Farm museums
		Visitor centres
		Riding
		Game shooting
		Other shooting
		Fishing
		Farmhouse catering
Adding value to conventional farm products	*Animal products*	Meat (direct sales, etc.)
		Skins/hides/wool
		Dairy products (direct sales/ processing)
	Crop products	Milled cereals
		PYO and direct sales of vegetables
Unconventional agricultural enterprises	*Animal products*	Sheep milk
		Rare breeds
		Fish
		Deer and goats
	Crop products	Linseed
		Evening primrose
		Teasels
	Organic production	
Use of ancillary buildings and resources	*Woodland products*	Fuel wood
		Craft timber products
	Redundant buildings	Industrial premises
		Accommodation
	Wetland	Fish
		Game
Public goods	*Wildlife*	ESA payments
	Landscape	Management agreements
	Historic sites	Heritage relief
	Access	Access agreements

Source: Slee (1989: 62).

'Public goods', as a form of diversification, are also produced unevenly within agriculture. 'Farming the environment' to produce environmental goods is being actively encouraged by state subsidies in several countries, but usually in closely defined areas. In the UK, state subsidies for farming in environmentally sensitive ways are at present restricted to defined Environmentally Sensitive Areas (ESAs) and Nitrate Sensitive Areas (NSAs). There is every prospect, however, of similar subsidies being made more widely

157

available to produce, for example, herb-rich pastures, flora-rich hedgerows, and landscapes containing well-maintained farm buildings, field boundaries and footpaths.

Conclusion

This chapter has emphasized individual farmer decision-making as the basis for understanding enterprise changes in agriculture. Farm businesses are in a constant state of flux, not least in recent years with the need to diversify the sources of income for the farm family. In aggregate, the individual decisions produce spatially uneven changes in the pattern of agriculture as summarized in type-of-farming regions. Some changes reflect the attempts by farmers to adjust their enterprise structures in the face of market prices; other changes are brought about by the diffusion of innovations through the farming community. Indeed farm diversification can be interpreted in this way. Yet other changes result from the role of external capitals in marketing new farm inputs or processing farm outputs, and these influences are examined in more detail in Chapter 7.

References

Blaikie, P. (1978) The theory of the spatial diffusion of innovations: a spacious cul-de-sac. *Progress in Human Geography* **2**: 268–95

Bowler, I. R. (1979) *Government and agriculture.* Longman, London

Bowler, I. R. (1981) Regional specialization in the agricultural industry. *Journal of Agricultural Economics* **32**: 43–54

Bowler, I. R. (1986) Intensification, concentration and specialization in agriculture: the case of the European Community. *Geography* **71**: 14–24

Britton, D. K. (1969) *Cereals in the United Kingdom: production, marketing and utilisation.* Pergamon, Oxford

Britton, D. K. (1977) Some explorations in the analysis of long-term changes in the structure of agriculture. *Journal of Agricultural Economics* **28**: 1–12

Britton, D. K. and **Ingersent** (1964) Trends in concentration in British agriculture. *Journal of Agricultural Economics* **16**: 25–52

Brown, L. A. (1981) *Innovation diffusion: a new perspective.* Methuen, London

Carruthers, S. P. (ed) (1986) *Alternative enterprises for agriculture in the UK.* Centre for Agricultural Strategy Report No. 11, Reading

Coppock, J. T. and **Gillmor, D. A.** (1967) The cattle trade between Ireland and Great Britain. *Irish Geography* **5**: 320–6

Dalrymple, D. J. (1979) The adoption of high yielding grain varieties in developing nations. *Agricultural History* **53**: 704–26

Dalton, G. and **Wilson, C.** (1989) *Farm diversification in Scotland*. Economic Report 12, Scottish Agricultural College, Aberdeen

Edwards, C. J. (1974) Farm enterprise systems in east Co. Londonderry. *Irish Geography* **7**: 29–52

Edwards, C. J. (1976) Short term changes in farm enterprise systems: a case study from Co. Londonderry 1972–1975. *Irish Journal of Agricultural Economics and Rural Sociology* **6**: 155–75

Edwards, C. J. (1980) Complexity and change in farm production systems: a Somerset case study. *Transactions of the Institute of British Geographers* **5**: 45–52

Edwards, C. J. (1987) The changing role of sheep production in Northern Ireland agriculture. *Irish Geography* **20**: 98–100

Evans, N. J. and **Ilbery, B. W.** (1989) A conceptual framework for investigating farm-based accommodation and tourism in Britain. *Journal of Rural Studies* **5**: 257–66

Fliegel, F. C. and **Kivlin, J. E.** (1962) Farm practice attributes and adoption rates. *Social Forces* **40**: 364–70

Fotheringham, A. S. and **Reeds, L. G.** (1979) An application of discriminant analysis to agricultural land use prediction. *Economic Geography* **55**: 114–22

Frawley, J, Bohlen, J. M. and **Breathnach, T.** (1975) Personal and social factors related to farming performance in Ireland. *Irish Journal of Agricultural Economics and Rural Sociology* **5**: 157–81

Gasson, R. (1973) Goals and values of farmers. *Journal of Agricultural Economics* **24**: 521–42

Gasson, R. (1988) *The economics of part-time farming*. Longman, London

Gillmor, D. A. (1969) Cattle movements in the Republic of Ireland. *Transactions of the Institute of British Geographers* **46**: 143–54

Gillmor, D. A. (1977) *Agriculture in the Republic of Ireland*. Akademiai Kiado, Budapest

Gillmor, D. A. (1987) Concentration of enterprises and spatial change in the agriculture of the Republic of Ireland. *Transactions of the Institute of British Geographer* **12**: 204–16

Gregor, H. F. (1974) *An agricultural typology of California*. Akademiai Kiado, Budapest

Griliches, Z. (1960) Hybrid corn and the economics of innovation. *Science* **132**: 275–80

Hägerstrand, T. (1953) (1967 in English translated by Pred, A. and Haag, G.) *Innovation diffusion as a spatial process*. University of Chicago Press, Chicago

Haines, M. (1982) *An introduction to farm systems*. Longman, London

Halliday, J. (1989) Attitudes towards farm diversification: results from a survey of Devon farms. *Journal of Agricultural Economics* **40**: 93–100

Hill, B. and **Ray, D.** (1987) *Economics for agriculture: food, farming and the rural economy*. Macmillan, London

159

Ilbery, B. W. (1979) Decision making in agriculture: a case study of NE Oxfordshire. *Regional Studies* **13**: 199–210

Ilbery, B. W. (1983) Goals and values of hop farmers. *Transactions of the Institute of British Geographers* **8**: 329–41

Ilbery, B. W. (1985) *Agriculture geography: a social and economic analysis.* Oxford University Press, Oxford

Ilbery, B. W. (1988) Farm diversification and the restructuring of agriculture. *Outlook on Agriculture* **17**: 35–9

Ilbery, B. W. (1990) Adoption of the arable set-aside scheme in England. *Geography* **75**: 69–73

Ilbery, B. W. (1991) Farm diversification as an adjustment strategy on the urban fringe of the West Midlands. *Journal of Rural Studies* **7**: 207–18

Ilbery, B. W. and **Stiell, B.** (1991) Uptake of the Farm Diversification Grant Scheme in England. *Geography* **76**: 259–63

Jones, A. R. (1989) The reform of the EEC's table wine sector: agricultural despecialization in the Languedoc. *Geography* **74**: 29–37

Jones, G. E. (1962) *Bulk milk handling: an investigation into the adoption of a new dairy technique in Lindsey.* Department of Agricultural Economics, University of Nottingham, Nottingham

Jones, G. E. (1963) The diffusion of agricultural innovations. *Journal of Agricultural Economics* **15** 387–405

Jones, G. E. (1967) Agricultural innovation and farmer decision making. *Open University Social Science Block III* **2**: 29–56

Joseph, A. E. and **Keddie, P. D.** (1981) The diffusion of grain production through southern Ontario, 1946–1971. *Canadian Geographer* **25**: 335–49

Katz, E. and **Lazarsfeld, P. F.** (1955) *Personal influences: the role played by people in the flow of mass communications.* Free Press, Glencoe

Lee, L. K. and **Stewart, W. H.** (1983) Land ownership and the adoption of minimum tillage. *American Journal of Agricultural Economics* **65**: 256–64

Lockeretz, W. (1988) Agricultural diversification by crop diversification: the US experience with the soyabean. *Food Policy* **13**: 154–66

Moore, R. C. (1978) Aspects of the innovation characteristics of dairy farmers in East Antrim. Unpublished MSc thesis, New University of Ulster, Coleraine

Morgan, W. B. and **Munton, R. J.** (1971) *Agricultural geography.* Methuen, London

Patrick, G. F. and **Eisgruber, L. M.** (1968) The impact of managerial ability and capital structure on growth of the farm firm. *American Journal of Agricultural Economics* **50**: 491–506

Rogers, E. M. (1958) Categorizing and adopters of agricultural practices. *Rural Sociology* **23**: 343–54

Rogers, E. M. and **Shoemaker, F. F.** (1971) *Communication of innovations: a cross cultural approach* 2nd edn. Free Press, New York

Scott, P. (1981) *Australian agriculture: resource development and spatial organization.* Akademiai Kaido, Budapest

Simon, H. A. (1952) A behavioural model of rational choice. *Quarterly Journal of Economics* **69**: 99–118

Slee, B. (1989) *Alternative farm enterprises* 2nd edn. Farming Press, London

Tarrant, J. R. (1975) Maize: a new United Kingdom agricultural crop. *Area* **7**: 175–9

Thornton, D. S. (1962) The study of decision making and its relevance to the study of farm management. *Farm Economist* **10**: 40–56

Vogeler, I. (1981) *The myth of the family farm: agribusiness dominance of US agriculture.* Westview, Colorado

Winsberg, M. D. (1970) The introduction and diffusion of the Aberdeen Angus in Argentina. *Geography* **55**: 187–95

Winsberg, M. D. (1980) Concentration and specialization in United States agriculture, 1939–1978. *Economic Geography* **56**: 183–9

Winsberg, M. D. (1982) Agricultural specialization in the United States since World War II. *Agricultural History* **56**: 692–701

Wolpert, J. (1964) The decision process in a spatial context. *Annals of the Association of American Geographers* **54**: 537–58

Wrathall, J. E. (1988) Recent changes in arable crop production in England and Wales. *Land Use Policy* **5**: 219–31

7

Marketing agricultural produce

Patrick Hart

This chapter advances the discussion of the food supply system beyond the farm gate and into the food-processing, distribution and consumption sectors (see Fig. 1.1). Particular emphasis is given to two features of contemporary agriculture that were introduced in Chapter 1, namely the internationalization of the food supply system (p. 26) and the real and formal subsumption of agriculture (pp. 24) by external capitals in the food-processing and retailing sectors.

In advanced Western economies most agricultural production is conducted on a commercial basis and the market is a powerful influence on farming decisions. It is not, however, just the market which is important but also the means by which the producer is linked to it. Therefore, it is useful to clarify the distinction between the concepts of the market and marketing. The 'market' is a term describing aspects of the demand situation for a commodity. It commonly equates with the level of aggregate demand, but can have wider connotations, being used to describe arrangements which exist to allow transactions to take place between buyers and sellers. These arrangements may be physical, for example provision of buildings, or organizational, for example those at national or international scales for trade in commodities such as grain or tea.

The term 'marketing' describes the sequence of processes which link producers and consumers (see Fig. 1.1). It involves the performance of all business activities involved in the flow of food products and services from the point of initial agricultural production until they are in the hands of consumers (Kohls and Uhl, 1985). This highlights the fact that marketing comprises more than just exchange transactions. The sequence which links producers and consumers is called the marketing chain. Figure 7.1 depicts the major participants in the UK food-marketing chain and exemplifies some of the processes they engage in. However, it must be stressed that the processes and their sequencing vary between products. Thus, farmers may sell to: livestock traders or produce merchants (either on the farm or at livestock or produce markets), food processors and retailers, as well as directly to the public. The commodities differ more widely. There is a basic distinction between plant

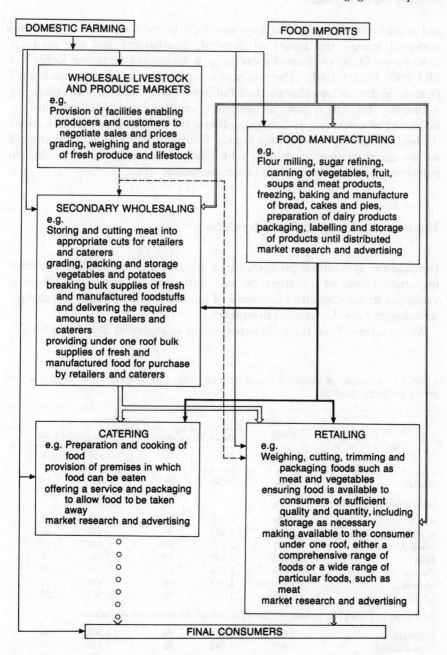

Fig. 7.1 Major participants, processes and distribution channels in the UK food marketing chain. Distribution channels of: (→) farm produce; (⇒) food imports; (-→) produce from wholesale livestock and produce markets; (⇒) produce from secondary wholesaling; (→) manufactured foodstuffs; (o o o) output of catering industry; (xxx) retailed food (*Source:* adapted from Tanbaum 1981)

163

and animal products and differences also occur in end use (including food and non-food items), the nature of demand, perishability and the need for processing. These are treated more fully in specialized texts (see Kohls and Uhl 1985; Barker 1981). The resultant complexity of marketing has limited farmers' ability to capitalize on their full potential in this area since familiarity with many different systems is required. Complexity and variety are thus the hallmarks of marketing and these confound the task of generalization. None the less, certain key characteristics can be identified which provide the context for an understanding of changes which have been and are taking place in marketing systems downstream of the production sector.

The demand for agricultural produce

Demand for agricultural products varies spatially and temporally. The most important causes of variations include differences in tastes and affluence, variations in the elasticity of demand of products, and technological change, including the development of substitutes.

At the international scale differences occur in aggregate demand as well as

Table 7.1 International variations in calorie supply per capita derived from vegetable and animal products, 1983/85

	Average daily intake	Intake from:			
		Vegetable products		Animal products	
		Total	%	Total	%
World	2 666	2 258	85	407	15
(a) Continental variations					
Africa	2 278	2 104	92	174	8
North Central America	3 378	2 382	71	996	29
South America	2 617	2 169	83	448	17
Asia	2:437	2 247	92	190	8
Europe	3 390	2 286	67	1 104	33
Oceania	3 133	2 143	68	990	32
USSR	3 403	2 536	75	867	25
(b) Countries having the greatest and least calorific intake from animal products					
Denmark	3 529	1 961	56	1 568	44
New Zealand	3 402	1 892	56	1 510	44
Belgium and Luxemburg	3 695	2 214	60	1 481	40
Burundi	2 217	2 167	98	50	2
Indonesia	2 504	2 450	98	54	2
Mozambique	1 664	1 608	97	56	3

Source: FAO (1986a).

in its composition (Table 7.1). People living in the developed countries of Europe, North America and Oceania consume more food and derive a significantly greater proportion of their diet from animal foodstuffs than do inhabitants of the Third World. Variations in living standards are the main cause of these differences, since animal products are more expensive than cereal or root crops. Demand for particular animal products may also be influenced by cultural factors such as the attitude of the Jewish and Muslim religions towards pigmeat. These variations in demand are matched by differences in agricultural production patterns between countries. For example, although many of Western Europe's needs are provided from other parts of the world, a significant proportion of European agriculture is geared towards livestock production in response to local demand.

More recently a reversal has occurred in the trend towards increased demand for animal products in more affluent countries. This illustrates the importance of variations in tastes. There has been a growing awareness of the health hazards of diets which are significantly based on animal foods. This has led to some shifts in production, including increased cultivation of crops yielding vegetable oils. Between 1975 and 1985 there were increases from 59 000 to 902 000 ha in the area under sunflowers in France and from 32 000 to 299 000 ha under oilseed rape in the UK. Such adjustment may not always be possible due to climatic factors. This has created problems in the Third World due to the process of dietary disjuncture (Mabogunje 1973). This occurs when tastes and demand develop for foods which cannot be produced locally. As a result, many developing countries are forced to import more of their food needs, with harmful implications for their balance of payments and development prospects.

From the perspective of marketing, it is temporal changes in demand which can create some of the greatest problems. Figure 7.2 depicts trends in consumption of certain food items in the UK from 1955 to 1985. Consumption of both butter and eggs shows an initial increase followed by a period of stability and then decline. The changes are more pronounced for butter. The initial upturn is due to the rise in household incomes of the late 1950s and the greater increase for butter results from its higher income elasticity of demand (see Appendix for explanation). The growing demand for substitute foods on health grounds accounts for the recent decline. The popularity of new brands of soft margarine has led to a marked decline in butter consumption. Demand for eggs is affected by changing tastes, notably a decline in the popularity of the cooked breakfast and this product is increasingly substituted by numerous convenience foods. Consumption of peas, by comparison, has been stable since 1960, but this situation is more complex since the product is marketed in different forms. The trend for fresh peas, and to a lesser extent the canned variety, has been downwards. They have been substituted by frozen peas, due to technological developments in food processing, increased affluence and rising demand for convenience foods. This structural shift in demand is important geographically, since different areas are used for the two types of

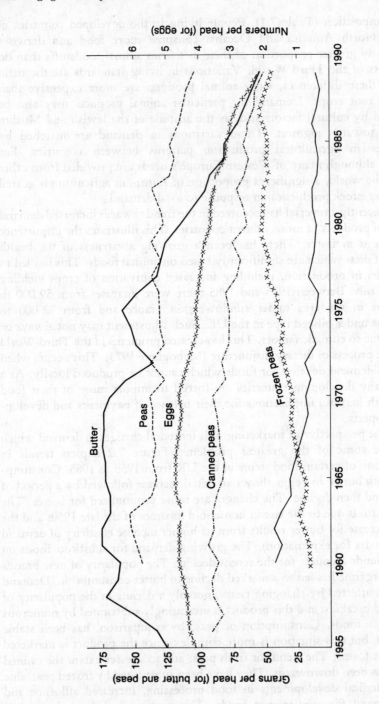

Fig. 7.2 Average weekly consumption per head of selected food items in the UK (*Source:* National Food Survey Committee, annual)

crop (fresh or for processing). The markets to which they are supplied also differ as do the most appropriate marketing methods (Coppock 1971).

Limitations of traditional geographical approaches to agricultural marketing

Geographical interest in the implications of demand traditionally focused on the location of markets; this derived from the ideas of von Thünen's isolated state model (Hall 1966). The value of the model, however, is reduced by its assumption of a single market and the partial equilibrium situation in relation to demand, price and output. It is rare today to find a situation where just one centre is the dominant market for a region's produce. For most products more than one market exists and markets for different products do not coincide. Hoggart (1979) has demonstrated the complex pattern of patronization of corn and grain markets by farmers in part of Huron County, Ontario (Fig. 7.3). Proximity is important, but farmers are also influenced in their choice of market by perceived price differences and personal factors, such as knowledge of personnel. Multiple markets produce an intricate pattern of land use zones, as was shown by Hoover (1948).

The realities of modern agricultural marketing systems include the imperfect matching of agricultural supply and demand. In addition to the influences on demand noted above, supply is subject to variation, and the lack of correspondence between these two economic influences provides a fundamental problem for marketing. Supply is unpredictable due to physical, structural and behavioural aspects of agricultural production. The level of supply at any time is, in part, a function of output levels over the preceding year, which are influenced by environmental factors, primarily the weather. Thus, irrespective of the land area and inputs applied, output is capable of marked fluctuations from year to year. One purpose of marketing is to reduce the disruptive impact of these, for example, through storage or price-fixing strategies. The structural characteristic which contributes to unpredictability is the large number of producers. The greater their number the more difficult it is to forecast their production decisions, since individuals will be less confident in predicting the actions of competitors. This ties in closely with behavioural factors in decision-making (see pp. 136–9). Farmers lack the ability to make decisions which would maximize returns because of this marketing uncertainty.

Shifts in demand tend to be more gradual than those in supply. This lack of synchronization is aggravated by the biological nature of agricultural production. Most production decisions have to be made at least a year in advance of any expected return from sales. In some cases, for example tree crops or other production systems involving considerable financial investment, the farmer needs to be confident about long-term price trends. The problem is thus not

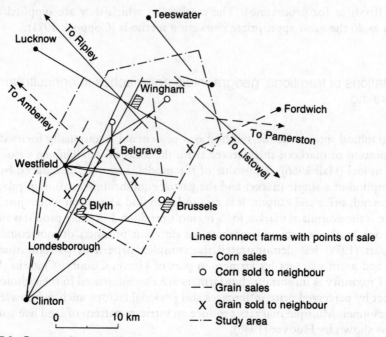

Fig. 7.3 Patterns of corn and grain marketing among farmers in part of Huron County, Ontario. Lines connect farms with point of sale. (——) corn sales; (o) corn sold to neighbour; (– – –) grain sales; (x) grain sold to neighbour; (– · –) study area (*Source:* Hoggart 1979)

one of reacting directly to current prices but of extrapolating from them. In an unregulated market this can lead to regular cyclical fluctuations in supply and prices, to the detriment of both producers and consumers (Hallberg 1982). Some farmers react by increasing their production when prices are low, in anticipation of a reduction by their competitors and a consequent price rise. This underlines the difficulties of predicting levels of supply.

The mismatch between supply and demand reveals one function of marketing to be a distinctly geographical one. That is the reallocation of commodities from areas where they are produced to those in which they are demanded. Trade is a manifestation of this. Trade may arise from the exploitation of comparative advantage in production of particular commodities, notably the exchange of tropical and temperate products. It is also used to overcome temporary shortfalls in supply. This is especially important to countries with less predictable climatic regimes. For example, the USSR, the world's leading wheat producer, has been more subject to output fluctuations than other leading producers (Fig. 7.4). This reflects difficult weather conditions as well as bottlenecks in production and distribution systems. Such fluctuations, coupled with rising living standards and domestic policies to encourage consumption of meat, have led to a greater reliance on wheat imports (Tarrant 1984). Therefore, trade is a useful marketing mechanism, but is subject to the

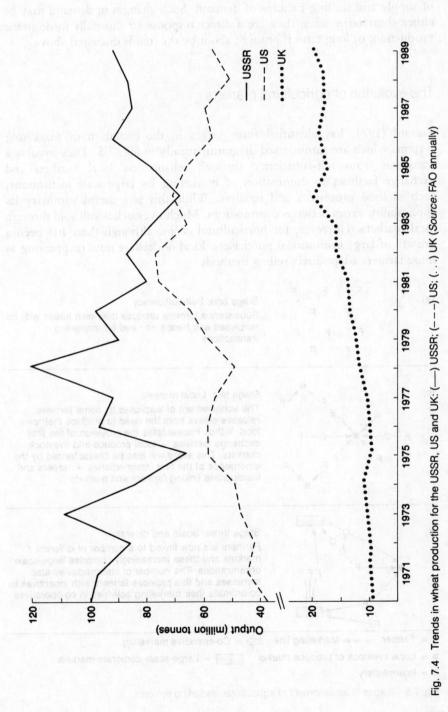

Fig. 7.4 Trends in wheat production for the USSR, US and UK: (——) USSR; (– – –) US; (· · ·) UK (*Source:* FAO annually)

same limitations as other forms of agricultural marketing, namely unreliability of supply and shifting patterns of demand. Such changes in demand may be either short term when they are a direct response to shortfalls in domestic production; or long term if brought about by the trends discussed above.

The evolution of agricultural markets

Tarrant (1974) has identified three stages in the evolution of marketing systems which are summarized diagrammatically in Fig. 7.5. They involve a transition from self-sufficiency through reliance on local markets and exchange facilities to domination of marketing by large-scale institutions, such as food processors and retailers. While this is a useful summary its applicability varies between commodities. Much livestock is still sold through local markets. However, for horticultural crops, although there has been a growth of large institutional purchasers, local outlets are now reappearing as more farmers adopt direct selling methods.

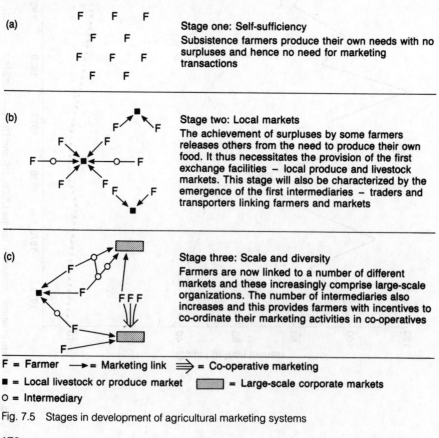

(a) Stage one: Self-sufficiency
Subsistence farmers produce their own needs with no surpluses and hence no need for marketing transactions

(b) Stage two: Local markets
The achievement of surpluses by some farmers releases others from the need to produce their own food. It thus necessitates the provision of the first exchange facilities – local produce and livestock markets. This stage will also be characterized by the emergence of the first intermediaries – traders and transporters linking farmers and markets

(c) Stage three: Scale and diversity
Farmers are now linked to a number of different markets and these increasingly comprise large-scale organizations. The number of intermediaries also increases and this provides farmers with incentives to co-ordinate their marketing activities in co-operatives

F = Farmer   ——▸ = Marketing link   ⟹ = Co-operative marketing

■ = Local livestock or produce market   ▭ = Large-scale corporate markets

o = Intermediary

Fig. 7.5 Stages in development of agricultural marketing systems

To understand recent changes it is helpful to examine the traditional systems, which equate with the 'local markets' stage in Tarrant's sequence (Fig. 7.5b). This form of market relies on the transport of produce by farmers to accessible locations. Prices are determined by face-to-face contact between buyers and sellers. However, such systems are not suited to deal with high levels of demand and do not give consumers access to products of more distant regions. Thus interregional trade develops. Initially this tends to be grafted on to the pattern of local markets, but the customers increasingly comprise specialized traders. Benefits accrue to both producers and consumers in the form of expanded markets and a greater variety of products, but there is a limit to the volume of trade which can be transacted and in the case of some products, such as vegetables, quality may be impaired. Another limitation relates to information. As direct contact between producers and consumers is reduced they become more dependent on the middleman for information on prices. This enables traders to manipulate markets. Their activities have tended to arouse scepticism among producers and, in many developing countries, there continues to be a debate about the value of their role (Blain 1984).

Change thus reflects two pressures: the need for a further increase in the scale of marketing, and the desire for greater efficiency in linking producers and consumers. Many of the scale problems have been overcome by technological developments, notably in transport, including the expansion of networks, increased speeds and carrying capacities and the development of refrigeration. However, as the scale of marketing has increased, the potentially disruptive effect of gluts and shortfalls has been magnified. Rising demand has been accompanied by socio-economic progress which has helped to raise farmers' awareness of the importance of marketing. Most producers now appreciate that they can take actions to influence prices, albeit to a limited extent.

The marketing of many agricultural products has, therefore, progressed from the 'local markets' stage. One form of marketing which exemplifies the reasons for this change is the livestock auction market. Through time there has been a reduction in their number, frequency and the amount of trade transacted through them. Guellec (1983) has examined some of the reasons for their decline in the traditional livestock areas of Brittany. A major factor has been increasing livestock numbers. For example, in the 10 years up to 1981 the number of pigs in Brittany more than doubled to 7.5 million. There is a physical limit to the number of livestock that auction markets can handle. Farmers have also begun to appreciate the potential financial gains to be made from cutting out the middleman in marketing transactions. This has led to an increase in direct trading with abattoirs. A prominent role in this is played by co-operatives, many of which have set up their own slaughtering facilities. Abattoir locations are thus becoming a more significant influence on marketing patterns than are auction markets, although the latter have not yet disappeared completely. They retain significance for certain livestock, especially those, such as young calves, which occupy an intermediate position in the fattening cycle (Fig. 7.6).

Livestock auction markets have also proved to be a resilient feature in the UK (Fig. 7.7). Carlyle (1978b) has examined factors underlying their continuation in northern Scotland. For farmers, their proximity is a major advantage. This minimizes transport costs and reduces the risk of weight loss which could result from longer market journeys. It is also easier to withdraw stock if the price does not reach the expected level. Costs of returning unsold stock from more distant markets could seriously erode profits. The opportunity costs of marketing locally are also lower, as farmers are absent from their farms for a shorter time. A final advantage for many is the method of price determination. Reliance on the impartial role of the auctioneer is preferable to face-to-face negotiations with buyers. Reputations of market personnel are also important. Farmers often continue to patronize a market because of their faith in the auctioneer's ability to obtain the best prices.

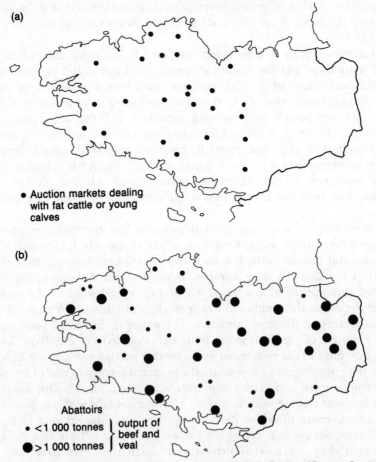

Fig. 7.6 Locations of (a) auction markets and (b) abattoirs in Brittany (*Source:* Guellec 1983)

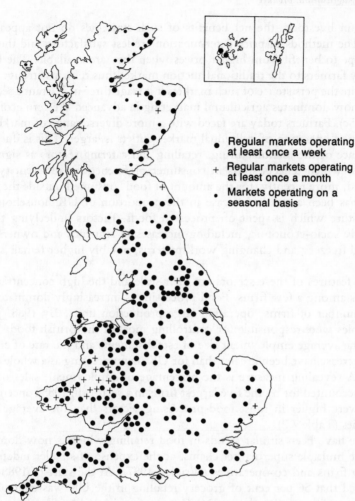

Regular markets operating
at least once a week

Regular markets operating
at least once a month

Markets operating on a
seasonal basis

Fig. 7.7 Livestock auction markets operating in Great Britain: (●) regular markets operating
at least once a week; (+) regular markets operating at least once a month; (.) markets
operating on a seasonal basis (*Source:* Livestock Auctioneers' Market Committee for England
and Wales and the Institute of Auctioneers and Appraisers in Scotland 1987)

There are also disadvantages associated with these markets, the effect of
which is to produce lower prices unless a market has a reputation for
specialized or high-quality stock. The small volume of business attracts fewer
traders and so there is less competitive bidding. Dealers will pass on their
travel costs in the form of lower bidded prices. Small markets also offer scope
for exploitation by traders through the processes of forestalling, blocking and
collusion (Curtin and Varley 1982). Alternative opportunities do exist, such
as selling directly from the farm or marketing via a co-operative, and these
options are discussed later. For smaller farmers, however, especially those

dealing in livestock, the net benefits of such strategies do not appear very great. The methods of price determination are less satisfactory and they give less scope to benefit from higher prices when they are available. The loyalty of many farmers to the traditional auction market thus remains strong.

Despite the persistence of such markets it is stage three of Tarrant's sequence which now dominates agricultural marketing in advanced Western economies (Fig. 7.5c). Farmers today are faced with a more diverse range of markets and the scale of operation of individual market outlets is larger. This is due to the emergence of the manufacturing, retailing and catering sectors as significant markets, which reflects shifts in consumer preference for different types of food and, more recently, for the amount of food consumed outside the home. There has been a steady increase in the proportion of UK household food expenditure which is spent on processed foods. Factors underlying this are primarily socio-economic, including increases in affluence and ownership of cars and freezers; and changing work patterns, notably higher female activity rates.

Two features of these sectors are their scale and the high concentration of business among a few firms. Food processing is increasingly dominated by a small number of firms, operating large production units. By 1988, just 25 companies were responsible for controlling most of the British food system, while the average employment size of establishments and the rate of employment increase have been greater than for UK manufacturing as a whole (Burns 1983). A revealing measure is the percentage of employment, sales or value added accounted for by the five largest firms in a sector. In 1985 concentration levels were higher in most food-processing sectors than the average for all industries (Table 7.2).

There have been similar trends in food retailing which is now dominated by large multiple supermarket chains, at the expense of smaller independent grocery firms and co-operatives (Wrigley 1987). At the start of 1988 it was estimated that 56 per cent of grocery retailing in the UK was controlled by five groups (Tesco, Sainsbury, Dee Corporation (including Fine Fare, Carrefour and Lennons), Asda and and Argyll Group (including Presto, Liptons, Safeway and Lo-Cost)). This has been paralleled by rationalization in shop numbers (Table 7.3). In particular there has been a growth of large supermarkets and superstores, the latter term describing units exceeding 2 300 m^2 of selling space and carrying a range of both food and non-food items. Although independent concerns still account for more outlets, their significance is reduced when figures for retail turnover are considered. The reduction in superstores operated by independents is due to take-overs by multiple concerns.

It is customary to subdivide catering establishments into three – the commercial, industrial and institutional sectors. In 1977 these respectively accounted for 56.1, 14.2 and 29.7 per cent of food bought by caterers (Tanbaum 1981). A trend within the sector is the growing importance of large contract catering firms which benefit from economies of scale. This is linked to developments in the institutional sector where government policy has encouraged competitive tendering, but contract caterers also play a prominent role in the other sectors, for example, serving airlines and industrial premises. There is a lack of data for catering although it is acknowledged to be a rapidly

Table 7.2   Five-firm concentration ratios in the UK, 1985

Minimum list heading	Percentage concentration in the five largest firms		
	Employment	Total sales and work done	Gross value added
411 Organic oils and fats	68	68	81
412 Slaughtering of animals and production of meat and by-products	22	17	18
413 Milk and milk products	46	55	54
414 Processing of fruit and vegetables	30	32	32
415 Fish processing	68	67	74
416 Grain milling	62	71	74
419 Bread, biscuits and flour confectionery	48	55	54
420 Sugar and sugar by-products	100	100	100
421 Ice cream, cocoa, chocolate and sugar confectionery	59	62	67
422 Animal feeding stuffs	40	44	55
423 Starch and miscellaneous foods	31	28	32
424 Spirit distilling and compounding	57	56	68
426 Wines, cider and perry	89	92	89
427 Brewing and malting	41	50	42
428 Soft drinks	44	45	45
Average for 106 industries listed	42	44	45

Source: Department of Trade and Industry Business Statistics Office (1988).

Table 7.3   Changes in grocery trading in the UK, 1982–86

	Numbers in:		Percentage change 1982–86	Turnover (£m. 1986)
	1982	1986*		
(a) All grocery outlets operated by:				
Multiple chains	4 599	4 418	−2.2	19 705.1
Co-operatives	4 096	2 969	−27.5	3 048.6
Independents	47 069†			4 696.7
(b) Superstores operated by:				
Multiple chains	268	393	+46.6	
Co-operatives	49	60	+22.4	
Independents	16	7	−56.2	
Total	333	460	+38.1	

* As at 1 Jan. 1987.
† Data on shops operated by independents available only for 1983.

Source: Nielsen Marketing Research (1982, 1987).

expanding sector, having been referred to as the 'sleeping giant' of the food system (Lang and Wiggins 1985). Some of the most rapid growth appears to be in the commercial sector, which is already the largest. This is especially due to the growth of take-away establishments, and specialist and fast-food chains (Tanbaum 1981). The expansion of these and of contract caterers again points to growing market dominance by larger purchasing units.

From the perspective of agricultural marketing it is the size of these sectors, and of the enterprises within them, which is important. Size brings economic power and this is particularly significant when set against the structure of the farming industry with its large number of relatively small producers. Some of the most dramatic recent changes in agricultural marketing reflect the power of these new markets to extract their requirements from the farming industry. In the 1970s the food-processing sector led the way in dictating terms to agriculture, but more recently retailers are acknowledged to have become more important. This is reflected in the tendency for large multiple chains to centralize their purchasing of meat and fresh produce through head offices. The scale of their requirements allows them to lay down detailed specifications and, if these cannot be met from domestic supplies, such firms turn to overseas producers. The changing relative power of retailers and processors is illustrated by the increased importance of retailers' own-brand lines, for example of frozen vegetables. These are produced for the retailers by the same processors with whose products they are in competition. Own-brand lines, however, usually retail at lower prices than the processors' own brands due to the scale economies in marketing available to retailers. This forces processors into a position in which they have to further reduce costs to remain competitive. Both these features of retailers' power have implications for farmers, being an important contributory factor to the cost–price squeeze which they have faced in the last 25 years.

Modern marketing structures

A useful method of classifying the ways in which marketing systems have evolved is according to the origin of change and this will form the framework for the following discussion. Change has emanated from both production and consumption sides as well as from governments, but these groups have different motives in pursuing change. Producers seek to maximize their returns; this can be achieved by increases in either prices or sales. It is also facilitated by having a guaranteed outlet for produce. For consumers, certainty of supply (in terms of quantity and quality) and low prices are the priorities. However, the consumer's objective to gain cheap supplies conflicts with that of the producer. In addition, lower prices may lead to market instability. The need to resolve some of these issues has prompted governments to intervene in order to safeguard the interests of both sides. Governments have also been involved in regulating international trade; in this context the scale of the problems and their implications for other policies preclude the independent involvement of producers or the market. However, the interests of producers, consumers, and the various organizations involved in processing and distribution are interlinked in a complex way with those of government. For example, although the incentive to form marketing co-operatives has been strongest among producers, this trend has also been encouraged by governments and food processors.

The role of governments

The motives behind government intervention in agricultural marketing cannot be separated from the broader agricultural policy goals discussed more fully in Chapter 9. For example, the desire to ensure consistently fair returns to farmers is commonly set against a commitment to maintaining reasonable food prices for consumers. These conflicting concerns are supplemented at the international scale by the strategic and economic considerations of maintaining adequate supplies and market outlets. Many governments also claim more altruistic reasons for their policies, including greater stability in world markets and the possibility of ensuring more equitable production allocations if international trade is regulated. However, attention is focused here on those policies explicitly geared towards marketing.

International trade policies

As agricultural trade expanded in the eighteenth and nineteenth centuries a disruptive effect was felt in many established producing areas in Europe. This led to the introduction of unilateral protectionist policies, including import tariffs and, in extreme cases, such as the UK's Corn Laws, bans on imports. These were intended to benefit producers by reducing competition and guaranteeing domestic markets. Consumers gained few immediate benefits as higher prices inevitably resulted. In the long term, however, it was argued that strategic benefits would accrue from the existence of a prosperous home agricultural sector, which the policies encouraged. As economic development has proceeded and demand has diversified, few countries have been able to achieve self-sufficiency in food. World trade has thus expanded, and although many countries maintain domestic restrictions, these have been supplemented by numerous multilateral agreements.

More recently international trade negotiations have had to take account of new issues, due to the need to reconcile the interests of countries at different levels of development. Although the productive capacity of developed countries has significantly increased there has not been a commensurate growth in the ability of developing countries to pay for their growing food needs. Consequently, much surplus production can only be disbursed as aid, rather than through conventional trade channels, and this can be disruptive for the recipients (Tarrant 1980). Furthermore, these surpluses in part result from the subsidization of production in developed countries. This is detrimental to the interests of Third World exporters, as it limits their competitiveness. The maintenance of reserve stocks, their equitable disbursement and provision of incentives to developing countries' domestic production are thus difficult to achieve simultaneously.

Table 7.4 depicts the major donors and recipients of two elements of food aid in 1988/89 – cereals and dairy produce. The list of donors is dominated by the advanced Western nations. Members of the Organization for Economic Co-operation and Development (OECD) account for approximately 95 per cent of all food aid (FAO 1989b) and the US alone is responsible for more than 50 per cent of the two items depicted in Table 7.4. These countries look to aid as a means of distributing their domestic surpluses, a policy which has

Table 7.4 Breakdown of major international food aid flows by donors and recipients, 1988/89

(a) Major donors

Country	Disbursements of:	
	Cereals (thousand tonnes)	Skimmed milk powder (tons)
Argentina	21.0	—
Australia	347.9	200
Austria	20.8	—
Belgium and Luxemburg	32.0	—
Canada	1 169.9	6 770
Denmark	74.7	—
EC*	882.4	108 834
Finland	25.0	2 395
France	261.7	—
Greece	5.3	—
Ireland	4.6	—
Italy	168.2	—
Japan	440.8	—
The Netherlands	78.9	7 001
Norway	31.6	—
Saudi Arabia	10.0	—
Spain	39.1	—
Sweden	132.4	—
Switzerland	64.5	1 363
UK	161.0	—
US	5 286.1	114 241
West Germany	272.3	3 852
Others	492.9	—
Total	10 043.7	245 008

* Figures exclude disbursements by individual EC member states shown elsewhere in the table.

Table 7.4 (continued)

(b) Major recipients

20 leading recipients of food aid in:			
Cereals (thousand tonnes)		Skimmed milk powder (tons)	
1. Egypt	1 426.7	1. Philippines	20 330
2. Bangladesh	1 161.0	2. India	20 192
3. Ethiopia	572.8	3. Mexico	11 510
4. Mozambique	424.1	4. Peru	10 258
5. Pakistan	416.3	5. Sri Lanka	9 185
6. Jamaica	364.7	6. Ethiopia	8 927
7. Mexico	290.5	7. China	7 255
8. Tunisia	284.2	8. Egypt	6 900
9. Guatemala	276.6	9. Bolivia	6 083
10. Sri Lanka	272.1	10. Brazil	5 381
11. Morocco	237.4	11. Nicaragua	5 110
12. Dominican Republic	228.0	12. El Salvador	4 964
13. China	223.0	13. Guatemala	4 210
14. Malawi	216.5	14. Tanzania	4 150
15. Sudan	198.1	15. Panama	4 086
16. El Salvador	197.4	16. Pakistan	4 086
17. Peru	145.6	17. Honduras	3 858
18. Philippines	134.9	18. Sudan	3 493
19. Kenya	111.8	19. Haiti	3 287
20. Vietnam	99.9	20. Mozambique	3 164

Source: FAO (1989).

both immediate and potential long-term pay-offs. For example, Tarrant (1985) has described the use of aid by countries such as the US as 'priming the market', for example, creating demand for bread in non-wheat-producing countries, such as Sri Lanka. The need to dispose of domestic surpluses is also reflected in the lists of recipients. The leading recipients of cereals form two main groups: countries of Africa which have been hard hit by recent natural disasters; and those in Asia where domestic output cannot provide adequately for local needs due to high population densities. However, not all dairy products are relevant to the needs of countries which face severe food shortages, such as Ethiopia, and so Western producers have sought alternative outlets for their surpluses. Thus, although some of the major cereal recipients appear on the list for dairy products, this also includes more affluent Third World countries, such as Egypt and Brazil. Political factors, too, are important influences on the allocation of food aid. The prominence of Egypt as a recipient for cereals is, in part, accounted for by the favourable treatment she has received from the US since the breakdown of her relations with the USSR.

Much government policy on international trade is via multilateral agreements. Of these the most influential has been the General Agreement on Tariffs and Trade (GATT). This originated in a committee established in 1947 by the Economic and Social Council of the UN to prepare a treaty on the organization of world trade. Although this was never ratified, GATT has remained in existence and still constitutes a focus for the negotiation of rules for the conduct of trade. Seven rounds of negotiations have been completed and the eighth, the Uruguay Round which continues at the time of writing, was launched in 1986 when 95 countries formally subscribed to GATT. The 23 original signatories included the world's major industrialized nations, whose aims were to expand the trading system through liberalization and removal of barriers. This set the tone for later negotiations. However, from the late 1950s many Third World nations began to express dissatisfaction with the working of the world trade system and its implications for their economies. As exporters of primary products and importers of manufactured goods they were faced with worsening terms of trade. In 1962, 77 Third World countries called for a UN Conference on Trade and Development (UNCTAD). The title of 'Group of 77' is still used, even though the membership has now grown and been constituted as a permanent conference. UNCTAD has a different emphasis from GATT, with more interest being shown in regulation so as to protect Third World economies from price fluctuations, which are particularly marked for primary products. For example, at UNCTAD IV, held in Nairobi in 1976, the Group of 77 proposed the creation of a common fund of $6 bn to regulate prices of 17 major mineral and agricultural commodities. Although agreed in principle this has never been ratified.

Generally GATT has been more successful in liberalizing trade in industrial rather than agricultural products. Nagle (1976) attributes this to the reluctance of governments to agree to measures which might undermine their farm income support policies. It is felt that sharp price fluctuations may occur which would totally contradict the objectives of domestic policies. UNCTAD has been concerned with agricultural aid and trade within the broad context of North–South relationships. It has had some minor successes. For example, after Nicaragua objected to the US decision to reduce her sugar quota, a vote was passed at UNCTAD VI in Belgrade requiring developed countries to refrain from applying such restrictive measures. This was deemed to conflict with the UN Charter. However, major reforms of world trade, in particular in food products, have been more elusive. The Group of 77 attribute this to blocking by industrialized countries, especially the US.

Similar issues are apparent when individual commodities are considered. The first International Wheat Agreement was concluded in 1933 but broke down after one year. Further negotiations brought the International Wheat Council into existence in 1942 but not until 1949 was a new agreement negotiated. The signatories of this second agreement accounted for a major proportion of world trade, although Argentina and the USSR declined to take part. Its main provisions were the definition of maximum and minimum prices and guarantees by the signatories to trade specified minimum amounts within that range. This succeeded largely due to high prices on the open market; it encouraged importers to maximize their trade transacted under the terms of the agreement. Although exporters gained a lower return this was accepted as a safeguard against a possible drop in prices. Six further agreements have been negotiated. Increasingly they have been based on broader objectives, including trade expansion, market stabilization and the furthering of co-operation in wheat matters generally. In 1971, agreement, for example, included no specific price or supply provisions, primarily due to the prevailing low prices on the free market. Instead it comprised two legal instruments: the Wheat Trade Convention, which sought to promote the objectives outlined above, and the Food Aid Convention, which committed exporters to minimum annual contributions of grain as aid. The International Wheat Council remains in existence and is a useful holding body, allowing negotiations to continue and specific issues to be discussed by importing and exporting countries. The remit of the Council has now been extended to all cereals.

In recent years international market conditions for cereals have changed markedly. The USSR has altered its agricultural policy to become a significant importer in years of domestic crop failure. North American producers have reduced their acreage levels and reserve stocks whereas production increases in Western European countries, especially the EC, have transformed the region into an exporter of many cereals.

Despite these changes the pattern of world exports of wheat is still dominated by a relatively small number of countries (Table 7.5). The two main North American producers account for almost half the total, with EC countries, Australia and Argentina being the other main exporters. The degree of concentration is noticeably lower among importing countries. These are

Table 7.5 The leading exporters and importers of wheat and flour, 1988

	Export			Imports	
	100 million tonnes	Percentage of world exports		100 million tonnes	Percentage of world imports
World	1 067 687	100.0	World	1 077 430	100.0
US	382 700	35.8	China	153 352	14.2
France	170 842	16.0	USSR	144 357	13.4
Canada	114 869	10.8	Egypt	69 970	6.5
Australia	105 858	9.9	Italy	56 481	5.2
West Germany	46 303	4.3	Japan	55 786	5.2
Argentina	43 742	4.1	Algeria	53 668	5.0
UK	33 890	3.2	Iran	45 000	4.2
Italy	27 514	2.6	Iraq	34 389	3.2
Hungary	14 312	1.3	South Korea	22 883	2.1
USSR	13 437	1.3	West Germany	21 841	2.0
Percentage of total accounted for by 10 leading exporters	89.3		Percentage of total accounted for by 10 leading importers	61.0	

Source: FAO (1989).

mainly ones with large populations which have production difficulties due to low or unreliable yields, for example the USSR, or limited suitable land resources, such as Japan and South Korea. Some European countries feature both as leading exporters and importers. This may be accounted for by their inability to produce sufficient quantities of certain kinds of wheat or the practice of exporting flour milled from imports. For example, in Italy there is considerable demand for hard wheat, the basis of pasta, but conditions are only suitable for its growth in the south of the country. Large wheat-growing areas in the north can only produce soft varieties.

Given the reductions in output levels among the main exporters, the degree of export concentration has potentially harmful implications for those countries which have to rely on imports. Although European production and exports have increased this has not compensated for the North American reductions. Thus at a time when the International Wheat Agreement included no price provisions, the free market has been characterized by increasing price instability (Marticorena 1982). One feature to emerge is the desirability of linking international trade agreements with national policies. The total world supply situation has been a more potent influence on prices than any agreement provisions. Blandford (1963) has examined the links between variability in national grain markets and international price instability. Domestic market variability is unavoidable due to fluctuations in yields, for example as a result of variable weather conditions, but it is possible to minimize the extent to which this is transmitted to the international market. Similarly, individual countries can adopt policies which enable them to absorb some of the variability emanating from other market participants. A key policy will be the holding of reserves, but this is one of the issues which has proved a stumbling block in negotiations of a new agreement. Consensus has not been possible on three issues: volumes of reserve stocks, mechanisms which should operate to allow the release of reserves, and provisions to enable developing countries to meet their agreement obligations (primarily the financing of imports).

Marketing boards

These are characterized by their statutory responsibility for marketing or some aspect of its regulation. They were first established in the 1920s in Canada and Australia and have subsequently spread to many developed Western nations. The enabling legislation usually requires proof of majority producer support for a board's establishment. They are democratic producer organizations but also include government representation. In certain countries, including the UK, boards are mainly concerned with domestic marketing. Elsewhere separate boards exist for export commodities. European countries also established boards in former colonial territories to organize the purchasing of export crops, for example cocoa and groundnuts in West Africa (Bauer 1963).

The major concerns of marketing boards are market stability and producer incomes. A common method of stabilization is through payment of a pool price. This involves pooling the revenue from a product and dividing it uniformly among producers. This has been operated by the New Zealand Apple and Pear, and Dairy Boards so as to reduce uncertainties deriving from short-term price fluctuations and, in the case of dairy products, from the variety of end uses. Similar considerations underlie the price scheme operated by the British Milk Marketing Board (MMB). Other methods of stabilization include control over supplies, such as the British Potato Marketing Board's quota system, or the holding of reserve or buffer stocks. These are stocks withheld from the market when there is excess supply, to be released at times of shortage, for example, the New Zealand Wool Board's national buffer stock scheme. Buffer stocks tend to be more successful in balancing the market when production is of a seasonal nature. They are less effective at redressing longer-term imbalances unless accompanied by a quota system.

Some stablization strategies, such as supply control, may also increase producer incomes. A range of other measures can be used to reduce costs or increase revenues, including market and product development, joint promotion schemes and provision of specialized services. These are exemplified by the work of the British Milk and Wool Marketing Boards. In the UK, milk production expanded until the mid-1980s in the face of stagnant or even declining demand for dairy products (during the 1980s household consumption of milk and cream declined by 19 and 49 per cent respectively). To compensate, the MMB has been active in promoting new lines, such as Lymeswold cheese, and in more innovative strategies, for example research into the potential of milk as a petfood ingredient. The Wool Marketing Board, faced with declining demand, has promoted the quality aspects of wool and sought to encourage manufacturers to use the British product. Both boards provide a variety of producer services. For dairy farmers these include management advice, artificial insemination and veterinary services. The Wool Board provides assessment of ram fleeces, which acts as a guideline for selective breeding. It has a long history of co-operation with its New Zealand counterpart in training shearers. Such items all represent major expenditure on behalf of producers. Export boards may use similar strategies and can also increase the bargaining power of their membership in international markets. The promotion of products such as New Zealand and Irish butter and Danish bacon provide examples of their effectiveness.

The policies of the MMB also provide one of the clearest illustrations of the potential geographical impact of marketing board policies. This has been fully covered in standard texts (Morgan and Munton 1971; Tarrant 1974; Ilbery 1985) and only an outline account will be provided here. Prior to the Board's establishment in 1933, the location of dairying in England and Wales was influenced by classical distance-from-the-market factors. Regions with large population concentrations, such as the South-east and North-west, were major markets for milk and this conferred locational advantages on producers in those areas. This situation changed with the advent of the Board and its

183

system of pool prices and collection charges. England and Wales were divided into 11 regions. Within each region producers received a fixed pool price, the exact level of which reflected the returns from all milk sold in that region. Regions with high proportions sold as liquid milk, such as the South-east, had higher pool prices. However, to ensure that all producers were equitably treated, differences between regional pool prices were kept to a minimum. Similarly, the Board levied a uniform collection charge within regions, regardless of the distance between a producer's farm and the nearest dairy. Regional variations in the collection charges were also minimized.

More recently the Board has operated a standard pool price in all regions. This does not mean that all producers receive identical prices for their milk, but they are all paid on the same basis, which takes into account the quality of their milk. Since April 1984 a centralized milk testing scheme has operated. Farmers are paid according to the butterfat, protein and lactose contents. Regional price variations are still discernible, but these are small. For example, in 1986–87 the annual average price paid per litre varied from 16.022p in

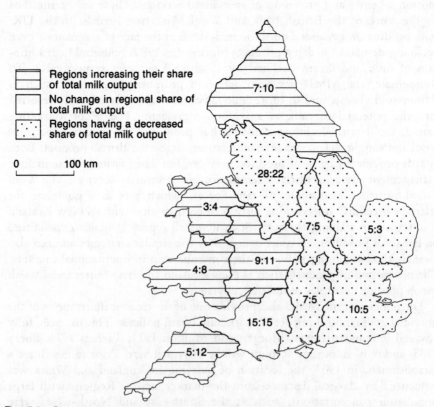

Regions increasing their share of total milk output

No change in regional share of total milk output

Regions having a decreased share of total milk output

0   100 km

7:10

28:22

3:4

7:5   5:3

9:11

4:8

7:5

15:15   10:5

5:12

Fig. 7.8 Changes in milk production between MMB regions, 1937–87. Figures indicate percentage of milk produced in a region in 1937 and 1987 (*Sources:* Morgan and Munton 1971; Federation of United Kingdom Milk Marketing Boards 1987)

North Wales to 16.232p in the South-east. The differences now reflect variations in milk quality, rather than locational factors. The effect of these policies has been to erode the former advantage of producers in regions of high demand and to bring about changes in the geography of milk production (Fig. 7.8). Notably, there has been a shift in production away from the South-east to less accessible western and northern regions.

A further aspect of the role of marketing boards is their use by governments to implement other policies. For example, in the UK the MMB administered the quota system introduced in the EC in 1984 to curb overproduction of milk. This was accompanied by the Milk Supplementary Levy (Outgoers) Scheme, which was designed to release quota for redistribution to farmers facing hardships. The Board's existence clearly facilitated the task of implementing the system, although it was required to act in a way which many producers saw as detrimental to their interests.

Other government policies

Governments have established various bodies to promote production and marketing efficiency without having the statutory role of boards. In the UK these include the Home Grown Cereals Authority, the Meat and Livestock Commission and Food from Britain. The latter was formed under the Agricultural Marketing Act of 1983 and replaced the Central Council for Agricultural and Horticultural Co-operation. Its objectives are the promotion of British food in domestic and overseas markets and the improvement of marketing structures for farmers. Its work in the export field focuses on advertising and public relations, for example, arranging special promotions of British foods in foreign retail establishments and the organization of visits to the UK by foreign buyers. Four areas have been targeted as priority markets: France, Germany, the Benelux countries and the US. At home, efforts have concentrated on quality improvement. The Foodmark scheme has been developed to enable consumers to identify food which has reached set quality standards. This is intended to encourage farmers, processors and distributors to achieve those standards, thereby increasing the market for domestically produced food.

Governments also provide the legislative framework and, in some circumstances, financial assistance for other marketing ventures, such as co-operative developments. Food from Britain is active in promoting co-operative marketing and its annual grant expenditure in this area now exceeeds £1 m. In various parts of the world governments have been involved in modernizing physical market structures. The intention is to enhance the attractiveness of markets and streamline marketing processes, thereby increasing the volume of trade and reducing marketing costs. This type of policy is common in developing countries where infrastructure is often lacking. A recent example of such developments in Europe is the post-war creation, in France, of a network of new centres – Marchés d'Intérêt National. Dealing primarily with

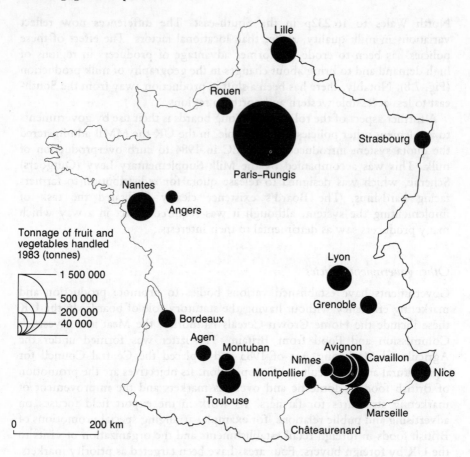

Fig. 7.9 Marchés d'Intérêt National: throughput of fruit and vegetables

horticultural produce, they are equipped with modern facilities and have
excellent access to interregional transport routes. They have brought some
rationalization to the formerly chaotic distribution system which often
involved moving produce over unnecessarily long distances. By 1983, 19 were
in existence (Fig. 7.9). The intention behind their creation was that they
should perform two roles. One type acts as collection points for produce from
traditional horticultural regions, such as Provence and the Agenais. These
include Cavaillon, Châteaurenard and Montauban. Complementary markets
serve large population centres, including Paris (Rungis), Lille and Strasbourg.
The markets have experienced differential growth rates. With the exception
of Cavaillon and Châteaurenard it is those in major urban centres, especially
Rungis, which have grown most rapidly. One reason for the slow growth of
the markets in horticultural areas has been the emergence of alternative forms
of marketing at about the same time, including co-operatives and direct links

between growers and customers. However, the 19 markets now handle more than 4.5 million tonnes of produce annually, which represents a significant contribution to the total pattern of marketing in France.

The role of producers

Barker (1981) suggests that the business orientation of farmers is primarily towards production. Consequently, many have failed to appreciate the benefits to be gained from greater involvement in marketing. However, more farmers are now starting to take account of factors beyond the farm gate. Two main initiatives are apparent.

Co-operative marketing

The essence of agricultural co-operation is that farmers can achieve some objectives more successfully through mutual action than as individuals. Marketing is one of three main areas of agricultural activity in which co-operatives have intervened, the others being supply purchase and production, although many societies have multiple functions (Bowler 1972). A universally acceptable definition of co-operative societies has proved elusive because of the variety of forms they make take. The International Co-operative Alliance has approved a set of fundamental principles of co-operation (Sargent 1982), but individual groups differ in their adherence to these. Essential character-istics include voluntary and unrestricted membership, democratic control and the allocation of profit either to benefit the members on an equal basis or in proportion to the amount of trade they conduct through the society. This latter characteristic is a crucial distinction between a co-operative and other forms of business organization in which profits are allocated in proportion to the amount of capital invested (Levay 1983).

Marketing co-operatives aim to increase the returns of their members from sales of produce or provide them with greater competitive security. This can be accomplished by three interrelated means.

Scale economies. Individual farmers may be unable to afford many marketing facilities such as for washing, grading, packaging, transport and storage. However, a group purchase will grant individuals access to the facility at a lower rate than commercial hire charges. A particularly important scale economy is access to specialized management skills. Sargent (1975) refers to the appointment of a general manager as 'unquestionably the most important decision to be taken by a marketing co-operative in determining its success or failure'. Co-operation has thus allowed farmers to tap specialized marketing expertise, including, in some instances, the appointment of well-qualified personnel from industry. Even if a group is unable to purchase its own

187

facilities or employ specialist personnel, it may be able to negotiate more favourable rates for members than would be charged to individuals. This leads to the second source of benefits.

Countervailing power. A classic cause of the individual farmer's weakness in the market is the inability to influence prices due to the small proportion of total output for which the farmer is responsible. Concerted action by a number of producers may allow them to exert more power over prices, especially if it is possible to withhold a significant proportion of a product from the market. In practice, many co-operatives have found it difficult to achieve this degree of power. This is often due to the fact that non-members account for a significant proportion of sales of a product and societies have no control over them. Even where a co-operative does control a significant proportion of production, it needs to ensure the commitment of its members to selling through the society. This can prove difficult when members have confidence, however misplaced, in their own marketing ability.

Vertical integration. Investments made by societies may allow them to extend their operations beyond those normally associated with individual producers. This can bring about a gradual take-over of some of the roles of traders or manufacturers, thereby increasing the overall profits accruing to members.

Some societies have developed highly sophisticated strategies using all these devices. An example is the Kerisnel society in Brittany, whose 6 000 members are important producers of early vegetables and soft fruit. The society's size enables it to impose fair prices on traders. The head office is linked by telex to markets throughout Europe and prices are constantly monitored. When they drop below a given point produce is withheld. Vertical integration has occurred into canning and freezing and the society has an interest in the Plymouth–Roscoff ferry service, which was developed in part to open up UK markets (Ardagh 1982).

As the scale of co-operation has increased, hierarchical systems of organization have developed. Secondary co-operatives, usually referred to as co-operative unions, are groupings of individual primary societies. The theory behind their formation is that they enable even greater scale economies and give more scope for vertical integration and the use of countervailing power than do the primary societies. In many countries a third tier now exists in the hierarchy with a small number of tertiary bodies co-ordinating the secondary unions at national level. In France, for example, the Confédération Française de la Co-opération Agricole is the national representative body, to which are affiliated 19 national co-operative federations (organized on a commodity basis) and 17 regional federations. Thus a pyramidal structure exists with each organization affiliated to one or more others in the next tier. This has brought further advantages, notably the potential for integration with other bodies providing specialist services. For example, financial institutions have been developed in many countries to service the credit needs of the co-operative movement. These include the French Caisse National du Crédit Agricole and the Dutch Rabobank.

Although the increasing scale of the movement has brought obvious advantages, this has not been without costs. The philosophy of co-operation is based on democratic principles: members should have the power to influence the policies of their societies. However, with the increasing importance of professional managers and the tendency for primary societies to become subservient to the secondary and tertiary bodies, some of the original ideals of the movement may be lost, especially if divergence occurs in the objectives of different types of member (Levay 1983).

Table 7.6 Percentage of selected products sold through co-operatives in EC countries,* 1984

Product	West Germany	France†	Nether-lands	Belgium	Luxem-burg‡	UK‡	Ireland	Denmark†	Greece
Pigmeat	n.a.	64	26	10	±25	13	23	92	4
Beef and veal	n.a.	21	15	<1	±25	10	14	71	7
Poultrymeat	n.a.	45	27	<1	n.a.	2	10	50	23
Eggs	n.a.	25	19	<1	±10	28	2	68	15
Milk	n.a.	52	87	65	86	0	94	88	21
Sugar-beet	0	17	63	<1	n.a.	0	0	14	0
Cereals	50	67	55	20	n.a.	20	33	47	0
All fruits	n.a.	42	78	55	20	33	0	50	39
All vegetables	n.a.	30	80	55	n.a.	17	0	50	14

n.a. Not available.
* Data unavailable for Italy.
† Data relate to 1979
‡ Data relate to 1982

Source: Adapted from Commission of the European Communities (1986).

Despite these comments, Foxall (1986) suggests that a strong formal organization is important in encouraging and sustaining a high level of participation in co-operatives. This is one of the factors which explain geographical variations in co-operation. Table 7.6 depicts the co-operative share of the market for various commodities in countries of the EC in 1984. This largely confirms the classification of these countries into three groups by Foxall (1986). In group 1, comprising the Netherlands, Denmark and Lux-emburg, co-operatives play a crucial role in agriculture and market shares are generally high for major production sectors. Group 3 includes the UK, Greece and Italy. Here the level of co-operative involvement is much lower and this is linked to a poorly developed structure at secondary and tertiary levels. Group 2, made up of France, Belgium, Ireland and Germany, is less homogeneous but lacks the extremes of the other two. These countries have a well-developed organizational hierarchy, but the bodies at the top of that hierarchy exert less overt pressure on ones lower down than do those in group 1. This is matched by a more ambivalent attitude towards co-operative marketing on the part of farmers.

Social, economic and political factors also help to explain international differences in the extent of co-operative development. The strong co-operative movement in Denmark developed as a response to economic adversity arising from depressed prices in the nineteenth century. Favourable domestic conditions included a high level of education among farmers and a sophisticated local democratic system, out of which co-operative societies emerged quite spontaneously (Foxall 1982). National governments differ in their commitment to co-operation. French governments have been particularly supportive, especially because of the role they see for co-operative societies in regional development. An important stimulus was the legislation of the early 1960s enabling the formation of a new type of society, Sociétés d'Intérêt Collectif Agricole (SICA). These were released from the restriction of having to conduct business exclusively with their own members, which previously applied to all societies. The slow growth of co-operation in the UK has been attributed to various factors, including: the lack of a well-developed organizational structure, poor government support and the range of alternative devices set up by governments to improve marketing (such as marketing boards and price supports), structural factors (fewer small farms and a low contribution of agriculture to GDP and exports), the geographical diversity of agriculture and behavioural factors, notably the independent outlook of many farmers.

It should also be noted from Table 7.6 that co-operative market shares vary between agricultural commodities. The types of product for which co-operative marketing is significant include those which have high marketing costs, such as dairy produce, and those which have been subject to rapid marketing change or offer more scope for added value. Both the latter features are characteristic of horticultural products. These characteristics also help to explain distributions of co-operative societies within countries. Figure 7.10(a) illustrates their distribution in England and Wales. A concentration on the eastern side of the country is apparent. However, it needs to be appreciated that this represents the aggregation of a number of distributions of different types of co-operative society. Three individual patterns are also depicted. For grain and seed co-operatives a generally even pattern across the country is revealed (Fig. 7.10b). The societies involved in marketing peas and beans are strongly concentrated in the east (Fig. 7.10c) but those concerned with livestock are found mainly in the wetter western areas (Fig. 7.10d). The distribution patterns of these three types of marketing co-operative are thus accounted for by the underlying agricultural geography rather than by any differences in attitude on the part of farmers.

Direct marketing

This involves direct transactions between producers and consumers, thus bypassing the middleman. It is not new. Traditional markets involved this type of contact, and farmers, in many parts of the world, have at various times delivered produce to consumers. Some meat, dairy and horticultural

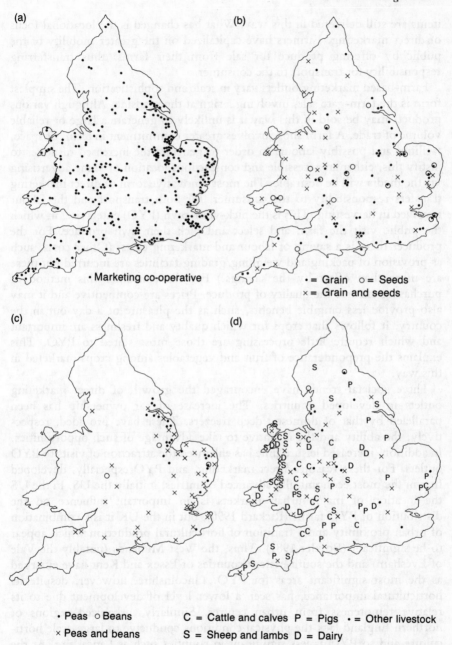

(a)

(b)

• Marketing co-operative

• = Grain   o = Seeds
× = Grain and seeds

(c)

(d)

• Peas o Beans   C = Cattle and calves P = Pigs • = Other livestock
× Peas and beans   S = Sheep and lambs D = Dairy
                              × = Combination of two or more forms of livestock

Fig. 7.10 Marketing co-operatives in England and Wales. (a): (.) marketing co-operative; (b): (.) grain; (o) seeds; (x) grain and seeds; (c): (.) peas; (o) beans; (x) peas and beans; (d): (C) cattle and calves; (S) sheep and lambs; (P) pigs; (D) dairy; (.) other livestock; (x) combination of two or more forms of livestock (*Sources:* Based on data from the Plunkett Foundation for Co-operative Studies 1987)

items are still delivered in this way. What has changed is the locational focus of direct marketing. Farmers have capitalized on the greater mobility of the public by offering produce for sale from their farms, thus transferring responsibility for transport to the consumer.

Farm-based marketing outlets vary in scale and sophistication. The simplest form is the farm-gate sale, involving a sign at the roadside. Although various products may be sold in this way it is unlikely to generate a large or reliable volume of trade. A farm shop involves greater commitment in terms of space, facilities and possibly labour. In order to generate the increased turnover to justify this, either an accessible and conspicuous location or some advertising via the media will be desirable. The most innovative form of direct marketing transfers responsibility to the consumer, for both transport and the labour involved in harvesting. This is the pick-your-own (PYO) enterprise, in which the public visit the farm and select and pick their own produce. For the producer there is a saving of labour and marketing costs. (Other costs, such as provision of packing and weighing/grading facilities are incurred but these are more than offset by the savings.) For the consumer this method of purchasing ensures the quality of produce. Prices are competitive and it may also provide less tangible benefits, such as the pleasure of a day out in the country. It follows that crops for which quality and freshness are important and which require little processing are those most suited to PYO. This explains the preponderance of fruit and vegetables among crops marketed in this way.

Three societal trends have encouraged the growth of direct marketing outlets in developed countries. The increase of car ownership has been paralleled by that of domestic deep freezers. These have provided, respectively, the ability and the incentive to take advantage of such opportunities. In addition, increased leisure time has enhanced the attraction of visits to PYO outlets. For these reasons direct marketing, and PYO especially, developed first in the most economically advanced countries, initially the US. In the US the location of major urban markets is an important influence on the distribution of PYO farms (Rickard 1990), but in the UK it is a combination of urban proximity and a tradition of horticultural production which appear to be significant (Bowler 1981). Thus, the West Midlands (notably the Vale of Evesham) and the south-eastern counties of Essex and Kent have emerged as the most significant areas for PYO. Lincolnshire, however, despite its horticultural importance, has seen a lower level of development due to its relative remoteness from urban centres. Similarly, urbanized regions of northern England lack the physical conditions conducive to large-scale horticulture and so PYO is less significant in counties such as Lancashire. At the local scale a site on a main transport route is of prime importance in the US but less so in the UK. This may be attributed to the relatively recent development of PYO in the UK. There has been a steady growth in the number of outlets, with indications that some areas may now be reaching saturation point. This, combined with other influences such as the rising costs of motoring, may lead to rationalization in the pattern and, in a more competitive environment, location alongside major roads could increase in importance.

The role of the market

In the last 30 years farming has become increasingly associated with and dependent upon the food-processing and distribution sectors, a trend most advanced in the US. This is part of the process referred to by political economists as capital penetration, whereby many contemporary rural trends are explained in terms of capital seeking out more profitable activities and locations. The ruralization of industry, institutional investment in farmland and the capital intensification of agriculture, with its attendant dependence on industrial inputs, are all part of the same process (see Ch. 1).

A Marxist view is that these trends are a response to the need for capital accumulation which is the driving force of capitalist societies. Some of the changes described above, which have led to increased demand for processed and convenience foods and innovative retailing methods, have provided new opportunities for capital accumulation. It is possible to identify the advantages the food-processing industry may derive from integration with agriculture; in particular it can gain greater control over material supplies. This is especially important when some of the vagaries of agricultural production and marketing systems are considered. Without control it is difficult for the processor's precise requirements of price, quantity, quality and delivery schedules to be met. Failure to satisfy these leads to reduced profits, for example, through machinery lying idle due to irregular deliveries, or loss of markets due to a decline in quality standards.

This integration between agriculture and industry has also been called agribusiness, a term originally used by Davis and Goldberg (1957) and subsequently applied in a range of different situations. Wallace (1985) suggests that it best refers to the whole system of agriculturally related activities upstream and downstream in the food chain. The growth of agribusiness has caused concern in developed countries, where the autonomy of the family farm is threatened, and in the Third World due to its neo-colonialist overtones (Burbach and Flynn 1980). Attention here is confined to agribusiness in the West, but these broader issues should not be ignored. An important focus of interest is the nature of the links between the elements within the system; for downstream activities, some of the most important of these will be marketing links.

Vertical integration

Although this term can describe any integration of successive stages of production, it is restricted here to situations in which it occurs through direct ownership of property within one firm (Cramer and Jensen 1982). Vertical integration can occur from the farm sector onwards when it is instigated by farmers, usually via co-operatives (see p. 188), but it more commonly occurs from the food-processing and retailing sectors. This represents an obvious method by which firms can extend control to their sources of supply (i.e. by the ownership of farmland). In practice, vertical integration of this nature (real subsumption) has not proceeded very rapidly. In the US, where it is most widely developed, only 1 per cent of farms are corporately owned. However, this figure needs to be treated with caution due to the large size of these farms. Hence this 1 per cent account for 15 per cent of cash receipts (Burbach and Flynn 1980).

Various reasons account for the limited development of vertical integration. Firms wishing to exert more control over supplies will be interested in specialist farming units. This avoids the need to diversify into activities which require new forms of expertise in production or marketing. The mixed structure of many British farms has thus been a disincentive to full integration. Similar factors have been cited as limiting vertical integration in other areas, including Quebec and New Zealand (Smith 1984; Le Heron and Warr 1976; Robinson 1988). Food companies in New Zealand rely on their own holdings for up to 10 per cent of supplies. These serve as demonstration farms and have played a role in introducing new crops, such as asparagus. As elsewhere, however, New Zealand firms prefer to avoid the direct ownership of farmland.

The preference for specialist holdings also helps to explain the distribution of corporate farming within countries. In the US this is related to the underlying agricultural and physical geography (Vogeler 1981). Corporate farms have developed more rapidly in areas of uniform soil and climate, for example the irrigated West and humid South. They are also strongly represented in areas producing commodities for processing, such as poultry, fruit and vegetables (Broadway and Ward 1990). This further increases the attractiveness of states such as Florida and California for investment. Finally, legal factors can have an influence, as in North Dakota, where state law prohibits farm corporations.

Contract farming

In the absence of vertical integration, food-processing and distribution firms still require a means of exerting control over their agricultural supplies (formal subsumption). Contract farming was defined by the Barker Committee as: 'the production and supply of agricultural or horticultural produce under forward contracts, the essence of such arrangements being a commitment to provide an agricultural commodity, at a time and in the quantity required by a known buyer' (Barker 1972: 2). Contracts have proved to be a very flexible marketing method. They are used by a variety of markets, including processors, agricultural merchants, retailers and wholesalers and in the purchase of many different products (Marsh 1965). The form of contract also varies. At one extreme is the forward sales contract, which is a commitment to sell an already harvested crop at a specified later date. These are normally entered into towards the end of the production process, and carry few management implications for the farmer. At the other end of the spectrum are contracts under which the purchaser is fully involved in farm management, including input provision. Under these the price, or the method by which it is to be calculated, will normally be agreed in advance of delivery. Some farmers have considered that this puts them at a disadvantage, especially in times of rising prices, since there is a tendency for contract terms to make over-cautious estimates of price changes through time. This was especially felt to be a problem among growers of vegetables for the freezing industry. As a result, a new form of contract, the joint venture contract, has been developed (Malcolm 1983). Under these, sales receipts are shared between growers and processors in proportion to their costs. Ironically, this type of contract is less advantageous to farmers in a period of stagnant demand than it would have been in the market conditions of the early 1970s. For this reason, in 1983, the NFU recommended growers to limit the proportion of their crop sold under such contracts.

Both parties face a mix of advantages and disadvantages in contract farming. For the processor there is a balance of advantages. Firstly, there is the ability to control the quantity, quality, price and timing of delivery of supplies, the importance of which has already been stressed. This is achieved via contract terms which may cover: the type of seed and fertilizer to be used; husbandry practices, including agreed access for the firm's fieldsmen for inspection purposes; arrangements for harvesting and transport; and the means of price determination. Closely linked to greater control, is greater certainty of supply. This is particularly important to the large industrial firm in its planning and budgeting, for example, when investing in additional processing capacity. More recently, a third advantage has become apparent for the processor as firms' interests have diversified. For example, it is not uncommon for firms to combine interests in processing with the manufacture of machinery or other farm inputs. Contracts provide scope to enlarge markets for these products. Where contracts require investment, for example in harvesting machinery, the firm may provide loans and subsequently benefit from farmers' interest payments. These examples provide further evidence of the emergence of agribusiness and the use of contracts to transfer value from agriculture in the capital accumulation process.

For the farmer the advantages and disadvantages are more finely balanced. Given the uncertainties of agricultural marketing, an assured outlet and price are very attractive. An associated gain is the higher prices which contracts often confer. Thirdly, the farmer may be granted privileged access to inputs and expertise, including the advice of agronomists working for the processor, subsidized supplies of seeds and fertilizers, and the ability to be at the forefront of technical developments, for example in harvesting machinery. However, realization of these advantages hinges on the exact wording of contract terms. This leads to a consideration of some of the disadvantages of contracting for the farmer.

A number of disadvantages stem from the unequal relationship between farmers and food-processing companies. In this respect contract farming reflects many of the problems of agricultural marketing in microcosm. Contracts are normally drawn up by processors and will include numerous clauses to safeguard their interests. For example, the processor's right to reject deliveries in certain situations is normally written into a contract and can sometimes be used unfairly in the farmer's view. The benefit of higher prices may be more illusory than real due to the early stage in the production process at which a contract is signed. This has been one of the factors underlying the development of joint venture contracts. However, the major disadvantages relate to the loss of independence on the part of the farmer. A contract transfers responsibility for many farm production decisions to the processor. The farmer is deprived of the opportunity to seek out more profitable markets at the end of the production cycle. There are also more subtle aspects of loss of independence. The farmer is described as being reduced to a 'propertied labourer' or caretaker on his own land (Davis 1980). Rural sociologists have focused on the class implications of the changed status of farmers in this situation. Capitalism transforms them from independent producers to wage earners on their own land. They can have more in common with hired labourers or piece-workers than private entrepreneurs. The significance of this can be both social and economic. Vogeler (1981) has described how, in 1978, seasonal tomato workers in Ohio demanded to be included in annual contract negotiations, recognizing that the contract prices of the canning companies had more influence on their wages than did the farmers who hired them. This was an acknowledgement of the changed class status of growers under a contract system.

Another socio-economic change which is linked to contract farming is increased farm size. Processors prefer to deal with large farmers so as to achieve scale economies. Through their power to award or refuse a contract, they have contributed to differences in profitability between small and large farmers and encouraged the process of farm concentration. Economically, contract farming, like vertical integration, has allowed capital to appropriate a greater proportion of profit from farming. In addition to the effects of processors' policies on prices and input costs this can be achieved more indirectly. For example, it has been suggested that industrial firms treat farmers as employees, without having some of the associated overheads, such as social security payments, paid holidays or the need to recognize minimum wages legislation (Vogeler 1981).

Despite the concern expressed about the implications of contract farming, a growing amount of produce is marketed by this means. It has not developed uniformly in geographical space for structural reasons, since it is more suited to certain products than others. Commodities most likely to be purchased by

Table 7.7   Percentage of selected products sold under previously concluded contracts in EC countries,* 1984

Product	West Germany	France	Nether-lands	Belgium	Luxem-burg	UK	Ireland	Denmark	Greece
Pigmeat	14	25	50	55	10	50	n.a.	0	n.a.
Calves	14	25	85	90	n.a.	<1	n.a.	5	n.a.
Poultrymeat	73	45	90	95	n.a.	95	90	70	n.a.
Eggs	20	15	50	70	n.a.	65	10	75	n.a.
Milk	27	1	90	0	n.a.	0	10	0	7
Sugar-beet	100	100	100	100	n.a.	100	100	99	100
Potatoes	0	8	70	15	n.a.	13	4	40	2
Peas	95	90	95	98	n.a.	95	100	100	9

n.a. Not available. Data unavailable for Italy.

Source: Adapted from Commission of the European Communities (1986).

contract include: those for which quality is important; new products for which there may be market uncertainty; products for which it is important to restrict supply; and products used for processing (Marsh 1965). Data on contract farming in the EC generally confirm this (Table 7.7). Quality is important for many products which are widely sold under contract, including pigs, calves and peas. Processing crops are also strongly represented on the list and, in the case of sugar-beet, there is the additional need to restrict supplies in accordance with international quotas. With some commodities there are variations between states. A notable example is milk. This is a product for which contract farming would seem to be particularly relevant at present, due to the need to limit production. However, in some countries it is considered unnecessary because of established alternative marketing systems, such as the MMB in the UK. Contract farming is significant for a similar list of crops in the US, including sugar-beet, fruit and vegetables for processing, seed crops, milk, poultrymeat and eggs (Vogeler 1981).

Geographically, contract farming has had a distinct impact at the local scale since it differs from other marketing forms in one crucial characteristic. It is the market rather than the farmer which determines where a commodity is produced. In the canning and freezing industries, for example, quality requirements dictate that vegetables are produced close to the processing plants. Freezing companies have commonly operated a rule under which more perishable crops must reach the plant within a given time from harvesting. For example, a 90-minute rule is operated by many companies contracting peas for freezing. This imposes a distance limit on production, which has led to well-defined supply areas focused closely on the plants. The significance of local supply areas for more perishable crops is reflected in the locations of plants which offer contracts for them. In the UK, the markets for vining peas are clustered in the major arable production areas of East Anglia, Lincolnshire and east Yorkshire, with a few in secondary areas having a horticultural tradition, such as the Vale of Evesham, Kent and Tayside. The pattern of plants contracting for beetroot, a less perishable crop, is more dispersed. Although they are found in the main producing areas of East Anglia and Lincolnshire, they have less need to be close to supplies. More are located in the industrial areas of northern England and other major cities, reflecting the influence of the market as well as historical factors. Many of these plants, involved in the manufacture of pickles, were established due to the availability of female labour in towns such as Glasgow, Newcastle and Rotherham (Hart 1976).

The ability of firms to influence agricultural geography via contracts also operates in a negative fashion. Firms have the power to determine where a commodity is *not* produced by refusing to grant a contract. Geographical patterns of production thus differ from those which would occur if the commodity were marketed by more traditional methods. This negative aspect of contract farming is illustrated by changes in Lincolnshire and East Anglia following the decision by Bird's Eye to stop processing peas at their Grimsby and Yarmouth plants in the late 1970s. In Norfolk 120 farmers who had formerly grown peas could no longer obtain a contract. They were forced to alter their production systems and many took up oilseed rape as an alternative. This example puts the geographical implications of contract farming very sharply into focus.

Policies of firms offering contracts are most likely to influence geographical patterns at a local scale within the supply area of a plant. It has been suggested that these patterns can be understood by means of a two–stage descriptive model (Hart 1978). Stage one (Fig. 7.11a) closely follows the single market (von Thünen) approach. Given the assumptions of location theory concerning physical homogeneity and transport costs directly related to distance, it is

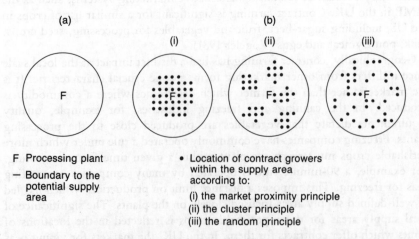

F Processing plant

— Boundary to the potential supply area

• Location of contract growers within the supply area according to:
(i) the market proximity principle
(ii) the cluster principle
(iii) the random principle

Fig. 7.11 A locational model of contract farming. (a): (F) processing plant; (—) boundary to the potential supply area; (b): (.) location of contract growers within the supply area according to: (i) the market proximity principle; (ii) the cluster principle; (iii) the random principle (*Source:* Hart 1978, *Tijdschrift voor Economische en Sociale Geografie* **69**: 205–15)

possible to define an area around any processing plant in which farmers can profitably produce crops for that plant. At the second stage, two decision-making influences are acknowledged. Firstly, decisions are made by farmers whether or not to seek contracts to produce the crop. These may be based on factors other than profitability and it has been found that the potential loss of independence is an important consideration deterring many farmers from seeking contracts (Hart 1980). If concern about loss of independence is randomly distributed this will bring about a random pattern of farmers who seek to obtain contracts. Secondly, the processing firm has to decide to which farmers it will offer contracts. Numerous factors may influence this. These include:

1. The desire to have production as close as possible to the plant, especially if the firm bears part of the transport costs or if the crop is highly perishable.
2. A policy of purchasing the crop through co-operative groups and/or maintaining close links with growers via company fieldsmen. Both of these may incline the company to offer contracts to farmers who operate in close proximity to each other.
3. Selection of growers on a first-come-first-served basis or on account of characteristics such as efficient husbandry practices, which may be assumed to be randomly distributed within the area.

Each of these may be translated into a different geographical pattern by applying one of three principles: market proximity, clustering or random (Fig. 7.11b). The resultant pattern of contract growing may thus be explained in terms of three factors:

(a) distance from the market;
(b) decision-making of farmers – a random influence;
(c) policies of the processors – one of the three principles or a mixed effect of two or more of these.

Futures markets

These first developed in North America in the mid-nineteenth century. They represent an attempt to reduce the risk deriving from temporal price fluctuations and may be used by any of the parties involved in agricultural marketing, including farmers, merchants and food processors. Their major function is to establish prices for particular commodities at some specified later date (Prior-Willeard 1984). Futures prices fluctuate in response to anticipated states of the market. Use is made of the markets by purchasing or selling a futures contract. Sale of a contract is a commitment to deliver a specified amount of a commodity at a future date and purchase of a contract is a commitment to take delivery. In practice, however, deliveries rarely take place. A commitment may be cancelled by making an opposite commitment to the original one. Thus a contract sale (a commitment to deliver) may be cancelled by purchasing a contract for the same commodity. This is termed closing out. This facility enables market participants to insure against disadvantageous price changes. For example, a pig farmer may sell a contract to deliver pigs at a future date. He receives the prevailing futures price. If, in the ensuing period, there is a reduction in the free market price of pigs the futures price will also drop. The farmer can thus be compensated for the lower free market price by purchasing a futures contract at a lower price than the one he sold. This purchase releases him from his commitment to deliver pigs on the futures market. Using the market as a form of insurance in this way is termed 'hedging'.

To date, futures markets have not been widely used by farmers. Sheldon (1986) attributes this to caution and a lack of knowledge about the markets and how to use them. Ironically, it appears that these markets are designed to deal with the issue of market uncertainty at a time when the major marketing problem facing farmers derives from their weakness in relation to large-scale agribusiness firms. Futures marketing does not appear to give farmers any additional protection against the power of such firms, or the cost–price squeeze to which their policies contribute.

Conclusion

Marketing systems are subject to continuous change, a feature which is of geographical interest at two levels. Firstly, marketing is itself a spatial activity, linking producers and consumers who normally occupy different locations. One purpose of marketing, therefore, may be seen as overcoming the friction of distance. The methods by which this and other marketing functions are achieved have been the concern of this chapter. A second level of interest centres on the relationship between marketing and the location of production. Three types of marketing effect can be suggested.

Locational effect

This has provided the traditional focus for agricultural geography and is still reflected in many studies. Plet (1983), for example, has demonstrated the spatial impact of the development of co-operative grain-processing industries in the Limagne region of France. The locational effects of markets may operate at different scales as is revealed by a study of sugar production in Queensland (Hoyle 1984). At the regional scale, production is influenced by the locations of ports which have bulk handling facilities and act as delivery points for sugar exports. At the local scale it is the location of sugar mills which exerts an influence, although now this operates through the Sugar Cane Prices Board, which allocates production areas to particular mills.

The influence of market location will be strongest where transport costs are high and farmers bear all or most of these. These factors help to explain why the town of Darwin continues to be supplied with most of its food needs by farmers from the surrounding Northern Territory of Australia (Cameron 1982). In this situation, isolation from the rest of the country, due to poor transport links, affords protection to local producers. Farmers in more distant regions are unable to compete because of the high costs of transport to the area. However, this situation will not last indefinitely and the completion of a rail link northwards from Alice Springs to Darwin is expected to bring about major changes in production patterns in northern Australia in the future.

Strategy effect

Marketing strategies may influence costs or involve physical intervention in production. One of the clearest examples of the financial impact of a marketing strategy was provided above in discussion of the MMB's pricing policy. The geography of production has changed as a consequence of this. Physical intervention is best exemplified by contract farming and the power of the market to dictate where production will take place, via contract terms. However, changes in marketing strategy do not always lead to new geo-graphical patterns. In some instances new strategies may be implemented to try to preserve existing patterns. Some co-operative ventures are of this type. In many instances these represent attempts to maintain farmers' strength in the market and are reactions to disadvantageous price trends, possibly due to the growing power of purchasers.

Behavioural effect

One consequence of the increasing emphasis on behavioural approaches to agricultural location has been the recognition that factors other than monetary returns influence farming decisions. Marketing methods differ in the combi-nation of costs and benefits they bring, while financial gains may be outweighed by less tangible considerations. Furthermore, marketing is an increasingly sophisticated process and farmers vary in their ability to under-stand and adapt to new methods and opportunities. The loss of independence associated with contract farming has been stressed. This also deters some farmers from joining co-operatives and it is probable that some have been dissuaded from developing direct marketing by their reluctance to allow large numbers of the public on to their farms. Most behavioural studies have dealt with production decisions but there is a need to focus on farmers' attitudes towards marketing. The importance of such attitudes has been revealed in a study of horticulture in the Vale of Evesham (Ilbery 1986). Horticultural production has declined in the Vale, partly due to inertia in marketing. Various factors account for this, including the lack of processing facilities in the area, but the independent outlook of farmers has also been important, especially with respect to co-operatives and contract farming.

These three effects will continue to be important focuses for geographical attention. It must also be stressed that studies of marketing, in common with the whole field of agricultural geography, will need to be based on a broader perspective then hitherto:

1. Contemporary marketing changes can only be understood if the position of farming in the food chain and the links with downstream activities are appreciated. The rapidly increasing influence, both directly and indirectly, of processors, retailers and caterers on farming patterns testifies to this (Smith 1984).
2. Changing marketing systems are one of the factors which are bringing about broader countryside changes, including negative environmental and social impacts (Hart 1987). This includes an altered decision-making context deriving from contract farming and agribusiness and changing opportunities for employment, for example, due to co-operative marketing, and for recreation as a result of the growth of PYO enterprises.
3. The theoretical insights provided by political economy have been shown to cast new light on marketing change and clarify many of the trends associated with the first two dimensions.

Appendix

Income elasticity of demand refers to the relative change in demand occurring as a result of changes in income. Demand is said to be elastic when an increase in income brings about a proportionally greater increase in demand. When the resultant increase is proportionally less than that in income, demand is inelastic. Given two products of differing elasticities, *ceteris paribus*, as incomes rise demand will increase more rapidly for that with the greater elasticity.

References

Ardaugh, J. (1982) *France in the 1980s*. Penguin, London
Barker, J. (1972) *Report of the committee of inquiry on contract farming*. HMSO, London
Barker, J. W. (1981) *Agricultural marketing*. Oxford University Press, Oxford

Bauer, P. T. (1963) *West African trade.* 2nd edn. Routledge & Kegan Paul, London

Blain, D. (1984) The stigma of the middleman – a closer look. *Ceres* **17**: 44–7

Blandford, D. (1983) Instability in world grain markets. *Journal of Agricultural Economics* **34**: 379–95

Bowler, I. R. (1972) Co-operation: a note on government promotion of change in agriculture. *Area* **4**: 169–73

Bowler, I. R. (1981) *Direct Marketing in British agriculture: a survey of pick-your-own farms in the English counties.* University of Leicester Department of Geography, Occasional Paper 6, Leicester

Broadway, M. J. and **Ward, T.** (1990) Recent changes in the structure and location of the US meatpacking industry. *Geography* **75**: 76–9

Burbach, R. and **Flynn, P.** (1980) *Agribusiness in the Americas.* Monthly Review Press, New York

Burns, J. A. (1983) The UK food chain with particular reference to the interrelations between manufacturers and distributors. *Journal of Agricultural Economics* **34**: 361–78

Cameron, B. J. (1982) The role of the Agricultural Development and Marketing Authority in sponsoring agriculture. *Journal of the Australian Institute of Agricultural Science* **37**: 97–9

Carlyle, W. J. (1978) Store stock marketing by small farmers in the crofting counties. *Scottish Geographical Magazine* **94**: 113–23

Commission of the European Communities (1986) *The agricultural situation in the Community 1985 Report.* Office for Official Publications of the European Communities, Luxemburg

Coppock, J. T. (1971) *An agricultural geography of Great Britain.* Bell and Sons, London

Cramer, G. L. and **Jensen, C. W.** (1982) *Agricultural economics and agribusiness* 2nd edn. John Wiley, New York

Curtin, C. and **Varley, A.** (1982) Collusion practices in a West of Ireland livestock mart. *Ethnology* **21**: 349–57

Davis, J. E. (1980) Capitalist agricultural development and the exploitation of the propertied labourer. In Buttel, F. H. and Newby, H. (eds) *The rural sociology of the advanced societies: critical perspectives.* Allenheld Osmun, Montclair, pp. 133–53

Davis, J. H. and **Goldberg, R. A.** (1957) *A concept of agribusiness.* Harvard Business School, Division of Research, Boston

Department of Trade and Industry Business Statistics Office (1988) *Business Monitor report on the census of production 1985* PA1002, HMSO, London

FAO (Food and Agriculture Organization of the United Nations) (1986a) *Food aid in figures. FAO, Rome*

FAO (1986b) *Food aid in figures. FAO, Rome*

FAO (1989) *Trade yearbook.* FAO, Rome

FAO (1990) *Annual production yearbook.* FAO, Rome

Federation of United Kingdom Milk Marketing Boards (1987) *United Kingdom dairy facts and figures 1987.* Federation of UK Milk Marketing Boards, London

Foxall, G. (1982) *Co-operative marketing in European agriculture.* Gower, Aldershot

Foxall, G. (1986) The structure of co-operative marketing in European agriculture. In Kaynak, E. (ed) *World food marketing systems.* Butterworths, London

Guellec, A. (1983) Tour d'horizon rapide sur les types de marchés aux bestiaux dans la région Bretagne.: *l'Information Géographique* **47**: 104–12

Hall, P. (ed) (1966). *von Thünen's isolated state.* Pergamon, London

Hallberg, M. C. (1982) Cyclical instability in the US dairy industry without government regulations. *Agricultural Economics Research* **34**: 1–11

Hart, P. W. (1976) Locational aspects of the supply of contract-grown crops to processing plants. Unpublished PhD thesis, University of London, London

Hart, P. W. (1978) Geographical aspects of contract farming, with special reference to the supply of crops to processing plants. *Tijdschrift voor Economische en Sociale Geografie* **69**: 205–15

Hart, P. W. (1980) Problems and potentialities of the behavioural approach to agricultural geography. *Geografiska Annaler* **62B**: 99–108

Hart, P. W. (1987) Economic processes and spatial change in agriculture. In Lockhart, D. G. and Ilbery, B. W. (eds). *The future of the British rural landscape.* Geo Books, Norwich, pp. 5–25

Hoggart, K. (1979) *The determinants of farm linkage patterns: an investigation of three townships in Huron County, Ontario.* University of London King's College Department of Geography, Occasional Paper 10, London

Hoover, E. M. (1948) *The location of economic activity.* McGraw-Hill Book Company Inc, New York

Hoyle, B. S. (1984) Ports and hinterlands in an agricultural economy: the case of the Australian sugar industry. *Geography* **69**: 303–16

Ilbery, B. W. (1985) *Agricultural geography. A social and economic analysis.* Oxford University Press, Oxford

Ilbery, B. W. (1986) Horticultural marketing: the case of the Vale of Evesham. *Transactions of the Institute of British Geographers* New Series **11**: 468–78

Kohls, R. L. and **Uhl, J. N.** (1985) *Marketing of agricultural products* 6th edn. Macmillan, New York

Lang, T. and **Wiggins, P.** (1985) The industrialization of the UK food system: from production to consumption. In Healey, M. J. and Ilbery, B. W. (eds.) *The industrialization of the countryside.* Geo Books, Norwich, pp. 45–56

Le Heron, R. B. and **Warr, E. C.** (1976) Corporate organization, corporate strategy and agribusiness development in New Zealand: an introductory study with particular reference to the fruit and vegetable processing industry. *New Zealand Geographer* **32**: 1–16

Levay, C. (1983) Agricultural co-operative theory: a review. *Journal of Agricultural Economics* **34**: 1–44

Livestock Auctioneers Market Committee for England and Wales and the Institute of Auctioneers and Appraisers in Scotland (1987) *A directory of livestock auction markets in England, Wales and Scotland.* LAMCEW/ IAAS, Basingstoke

Mabogunje, A. L. (1973) Manufacturing and the geography of development in tropical Africa. *Economic Geography* **49**: 1–21

Malcolm, J. (1983) Food and farming. In Burns, J. A., McInerney, J. and Swinbank, A. (eds.) *The food industry.* Heinemann, London, pp. 66–80

Marsh, J. S. (1965) *Contracts for farm products: examination of their use in SE England.* University of Reading Department of Agricultural Economics, Miscellaneous Studies 39, Reading

Marticorena, A. (1982) *A review of some contradictions existing in the international wheat market as a result of the different agricultural policies carried out by the main producers.* University College of Swansea Centre for Development Studies, Monograph 17, Geo Books, Norwich

Morgan, W. B. and **Munton, R. J.** (1971) *Agricultural geography.* Methuen, London

Nagle, J. C. (1976) *Agricultural trade policies.* Saxon House, Farnborough

National Food Survey Committee (annual) *Household food consumption and expenditure.* Annual report, HMSO, London

Nielsen Marketing Research (1982) *Grocery trading in 1982.* Nielsen Marketing Research, London

Nielsen Marketing Research (1987) *Grocery 1987 trade report.* Nielsen Marketing Research, London

Plet, F. (1983) Co-opération, productions contractuelles et transformations de l'espace agricole en Limagne. *L'Information Géograpahique* **47**: 12–22

Plunkett Foundation for Co-operative Studies (1987) *Directory of agricultural, horticultural and fishery co-operatives in the United Kingdom – 1987.* Plunkett Foundation, Oxford

Prior-Willeard, C. (1984) *Farming futures.* Woodhead-Faulkner, Cambridge

Rickard, T. J. (1990) Direct marketing as agricultural adaptation in Megalopolitan Connecticut. In van Oort, G. M., van den Berg, L. M., Groenendijk, J. G. and Kempers, A. H. (eds) *Limits to rural land use.* Pudoc, Wageningen, pp. 79–88

Robinson, G. M. (1988) Spatial change in New Zealand's food processing industry. *New Zealand Geographer* **44**: 69–79

Sargent, M. J. (1975) *Case studies in horticultural marketing co-operation.* University of Bath, Bath

Sargent, M. J. (1982) *Agricultural co-operation.* Gower, Aldershot

Sheldon, I. M. (1986) The early development of the London meat futures exchange: some interview results. *Journal of Agricultural Economics* **37**: 97–9

Smith, W. (1984) The 'vortex model' and the changing agricultural landscape of Quebec. *Canadian Geographer* **28**: 358–72

Tanbaum, J. (1981) *Food distribution: its impact on marketing in the '80s.* Central Council for Agricultural and Horticultural Co-operation, London

Tarrant, J. R. (1974) *Agricultural geography.* David and Charles, Newton Abbot

Tarrant, J. R. (1980) *Food policies.* John Wiley, London

Tarrant, J. R. (1984) The significance of variability in Soviet cereal production. *Transactions of the Institute of British Geographers* New Series **9**: 387–400

Tarrant, J. R. (1985) A review of international food trade. *Progress in Human Geography* **9**: 235–54

Vogeler, I. (1981) The myth of the family farm: agribusiness dominance of US agriculture. Westview Press, Boulder, Colorado

Wallace, I. (1985) Towards a geography of agribusiness. *Progress in Human Geography* **9**: 491–514

Wrigley, N. (1987) The concentration of capital in UK grocery retailing. *Environment and Planning A* **19**: 1283–8

8

Agriculture as a resource system

Martin Parry

Thus far the relationship between agriculture and its resource base (soil, climate and topography) has been treated in general terms. The quality of the resource base, particularly soil, has been shown to influence farm size (p. 88), farm type (p. 135) and the ownership pattern of farmland (p. 101), for example. Through these features, a varied pattern of opportunity is provided for capital accumulation by both farm and non-farm interests.

In this chapter we turn our attention to a more detailed analysis of the resource base and introduce the concept of the farm as a form of managed ecosystem (Tivy 1990). By this we mean the management of plant or animal communities to produce food. In some cases the degree to which the ecosystems are managed is small – in extensive ranching, for example. In others the ecosystems are so modified that they retain few recognizable natural elements, for example in intensive poultry, egg, veal and pig production. In general, the more intensive the input of management, labour and technology, the less obvious is the role of physical factors such as soils, weather and the requirements of plant genotypes. Yet, however intensive are the non-physical inputs, agriculture always involves both biophysical and human systems. For convenience, the natural resource base of agriculture is summarized by the term 'environment' in the following analysis in keeping with practice in the current literature.

The interaction of agriculture and the environment

This chapter concentrates on the biophysical variables within agriculture; it might even have been entitled 'the physical geography of agriculture'. Yet it would be inappropriate to treat these biophysical variables in isolation from human ones because it is precisely the interactions between them, that is, their 'management', which distinguish agriculture from many other economic activities. The interactions include three types of flow or exchange: flows of matter (e.g. inputs of fertilizer and outputs of farm produce); flows of energy (e.g. inputs of labour and outputs of calories); and flows of information (e.g. inputs of information on prices and outputs of information on yields which can, in turn, influence prices). We shall focus on the flows of matter and energy which largely comprise the biophysical inputs to farming systems, considering them under five main headings:

(a) the physical inputs to agriculture;
(b) the management of the resource base;
(c) externalities and the consequences of mismanagement;
(d) agriculture, land use and landscape;
(e) towards sustainable agriculture.

Under each of these headings it will be useful to view agriculture as an open system in which (a) the physical inputs that derive from the environment are, to varying degrees, limited in quality and quantity and thus place resource limitations on agriculture; and (b) some of the physical outputs from agriculture (such as animal waste, excess nitrogen, crop residues) need to be successfully absorbed and recycled by the environment if the system is to remain ecologically sustainable. The environment may thus be seen to place resource limitations on agriculture, while agricultural activities frequently have an impact on the environment.

Impact of environment on agriculture

With the gradual intensification of management over the centuries, the increased use of fertilizers to boost nutrients in the soil, drainage and irrigation to modify moisture conditions, buildings to adapt the microclimate, and plant and animal breeding to match a whole range of other physical inputs, one might expect that agriculture in the developed world was no longer much affected by the physical environment. This is true where soils, terrain and climate are all suited to both crop cultivation and grazing, but the area of land with no serious limitations for agriculture is itself quite limited, amounting to only 11 per cent of the global land area (FAO 1978). It is also distributed unevenly, with the largest proportion in Europe (36 per cent) and the smallest in north and central Asia (10 per cent) (Table 8.1).

Soils, terrain, water, weather and pests can be modified and many of the activities through the farming year, such as tillage and spraying, are directed towards this. But these activities must be cost–effective: the benefits of growing a particular crop, or increasing its yields by fertilizing, must exceed the costs of doing so. Often such practices are simply not economic, with the result that factors such as soil quality, terrain and climate continue to affect agriculture by limiting the range of crops and animals that can profitably be farmed. In this way the physical environment still effectively limits the *range* of agricultural activities open to the farmer at each location (see p. 136).

Table 8.1 Regional distribution of soils, with or without limitations for agriculture (% in each region)

	World	Europe	Central America	North America	South Asia	Africa	South America	Australia	South-east Asia	North and Central Asia
No serious limitation	11	36	25	22	18	16	15	15	14	10
Drought	29	8	32	20	43	44	17	55	2	17
Mineral stress	23	33	16	22	6	19	47	7	59	9
Shallow depth	21	12	17	10	22	12	11	7	6	39
Water excess	10	8	10	10	11	9	10	16	19	13
Permafrost	6	3	0	16	0	0	0	0	0	12

Source: Based on data in FAO (1978).

Moreover it limits them to a different degree, thus affecting the comparative advantage which one type of farming system may have over another in a given location. This can ultimately have an important influence on why some crops are grown in some places and not in others, leading in due course to a finer adjustment of land use to differences in the physical environment, and thus to more pronounced geographical patterns of agriculture. This trend was first observed by Baker (1926) when he described in detail the many ways in which land-use patterns in the US were becoming increasingly affected by variations in terrain, soils, temperatures and moisture.

Impact of agriculture on the environment

While environmental factors influence the range and profitability of different agricultural activities, agriculture frequently has an impact on the environment. In many respects this impact is adverse in the sense that environmental resources are frequently destroyed or depleted by agriculture. Possibly more serious has been the effect on soil resources, with agriculture causing world-wide denudation rates that are more than a thousand times the estimated pre-agricultural or natural rates of erosion (Huggett and Meyer 1980). Environmental degradation of this kind is an increasingly pressing issue both in tropical and temperate regions, and one of the most serious tasks facing researchers and planners today is somehow to optimize the use of environmental resources for agriculture while preserving their quality.

Other far-reaching effects stem from the replacement of natural vegetation by agricultural crops on a large scale. The greatest rates of change are at present occurring in the tropics, where rain forests are being felled and burned at an estimated rate of 11 million ha a year (Allen 1980). The consequent release of carbon into the atmosphere may account for, perhaps, one-third of the global warming that is expected to occur as a result of the so-called 'greenhouse effect' of radiatively active gases in the atmosphere (Crutzen and Gradel 1986).

Outside the tropics, and in regions of developed, market-oriented agriculture, forest areas have achieved a low-level equilibrium following centuries of contraction. For example, forest and woodland now covers only 9 per cent of the UK, and 22 per cent of the countries of the EC. Here the present-day impact of agriculture on the environment is felt mainly in the loss of habitat for wildlife, the reduction of the scenic quality of the rural landscape, extensive surface and groundwater pollution by agricultural chemicals, and the accumulation of poisons in higher organisms. These are also considered in greater detail later in this chapter.

The physical inputs to agriculture

The physical inputs to agriculture derive from processes of physical and chemical weathering of rocks (creating the mineral particles and nutrients of soils), from essentially atmospheric processes involving mainly the receipt by plants of solar radiation and water, and from the spatial transfer of materials from adjacent land surfaces by erosion and runoff. These terrestrial and atmospheric processes contain many feedback mechanisms whereby, for example, water and radiation are recycled between atmosphere and land, and between plants and their environment. The physical resources of agriculture can therefore usefully be considered as three resource cycles: energy, water and nutrients.

Energy

Almost all the energy which plants, and ultimately animals, use to grow comes from the sun. The process by which plants use light to fix energy in organic molecules is called photosynthesis. Generally speaking, the greater the intensity of sunlight the greater is the photosynthetic rate and thus the potential for growth, although there is a photosynthetic capacity beyond which plants become light-saturated. Depending on the cloudiness of the atmosphere, between one-quarter and three-quarters of the radiation at the edge of our atmosphere ultimately reaches the earth's surface. About half of this is formed into latent heat through the evaporation of water. Some is deployed in warming up the lower atmosphere, the earth's surface and its vegetation, and only a small amount (less than 1 per cent) is used by plants in photosynthesis. Interception of available radiation by plants is also determined by their leaf area, and the photosynthetic rate depends on resulting radiation receipt together with the photosynthetic efficiency of the plant.

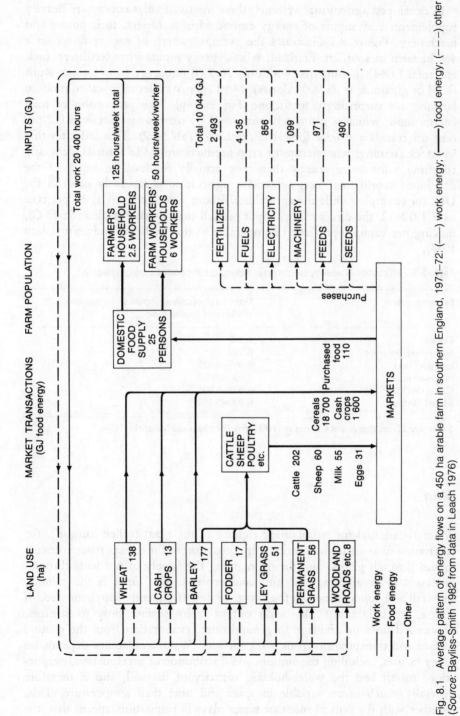

Fig. 8.1   Average pattern of energy flows on a 450 ha arable farm in southern England, 1971–72: (—) work energy; (——) food energy; (– – –) other
(*Source:* Bayliss-Smith 1982 from data in Leach 1976)

In developed agricultural systems, these inputs of solar energy are heavily supplemented by inputs of energy embodied in fertilizers, fuel, power and machinery. Figure 8.1 illustrates the average pattern of energy flows on a 460 ha farm in southern England, where energy inputs from fertilizers, fuel, etc. total 1 044 GJ compared with only 16.3 GJ of energy from human work (1 GJ or gigajoule = 23 900 calories). Most energy cycles in so-called modern farming are surprisingly inefficient. For example, the proportion of total energy input which is ultimately consumed by man ranges between 0.2 per cent for cereals and 0.02 for cattle ranching (Table 8.2). This indicates that livestock farming, where crops or crop products are fed to animals, generally consume a lot more energy than they actually produce. In much of the developed world this energy-intensive agriculture is relatively new: in the UK, for example, while the agricultural labour force declined by 50 per cent over 1952–72, the direct energy input per full-time worker tripled to 502 GJ, making per capita input levels comparable to that of heavy industry (Leach 1976).

Table 8.2   Efficiency of energy utilization in selected temperate farming systems

Farming system	Percentage of energy inputs (solar radiation) available as human food
Cereals	0.200
Sugar-beet/potatoes	0.250
Intensive beef	0.005–0.025
Intensive milk	0.030–0.080
Cattle ranching	0.002–0.004
Mixed farming	0.030–0.150

Source: Data from Briggs and Courtenay (1985) after Duckham and Masefield (1970).

Water

Water is essential for maintaining rigidity of the plant (called 'turgor'), for respiration to assist in the process of photosynthesis for transporting nutrients to and through plants, and for maintaining the supply of nutrients through processes of weathering, leaching and erosion. The input is via rainfall, snowfall and condensation in fog, mist, or dew (frequently supplemented, of course, by irrigation). The main output is by surface flow to streams, downward flow or drainage to groundwater, evaporation from the ground surface and transpiration from plant surfaces. Water availability depends on many factors, including the amount and distribution of precipitation, evaporation, runoff and the water-holding capacity of the soil, and is therefore generally much more variable in space and time than temperature. This, together with the critical role that water plays in respiration, means that it is the major factor limiting both the range of potential crops and the yields of crops throughout the world. Even under a maritime climate, such as that in north-west Europe, moisture deficiencies are one of the main limitations on crop yields (Briggs and Courtenay 1985).

Nutrients

A variety of mineral nutrients are necessary for healthy plant growth. Unlike radiation, temperature and rainfall, which are 'flow' resources in the sense that they are replenished continually (though often at variable rates), soil nutrients are not regularly self-renewing in the same way and thus require more careful management. This is particularly the case because the recycling of nutrients which would normally occur in a natural ecosystem is interrupted in agriculture by the removal of crops from the land. As a result, sustained cultivation is not possible without replacing the lost nutrients by fertilizing, manuring or mulching.

The nutrient needed in greatest quantity is nitrogen; this is required for the

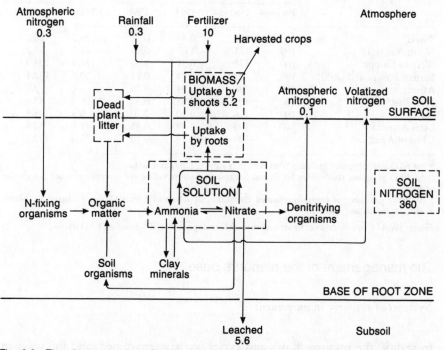

Fig. 8.2 The nitrogen cycle at Rothamsted, Hertfordshire, England, for the top 22 cm in a cultivated field. Quantities are in grams of nitrogen per square metre (*Source:* Bayliss-Smith 1982)

213

synthesis of amino acids, which then combine to form proteins. The flux of nitrogen between soil, plant and atmosphere is illustrated in Fig. 8.2. Its main source is the decomposition of organic matter in the soil. It is mainly either absorbed in solution through plant roots or leached by percolating rainfall out of the reach of plants. In the past nitrogen levels were maintained by growing leguminous plants (such as beans, peas and clover which have their own nitrogen-fixing soil organisms) in rotation with other crops. The modern use of artificial fertilizers is considerably more energy-intensive since 1.8 t of oil are generally used in the production of each tonne of nitrogen, and since only about half to three-quarters of the nitrogen added as fertilizers is recovered in the crop (Bayliss-Smith 1982).

In the past 20 years, large increases in the application of fertilizers have been needed to increase food production in line with population growth and a growing appetite for animal protein. Average world application rates per hectare almost trebled between 1964 and 1984, allowing a 12 per cent increase in per capita food production on less land per capita (Table 8.3).

Table 8.3   World per capita food production, cropped area and fertilizer use, 1964–84

Region	Per capita food production (1961–64 = 100)		Per capita gross cropped area (ha)		Per hectare fertilizer use (kg)	
	1961–64	1981–84	1964	1984	1964	1984
World	100	112	0.44	0.31	29.3	85.3
North America	100	121	1.05	0.90	47.3	93.2
Western Europe	100	131	0.31	0.25	124.4	224.3
Eastern Europe and USSR	100	128	0.84	0.71	30.4	122.1
Africa	100	88	0.74	0.35	1.8	9.7
Near East★	100	107	0.53	0.35	6.9	53.6
Far East†	100	116	0.30	0.20	6.4	45.8
Latin America	100	108	0.49	0.45	11.6	32.4
CPEs of Asia‡	100	135	0.17	0.10	15.8	170.3

★ An FAO grouping that includes West Asia plus Egypt, Libya and Sudan.
† An FAO grouping that covers South and South-east Asia excluding the centrally planned economies of Asia.
‡ An FAO grouping of centrally planned economies of Asia that covers China, Kampuchea, North Korea, Mongolia and Vietnam.

Source: World Commission on Environment and Development (1987) based on FAO data).

The management of the resource base

Systems of resource management

In reality, the resource flows and cycles we have examined, and the ways in which they are modified in agriculture, are integrated into *farming systems* (or types of farming). These embody many sets of interactions between matter, energy and information within the farming systems and between the systems and their 'environment'. In broad terms we can distinguish between *arable systems* and *grazing systems* because these involve sometimes quite different interactions of radiation, water and nutrients and therefore involve different forms of resource management. Since they are both production systems and yet managed ecosystems, they have a similar structure; their main output is to the economy and an important input is from the biosphere (Figs 8.3 and 8.4). Where the biotic rather than economic inputs appear dominant, for example in ranching or extensive cropping, the rhythm of climatic or biotic events may shape the system, as in the seasonal rhythm of annual cropping or of extensive livestock farming. By contrast, in economically intensive activities, such as modern poultry, pig or veal production, the rhythm of economic events may be more apparent, as in the seasonal market for turkeys. In all systems of agriculture, a central theme is the management of essential environmental inputs: water, nutrients, soil structure and biota.

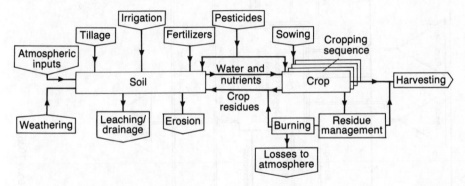

Fig. 8.3   The arable cropping system (*Source:* Briggs and Courtenay 1985)

The management of water

Irrigation

Three factors can limit the uptake of water by plants: the amount of precipitation, the nature of the soil and the structure of the plant. In global terms the unreliability of precipitation is probably the greatest constraint on agricultural productivity. The impact comes not simply from inadequate amounts of water but from the timing of spells with inadequate rainfall and from the uncertainty which stems from our general inability to predict and plan for these spells. The timing of moisture supply can be critical at certain stages in plant development, for example during the germination period and during flower formation and fertilization. Even in a wetter than average year, a few days of abnormally dry weather can seriously reduce yield.

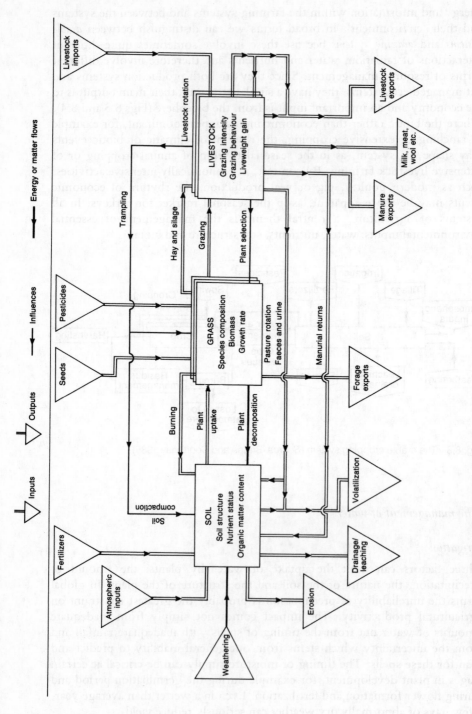

Fig. 8.4 The grassland system (*Source:* Briggs and Courtenay 1985)

The efficiency of irrigation in replenishing soil moisture depends on how and when it is applied, and in what quantities. Most irrigation is by surface sprinkling or spraying from static pipes or travelling booms. This system uses less space than surface furrows and canals, although it can cause structural damage to the soil. Subsurface irrigation, which directly replenishes soil moisture through underground ducts, reduces loss by evaporation from the surface and is thus a good deal more efficient in use of water, but has higher capital costs. Its use is generally restricted to high-value crops and areas where water is especially dear. Even where water for irrigation is plentiful and inexpensive, the cost-effectiveness of irrigation depends on applying just sufficient quantities of water at the right time. Excessive application not only wastes water, but may cause soil erosion, structural damage and leaching of nutrients (for details, see Briggs and Courtenay 1985).

Table 8.4 Growth in irrigated area, by continent, 1950–82

Region	Total irrigated area, 1982 (million ha)	Growth in irrigated area (%)		
		1950–60	1960–70	1970–80*
Africa	12	25	80	33
Asia†	177	52	32	34
Europe‡	28	50	67	40
North America	34	42	71	17
South America	8	67	20	33
Oceania	2	0	100	0
World	261	49	41	32

* Percentage increase between 1970 and 1982 prorated to 1970–80 to maintain comparison by decade.
† Includes the Asian portion of the USSR.
‡ Includes the European portion of the USSR.

Source: Worldwatch Institute (1985).

Since 1950 the irrigated area world-wide has increased from 94 million to 261 million ha, amounting to 17 per cent of the world's cropland and producing about one-third of the world's food production (Table 8.4).

Soil drainage

The oxygen requirement of plants for transpiration and other functions is taken from the atmosphere and the air in the soil. Because air and water compete for space in the soil, the relative amounts are crucial for healthy plant development. Excess water can frequently mean insufficient oxygen supply, while the anaerobic conditions can lead to a build-up of toxic gases such as ethylene. The processes are complex and not yet fully understood but it is clear that, even over short periods, waterlogging can seriously reduce crop yields. For example, waterlogging for one week in the middle of the growing season in the UK can reduce yields of winter wheat and sugar-beet by between

40 and 50 per cent, and can eliminate a potato crop entirely (Trafford 1970). Waterlogging generally occurs as a result of inadequate drainage resulting from finely grained or compacted soils, which hinder water filtration, or from impedance of surface drainage by flat terrain. Where this occurs there is a need to improve the vertical or lateral movement of water by soil drainage. Figure 8.5 illustrates that the extent of soil drainage varies greatly, from perhaps 10 per cent of agricultural land in France to 37 per cent in West Germany (FDR)

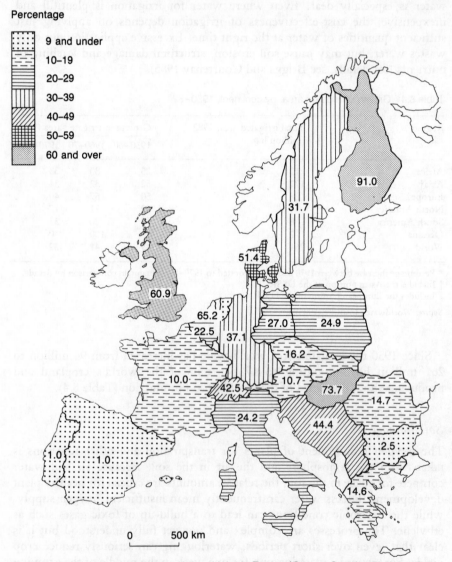

Percentage

	9 and under
	10–19
	20–29
	30–39
	40–49
	50–59
	60 and over

Fig. 8.5 Percentage of drained agricultural land in Europe (*Source:* Goudie 1981 after Green 1980)

and over 65 per cent in the Netherlands (Baldock 1984). Variations in terrain and soil explain some of this discrepancy, but it is largely the result of institutional factors. Drainage is only effective if organized on a large scale so the water-levels in adjacent fields are compatible and there are appropriate channels for removal of the surplus water. The small and fragmented nature of many French farms has been the main restraint on drainage in the past.

In contrast, large areas of land in England and Wales were drained in the period of 'high farming' (and cheap labour) in the mid-nineteenth century (Fig. 8.5). More recently, drainage grants introduced by the Agriculture Act 1947 are responsible for the annual rate of new soil drainage in England and Wales being one of the highest in Europe over the past 20 years, with more than 1 per cent of all agriculture being drained annually during the 1970s (Green 1980). Most of this is in the form of mole drains or pipe drains. The gains from drainage vary according to local conditions. Quoting data from surveys of yield changes on experimental plots, Briggs and Courtenay (1985) note increases of more than 30 per cent in some instances, and frequently of around 10 per cent.

The management of nutrients

The loss of nutrients as a result of harvesting of the crop is a major difference between a natural ecosystem and agriculture. One of the most important items of management in agriculture, therefore, is the artificial replenishment of nutrients by fertilizing or manuring. The quantities of inorganic fertilizer used in modern farming can be very high: in the UK their cost amounts on average to more than 16 per cent of the total on-farm cost of sugar-beet production, 12 per cent of winter wheat and 9 per cent of main crop potatoes (Eddowes 1976). Application rates range from about 250 to 700 kg/ha annually, depending on the type of crop and its place in the crop cycle.

Most inorganic fertilizers are applied as granules or pellets, but they may come in quite different forms (Table 8.5). Increasing use is being made of mixes or compound fertilizers, particularly of nitrogen, phosphorus and potassium, which are conveniently provided in appropriate proportions for application.

The convenience of inorganic fertilizers, particularly their guaranteed composition and their cost-effectiveness, have meant that they have far outstripped the use of organic manures. However, as indicated in Table 8.6, their price depends greatly on the price of oil (of which nitrogen fertilizers are a by-product) and it is therefore quite possible, if oil prices increase, that the use of manure could increase again in the future. Even so, supplies would be far from sufficient for many grain-growing regions of the developed world where mixed farming of crops and livestock has long since given way wholly to arable farming. Regular applications of organic manure can, however, achieve yields similar to those from inorganic fertilizers (Briggs and Courtenay 1985).

Table 8.5 Common types of inorganic fertilizers

(a) Nitrogen

Fertilizer	Chemical composition	% N	Form	Comment
Ammonium nitrate	NH_4NO_3	34	Prilled or granular	Becoming more widely used; cheap
Ammonium sulphate	$(NH_4)_2SO_4$	21	Crystalline	Now rarely used
Sodium nitrate	$NaNO_3$	16	Granular	Expensive; rarely used
Calcium nitrate	$Ca(NO_3)_2$	15.5	Prilled	Widely used in continental Europe
Urea	$CO(NH_2)_2$	45	Prilled	Expensive; may cause scorch
Anhydrous ammonia	NH_3H_2O	82	Liquid	Cheap; readily lost by leaching
Aqueous ammonia	NH_4OH	21–29	Liquid	Cheap; less susceptible to loss

(b) Phosphorus

Fertilizer	Chemical composition	% P_2O_5	Form	Comment
Ground rock phosphate	Variable	25–30	Powder	Often highly insoluble
Superphosphate	$Ca(H_2PO_4)$	18–21	Powder	*c.* 30% of world P fertilizer
Triple superphosphate	CaH_2PO_4	47	Powder	*c.* 20% of world P fertilizer
Basic slag	Variable	8–22	Powder	By-product of steel-making
Dicalcium phosphate	$CaHPO_4$	40	Granular	Water-insoluble
Superphosphoric acid	$H_4P_2O_7$ or H_3PO_4	60	Liquid	Becoming more widely used

(c) Potassium

Fertilizer	Chemical composition	% K_2O	Form	Comment
Potassium chloride	KCl	60	Granular	Most widely used K fertilizer
Potassium sulphate	K_2SO_4	48–50	Powder	Expensive; sulphur has some fertilizer value
Kainit	Variable	12–16	Powder	Contains Mg or Na
Potassium nitrate	K_2NO_3	37	Powder	Expensive; mainly used for vegetables

Source: Briggs and Courtenay (1985).

Table 8.6 World grain production and fertilizer use at three oil price levels, 1950–84

Period	Oil price per barrel*	Annual growth (%)	
		Grain production	Fertilizer use
1950–73	2	3.1	7.5
1973–79	12	1.9	5.6
1979–84	28	2.0	2.5

* Current US dollars.
Source: Worldwatch Institute (1985).

The management of soil structure

In order to take up sufficient quantities of water, nutrients and air, plants need to develop a full set of roots that can absorb minerals in solution from the surrounding soil. This in turn requires a soil with a profile and structure that permits root growth and allows the retention and passage of water, air and nutrients. Tillage helps to create this type of soil environment, at the same time reducing competition from weeds and forming an appropriate seedbed. The latter is especially important in mechanized agriculture where seed drilling needs to be at a uniform depth and spacing. In addition, tillage serves to break up the compact layers in the soil, developing a more compressible and friable structure and opening up pore spaces into which the root tip will be pushed.

The main processes of tillage – ploughing, harrowing and rolling – can, over the years, lead to quite major changes in soil structure. In general we might expect tilled soils to be more porous and less dense, but there can occur, especially under moist conditions, the compaction of the soil under the weight of frequent implement use, particularly from the wheels of farm machinery. In some cases where tillage is intense, up to 90 per cent of the field surface can be covered by machinery 'wheelings' in a single year (Briggs and Courtenay 1985). This has been one reason for the increasing use of so-called minimal tillage, which aims to reduce the amount of wheeled traffic on the soil surface. One example of this is the direct drilling of seed in untilled or little-tilled soil, with the previous year's stubble perhaps only raked up rather than ploughed under. Direct drilling is most suited to light soils which already have a good structure, but this has not prevented its use being extended to almost 250 000 ha in the UK, amounting to about 5 per cent of the arable area.

The management of biota

Plant and animal breeding

Improvements in varieties of plants and animals have made the greatest contribution to recent increases in agricultural output. In the US, for example, yields of maize and sorghum nearly tripled between 1950 and 1980, largely as a result of the judicious combination of selected high-yielding hybrid varieties, increased applications of fertilizer and a rapid growth in the extent of irrigation (Fig. 8.6). However, as the cost of water and fertilizer has risen (relative to the price of grain) the rates of input have been restrained and yields have, more recently, levelled off. Interannual variability of yield has tended to increase both as a result of the greater inherent vulnerability of some new varieties to disease, or suboptimal climatic conditions such as drought and waterlogging (Worldwatch Institute 1985).

Increased attention is now being given to the development of cultivars that provide more stable yields under varying conditions and thus contribute as much to food security as to productivity. World-wide co-ordination of this activity is provided by the Consultative Group on International Agricultural Research (CGIAR) which supports a network of agricultural research institutes in the improvement of agricultural technology, particularly in plant and animal breeding. Advances in biotechnology – including recombinant DNA, tissue culture and cloning – have opened new frontiers in the management of biota for agriculture, but the greatest potential for boosting food output probably lies in one or two developing countries. In the 1980s Argentine maize yields were, for example, scarcely half those of the US suggesting that over time yields in Argentina, which has a similar soil and climate, could double as new technologies are introduced (Worldwatch Institute 1985).

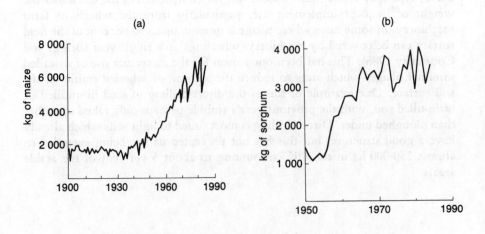

Fig. 8.6   Yields of (a) maize and (b) sorghum in the US (*Source:* Worldwatch Institute 1985)

Plant and animal protection

Nature has always 'taxed' the farmer: in the Middle Ages in Europe the loss of at least half the crop to pests and diseases was commonplace; even today up to one-third of the crop in Africa is lost or spoiled in this way. In the developed world, losses are now much lower but only because considerable sums are spent on pest control and crop protection. In the UK, for example, the costs of crop protection from spraying, weeding, etc. account for about 10 per cent of total costs of sugar-beet production, 4 per cent for potatoes and 3 per cent for winter wheat (Eddowes 1976).

As with the use of fertilizers, the cost-effectiveness of pest and disease control depends very much on rates of application and timing of treatment, and on the level of infestation which might otherwise occur as well as, of course, on the relative prices of chemicals and crops. Savings can vary greatly from, for example, doubled yields for effective protection against potato blight to increases of a few per cent for suppression of wild oats in cereal crops.

Briggs and Courtenay (1985) have identified five techniques of pest control and crop protection:

1. Direct control, by the weeding of plants and hunting of vermin, such as pigeons and rabbits. Especially in less-developed farming economies, the time spent of snaring rabbits and scaring birds, largely by the children of the farming household, can be considerable.
2. Chemical control, by the application of insecticides, herbicides, fungicides and nematicides (those compounds which attack worms).
3. Biological control, which is generally achieved by the introduction of a species that will predate on the pest or feed on the weed, for example the introduction of myxomatosis into the UK and Australia in the 1950s to control rabbits.
4. Cultural methods, that is, the use of tillage to control weeds and reduce soil-borne diseases, often generally encouraged by moist conditions, by improving drainage.
5. Habitat removal, such as of moorland, coppices and hedgerows, which may be refuges for pests. They may, of course, also be important refuges for predators of pests and for wildlife valued for its own sake. Frequently the benefits of pest removal must therefore be 'traded off' against other types of loss. Indeed, it is probable that all these types of pest control and crop protection inadvertently affect inoffensive wildlife and their habitat. These impacts are discussed in greater detail below.

Externalities and the consequences of mismanagement

Where inappropriate agricultural management is practised, natural resource inputs can become progressively degraded, leading to the depletion of soil resources through erosion, desertification and salinization, the pollution of water supplies and the destruction of wildlife and landscapes. These impacts are not new – they have occurred since agriculture began – but there is a growing sense in the developed world that they have been neglected as an issue and that resource depletion is an avoidable waste. Moreover, some increasingly intensive forms of farming have begun to create animal and plant waste, such as manure and straw, in such quantities that there are problems in their disposal without overloading natural systems of decomposition.

Soil erosion and desertification

Rates of development of topsoil are slow when compared with rates of technological change in agriculture: it may take 100–500 years to generate 10 mm of topsoil under natural conditions of vegetation. Soils are thus essentially a fixed rather than renewable resource for agriculture; once fully removed they are, for all practical purposes, destroyed for ever. Yet most forms of agriculture tend to cause striking increases in natural rates of erosion: on the world scale it is estimated that man has increased the world denudation rate from 20 million to 54 000 million tonnes per year (Huggett and Meyer 1980). This is largely the result of increased exposure of the bare soil to wind and rain by cultivation, and reduced root and other organic matter in the soil, which otherwise serves to reduce the rate of dislodgement by raindrops and increases infiltration capacity. Consequent increases in rain spatter on bare soil, and in rates of overland flow (known as sheet wash) removes topsoil on slopes and creates small channels (or rills up to 50 cm deep) along plough furrows; these may later develop into gullies (usually 1–15 m deep) that erode back upslope through all soil horizons.

Land loss from water erosion is probably more extensive in developing countries, where soil conservation services are still inadequate. But even in the US, with the world's largest conservation service, more than 1 per cent of the farmed area is estimated to be seriously degraded each year by soil loss (Allen 1980). The areas of most severe topsoil loss include the intensively farmed cotton and soyabean region east of the lower Mississippi, the wheat belt in the central and southern Great Plains, and the Appalachian foothills in Georgia and the Carolinas. In the UK, loss of soil results largely from wind erosion, the areas most susceptible being the light, sandy soils of East Anglia, Lincolnshire and Yorkshire, and the light peats of the Fens (Huggett and Meyer 1980).

In semi-arid regions mismanagement of soil can lead to extensive and largely permanent devastation of soil and vegetation. This process, now widely referred to as *desertification*, combines a syndrome of conditions which may include, for example, deterioration of rangelands, forest depletion, and dune encroachment, wind erosion of topsoil, and deterioration of irrigation systems. It was estimated in 1978 that 20 million km² – an eighth of the world's land surface – was already desertified or was at high risk (UN Conference on Desertification 1978). Reassessments in the 1980s suggest both that desertification is spreading, and that it is quite extensive in some developed countries, particularly in southern Europe. The areas particularly affected are Calabria, Puglia and Sicily in southern Italy, south–central Spain, and upland regions in Greece and Turkey (Worldwatch Institute 1985; Fantechi and Margaris 1986).

Depletion and pollution of water

Depletion

Substantial areas of irrigation, for example in the western US, northern India and northern China, are based on the withdrawal of groundwater from aquifers which receive little annual replenishment from rainfall. Even where recharge does occur, groundwater is often pumped at rates that exceed replenishment, causing water-tables to fall, and increasing the cost of pumping if not exhausting supplies altogether. In the Great Plains of the US the area under irrigation from the vast underground reserve of the Ogallala aquifer increased over 1944–78 from 2 to 8 million ha (equivalent to 5 per cent of the US farmed area). Faced with diminishing well yields and rising pumping

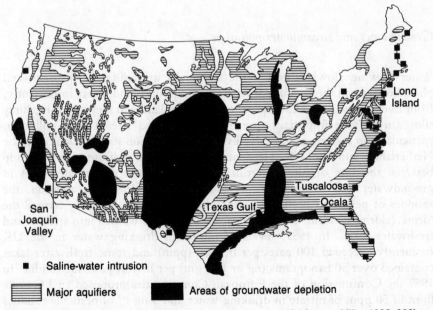

Fig. 8.7 Depletion of the major underground aquifers in the US (*Source:* Miller 1982: 366)

costs, however, farmers have more recently been taking land out of irrigation – 600 000 ha between 1978 and 1982 (Frederick and Hanson 1982). In the US groundwater depletion is most serious on the Great Plains (the Ogallala aquifer), the Gulf coast (the Texas Gulf aquifer) and central California (the San Joaquin Valley aquifer) (Fig. 8.7). On the western and higher parts of the Great Plains rising pumping costs and low commodity prices led to a 10 per cent reduction in the area of irrigated cropland in the early 1980s (Worldwatch Institute 1985). In coastal regions, the overpumping of groundwater has caused the inland intrusion of saline groundwater that is unusable for agriculture, industrial or domestic purposes.

Salinization

The depletion of aquifers may increase the salt content of groundwater in continental regions (as in the case of the Ogallala), as well as increase the risk of ocean salt-water intrusion in coastal areas. In parts of Israel, Syria and the Arabian Gulf states, for example, recent pumping has pulled the water-table below sea-level.

More extensive salinization can occur, particularly in dry climates, where waterlogging due to irrigation is accompanied by high rates of evaporation, leaving a damaging residue of salt. The problem is not restricted to developing countries nor to profoundly semi-arid regions. It is extensive in the San Joaquin Valley in California, the North Plain of China and Soviet central Asia.

Groundwater and streamwater pollution

Water flowing slowly through the soil picks up solutes of nitrogen and phosphorus from the residues of fertilizer compounds and, more importantly, from livestock waste. The disposal of slurry on land and seepage from slurry silos can thus be important source of ground and streamwater pollution, particularly in regions of intensive livestock production such as Denmark, the Netherlands and the US. Here numerous studies indicate a relationship between the rate of nitrate fertilizer application and the nitrate levels in groundwater and streams (Briggs and Courtenay 1985). In addition, the residues of persistent pesticides, which have become collectively termed the 'drins' (aldrin, dieldrin, endrin and isodrin), may be washed into streams and freshwater lakes. In 1987, levels of nitrate in drinking water in the UK frequently exceeded 100 parts per million (ppm) and some freshwater lakes contained over 50 nanograms (ng or 10^{-9} gm) per litre of pesticide residue. In 1987 the Commission of the European Communities proposed an EC-wide limit of 50 ppm of nitrate in drinking water and 5 ng l^{-1} of pesticide residues in freshwater lakes.

Effects on plants and animals

Habitat destruction

Through the 1950s the cropland area in the world grew at an annual rate of about 1 per cent (Urban and Vollrath 1984). By the 1980s this had slowed to perhaps 0.2 per cent, partly as a result of increases in productivity and partly because of the shortage of unreclaimed yet cultivable land. In the tropics, however, it has been estimated that during the period 1950–70 rain forests were being felled at the rate of 110 000 km² each year (Sommer 1976). The destruction of wildlife habitats on this scale has led to concern over the possibility of substantial increases in the rate of species extinction: one estimate is 1 million species (or one-tenth of the total) have become extinct during the last quarter of this century (Myers 1979). Estimates by the Nature Conservancy Council of habitat loss in the UK since 1949 reveal a 95 per cent loss of lowland herb-rich meadows, 30–50 per cent loss of ancient woodlands, and 40 per cent loss of lowland acidic heaths (Green 1986).

Destruction of genetic resources

The rate of species destruction may be important for agriculture itself, due to a diminishing pool of genetic resources. Experience shows that between 5 and 15 years after the introduction of a new crop strain with new resistance to diseases and pests, the resistance generally collapses in the face of newly evolved forms of disease or insect (Myers 1979). There is, therefore, a constant need to 'top up' a plant's genetic constitution with new germplasm. Substantial endemic diversity of wild grains and fruits, offering a broad genetic base for future crop development, is therefore important.

The risks attached to a narrow genetic base are illustrated by the spread of southern corn leaf blight in the US in 1970. At that time, when 70 per cent of US seed corn depended on five inbred lines of corn, the blight destroyed

Table 8.7 Some major US crops: extent to which they are dominated by a few varieties

Crop	Hectares (millions) 1976	Value ($m.) 1976	Total varieties	Major varieties	Hectarage, % of major varieties
Corn	33 664	14 742	197	6	71
Wheat	28 662	6 201	269	10	55
Soyabean	20 009	8 487	62	6	56
Cotton	4 411	3 350	50	3	53
Rice	1 012	770	14	4	65
Potato	556	1 182	82	4	72
Peanut	611	749	15	9	95
Peas	51	22	50	2	96

Source: Myers (1979).

one-seventh of the nation's crop, pushed corn prices up by 20 per cent and brought a bill to farmers and consumers of over £512bn (Myers 1979). Table 8.7 illustrates the continued narrow genetic base of US agriculture and its implied susceptibility to some new form of pathogen, insect pest or severe environmental stress such as unusual cold or drought.

Effects of pesticides

Since the publication of *Silent spring* highlighted the damaging effects of chemicals on wildlife, a good deal of attention has been paid to the effect of various pesticides on birds and mammals (Carson 1962). The accumulation of poisons in creatures near the top of various food chains is especially acute. This is illustrated in Table 8.8 which summarizes results from two surveys: of DDT in Lake Michigan (North America) and of DDD (a less toxic derivative of DDT) in Clear Lake, California. From these and other studies it is clear that organochlorines build up quite rapidly in some food chains and that a certain level of dose – different for each organism and each compound – is lethal (Briggs and Courtenay 1985).

Table 8.8 Concentration of pesticides in food chains

	Pesticide concentration (ppm)	
	DDT*	DDD†
Birds	3 177 (herring gulls)	Up to 1 600 (grebes)
Fish	3.0–8.0	40–2 500
Small invertebrates	0.41	(Not given)
General environment (lake bottom sediments)	0.0085	

* Lake Michigan (Hickey et al. 1966).
† Clear Lake, California (Hunt and Bischoff 1960).

Source: Briggs and Courtenay (1985) after Hickey et al. (1966) and Hunt and Bischoff (1960).

Inadvertent pest promotion

An unintentional result of heavy doses of pesticides, such as those frequently employed in the 1960s and 1970s, was the promotion of pesticide-resistant species. The number of resistant insects and mites is estimated to have doubled over 1965–77 (Allen 1980). Other pests have been promoted by the destruction of their predators.

While these forms of environmental degradation vary considerably from region to region, they are generally more apparent under intensive agricultural systems that incorporate high levels of resource input, such as fertilizers and pesticides, and strive for high levels of output, which frequently implies high levels of output waste. Increasing emphasis on this input-intensive farming has been most apparent over the past 20 years in Europe and North America (see Table 8.3). However, it is a recent and growing phenomenon in irrigated areas in the tropics, particularly southern India and South-east Asia. More extensive and 'resource-poor' forms of agriculture, for example dryland farming in the semi-arid tropics, have probably had less obviously marked environmental effects. But recent and large-scale extension of the cropped area, often into tropical forests or rangelands with unstable and erodible soils, has also led to widespread environmental damage (World Commission on Environment and Development 1987). Reduction of this damage through the development and use of sustainable agricultural systems is considered at the end of this chapter.

Agriculture, land use and landscape

The biophysical factors, such as plant genotype, weather and soils, considered in the preceding sections interact with other factors, such as technology, prices and the structure of landholding. Their interactions result in a continuously changing pattern of *land use* (the expression of agriculture on the land surface) and a continuously evolving *landscape* (the visible mix of terrain, natural vegetation, and the land uses of agriculture and other activities). The impact of land use on landscape has recently become a controversial issue, particularly in the EC and North America where intensive forms of agriculture, leading the the accumulation of large and expensive food surpluses, have been perceived as being unnecessarily destructive of traditional landscapes. The type and magnitude of these impacts have varied from country to country, though the overall trends have been common. The following section focuses on those occurring in the UK since about 1950.

Agriculture and land use

Of the total land area of 24 million ha in the UK, about 75.9 per cent is used for agriculture. Although this proportion has fluctuated by only 1 or 2 per cent over the past 30 years, there have been quite major changes of use between different agricultural land-use types which have led to shifts in the overall regional pattern of land use (Table 8.9).

Table 8.9   Changing land-use structure in the UK, 1935–87 (% of total)

	Rough grazings	Permanent pasture	Arable and other*	Woodland	Urban	Unaccounted for†
England and Wales						
1935	14.6	41.9	25.3	5.7	7.6	4.9
1939	14.9	42.3	24.1	6.2	8.6	3.9
1951	14.7	29.0	36.8	6.5	8.9	4.1
1961	13.4‡	29.0	36.7	6.9	9.9	4.1
1971	12.5	26.4	37.7	7.4	11.0	5.0
1980	11.4	26.4	36.3	8.0	11.6	6.3
1987	11.3	26.3	39.9	n.d.	n.d.	n.d.
% change in category 1935–61	−8.0	−31.0	+45.0	+21.0	+30.0	−16.0
% change in category 1961–87	−15.0	−31.0	+10.0	+16.0	+11.0	+54.0
Great Britain						
1931	26.4	30.7	22.5	5.6	5.0	9.8
1961	30.9‡	20.7	30.4	7.4	7.4	3.2
1971	28.2	19.3	30.5	8.2	8.2	5.6
1980	26.8	20.0	29.1	8.9	8.5	6.7
1987	25.3	20.1	31.7	10.1	n.d.	n.d.
UK						
1971	27.7	20.5	30.0	7.9	8.0	5.9
1980	25.9	21.3	29.0	8.7	8.4§	6.9
1987	24.7	21.2	31.3	9.8	n.d.	n.d.

* Includes fallow, temporary grass and orchards.

† Includes unutilized rural land and land which has escaped enumeration under other headings.

‡ In 1959 'rough grazing' was redefined and, as a consequence, the amount returned increased by about 600 000 ha.

§ Data for 1979.

Source: Parry (1991).

The 'loss' to urban use

In the UK, as in other developed countries, there has been a continued loss of farmland to urban use. The annual average rate over 1945–75 was 16 000 ha per year in England and Wales (about 0.2 per cent of the farmed area), although it was more marked in south-eastern England, where the rate may have exceeded 1 per cent each year (Parry 1991). In most developed countries such losses have been more than compensated by increases in agricultural productivity but, world-wide, there is some concern at the contraction of the cropland area in some developing countries, such as China, where agricultural output is quite closely matched to demand (Table 8.10).

Table 8.10   Declining cropland in selected countries

Country	Year of postwar peak in arable land area	Decline from peak year to 1980 (%)
China	1963	−5.1
France	1960	−13.3
Hungary	1955	−6.6
Ireland	1960	−29.4
Italy	1955	−4.8
Japan	1960	−19.6
The Netherlands	1955	−18.0
Poland	1955	−9.7
Portugal	1963	−18.1
South Korea	1968	−5.3
Sweden	1955	−21.0
West Germany	1955	−13.9
Yugoslavia	1969	−5.6

Source: Urban and Vollrath (1984).

Conversion to forestry

Even larger areas of farmland in Europe have been converted to forestry, mainly on poor-quality uplands. Over 50 000 ha were converted in England and Wales over 1946–80, the equivalent to a third of the area that was urbanized. However, continued excess food production in the EC has encouraged some progress in the reform of the CAP; as price supports for agriculture are reduced so the competitiveness of forestry has increased, especially in lowland areas. As a consequence, the rate of conversion to woodland is likely to accelerate in the 1990s, especially if favourable financial arrangements are offered to farmers under 'set aside' or 'farm extensification' schemes.

Increases in cropland

Despite the loss of good-quality farmland to urban development and poor-quality upland to afforestation, total agricultural production in Europe has increased substantially since the war. This has been the product not only of increases in yield but also of increases in the cropland area – particularly in the traditional grain-growing areas. These developments are linked to increasing specialization of production by farm and region due to the increasing scale of agricultural operations. For example, in the UK the area under crops has expanded in the eastern counties (Norfolk) and in the Midlands (Oxfordshire), but contracted in the west (Carmarthen, Northern Ireland) since 1950 (Fig. 8.8). Accompanying these changes, particularly in the east of England, has been an impact on the farming landscape through the enlargement of fields and removal of hedgerows.

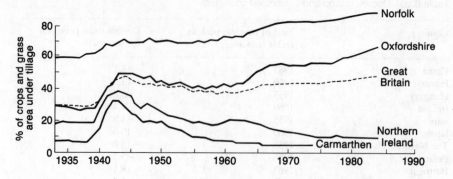

Fig. 8.8 Changes in the area under tillage in Great Britain and selected areas, 1933–87 (*Source:* Parry 1991)

A parallel response to the profitability of crop farming has been extensive reclamation of lowland heaths and upland roughland. Between 1933 and 1979 about 1 million ha in the UK (15 per cent of the nation's roughland) was converted to improved farmland (Parry 1991). In some gently sloping uplands reclamation has been particularly extensive, such as in the North Yorkshire Moors where 8 per cent of the roughland was ploughed up between 1950 and 1980 (Parry et al. 1982).

Agriculture and landscape

The land-use changes described above have frequently had an important effect on the appearance of the rural landscape. Indeed, the pace of the changes and the scale of the effects, particularly in Europe in the 1970s, placed a question mark over the assumption, widely held since perhaps the depressed years of the 1930s, that agriculture could reasonably be construed as the 'conservator' of the traditional rural landscape in the face of 'development' by industry and urbanization (Westmacott and Worthington 1974).

Lowland landscapes

The expansion of cropland and the increased specialization in arable farming has wrought the most substantial changes in lowland landscapes in Europe and North America: grassland and heath have been ploughed, coppices felled, wetlands drained, and hedgerows and field boundaries removed. Typical of the changes has been the cultivation of acid heaths in the Breckland of eastern England, which has been reduced in area by 87 per cent in the last 50 years (Briggs and Courtenay 1985). Similar areas in Dorset, which covered 40 000 ha in 1759, has been reduced to 6 000 ha by 1978 (Webb and Haskins

1980). Likewise, calcareous grassland, which covered millions of hectares in southern and eastern England in the early nineteenth century, had been reduced to less than 45 000 ha by the late 1960s. In addition, increases in field size, encouraged by the demands of mechanization, have been accommodated by the removal of hedgerows and the replacement of field ditches by pipe drains. Estimates of the rate of hedgerow destruction in eastern England in the 1960s have varied between 2 000 and 8 000 km yr^{-1} (Briggs and Courtenay 1985).

Wetlands

In England and Wales the annual area of new field drainage in the early 1980s was between 75 000 and 100 000 ha (Baldock 1984). Government grants for this work totalled £26m., a quarter of which was recoverable from the EC's agricultural fund. A consequence has been the extensive drainage of freshwater wetlands: about half of the UK's lowland fens and basin mires, and 60 per cent of the UK's lowland raised mires have been destroyed or significantly damaged since 1949 as a result of drainage operations and reclamation for agriculture; and about half of the area of raised bog in lowland Scotland has been destroyed by afforestation since 1930 (North 1983).

Uplands

Similar inducements to reclamation and afforestation, particularly by the EC's CAP in the 1970s and early 1980s, led to widespread reclamation and afforestation of uplands that previously had carried semi-natural vegetation under a regime of extensive grazing. In 7 out of a total of 10 National Parks in England and Wales, about 44 300 ha of roughland were ploughed and 24 100 ha were afforested between 1950 and 1980 (Parry et al. 1982). Over the 30 years, the total area of rough grazing in these areas fell by 7.5 per cent.

Towards sustainable agriculture

Understanding agriculture–environment interactions

We have seen that the productivity of both cropping and grazing systems in agriculture depend on the interaction of many factors. Knowledge of the effect of each factor on output would enable us, in theory, to allocate precisely the optimum amount of inputs, in the most appropriate mix, at the best time in the farming calendar. We could thus optimize the production system with respect to those factors over which we have some control; and when there is some uncontrollable variation, for example in sunshine, temperature or rainfall, we might at least be able to predict the outcome in terms of varying production and thus plan accordingly. Alternatively it might be possible to alter levels of other inputs to mitigate the variation and thus stabilize production. Attempts to stabilize sheep production in Iceland, for example, have recently been made by increasing the allocation of fertilizer in cool summers when the hay crop would otherwise be less than average (Bergthorsson et al. 1988). For this kind of response to uncertain weather to be worth while, we need to know the relative effects of temperature and fertilizer on grass growth, isolating the role of various factors (singly and in combination) and quantifying their effects. A fairly new branch of research is devoted to this kind of research – the formal mathematical modelling of production systems.

In general, the models fall into two broad classes: *empirical–statistical models* and *simulation models*. Empirical–statistical models are developed by taking a sample of annual yield data and input data for the same area and relating them through statistical techniques such as multiple regression analysis. They generally require only modest amounts of data and little computational time, but they only indicate the statistical association between variables and do not easily lead to a causal explanation of the relationships. In contrast, simulation models generally incorporate, through a set of mathematical expressions, those mechanisms or interactions that are understood to be important for plant growth, such as photosynthesis and transpiration. Some models are therefore quite detailed, using hourly weather data to simulate the growth of a single crop type on a particular soil. Others may use monthly average data and thus be more readily applicable to a variety of locations, including the Third World where reliable data are often scarce. One such set of simple models has been used by the Food and Agriculture Organization of the United Nations (FAO) to produce 1 : 5 million scale maps of the potential for crop production in Africa, Latin America and South-east Asia based upon climate, soils and terrain data (FAO 1984).

Protecting the environment

The expansion and specialization of agriculture, particularly in developed countries in the past 30 years, have resulted in damage to the rural environment and conflict between farmers and other users of the countryside. Much of this expansion has been fuelled by prices being maintained artificially high by national governments or supranational agencies such as the Commission for the European Communities. In the UK direct subsidies for land drainage and upland 'improvement' have been most damaging because they have often led to massive landscape changes in areas of great landscape or ecological value, in return for new agricultural land of somewhat marginal value.

Since about 1980, however, a number of developments have served to reverse these trends (O'Riordan 1987). First, and most important, the excess of food supply in the EC, the high cost of storage of surpluses, and the spiralling costs of the CAP have led to serious attempts both to control CAP spending and reduce production. Second, the introduction of subsidies for environmentally sound farming techniques has recently gained ground. In England and Wales, the Wildlife and Countryside Act (1981) made provision for financial compensation to farmers where an application for capital grant aid is refused on conservation grounds in protected areas such as National Parks and Sites of Special Scientific Interest. It also enables management agreements to be negotiated between farmers and local authorities in order to maintain farm income while reducing environmental damage (Hodge 1989). In addition, landscape protection was extended after 1986 to cover 17 Environmentally Sensitive Areas (ESAs) in England, Wales and Scotland. Farmers in the ESAs will be encouraged to adopt farming practices which will safeguard landscape, wildlife and archaeological features of each area (Potter 1988).

Finally, recent attempts have been made to extend the range of potential sources of income in rural areas, both in the variety of crops and livestock farmed and in the array of non-farm activities such as forestry, recreation and light industry, with the aim of reducing dependence on intensive cereal and livestock production (see pp. 156–7).

Research capacity to develop new technology

The maintenance of sustainable agricultural systems requires a steady flow of appropriate new technology. Crosson (1986) has emphasized that the inter-national network of agricultural research institutes supported by the CGIAR, and the related national research agencies, 'must be regarded as one of the major institutional innovations of modern times. Its positive impact on the welfare of people involved, both as producers and consumers of agricultural commodities, has already been enormous'. But the development of technologies that are suitable for the highly diverse soil and climatic conditions of farming systems in many different countries requires a decentralized research capability. There is a need to study the processes of institutional change in agriculture, such as changes in landownership, credit systems and marketing, since the long-term sustainability of agricultural systems depends as much on institutional flexibility and resilience as on the successful adaptation of technology.

Conclusion

This analysis of the resource base of agriculture has highlighted the unsustainability of contemporary, industrialized farming systems when viewed from a managed ecosystem perspective. Even so, many references have been made to show that agriculture–environment problems are not confined to the industrial model of agricultural development – they can be observed in many other parts of the world, including developing countries.

However, there is evidence of an increasing willingness among policy-makers and the farming community to manage agro–ecosystems so as to ameliorate, if not avoid, the worst environmental impacts of modern farming. A growing academic literature has explored this theme (e.g. Jenkins 1984), and a range of management policies have been put in place in several countries (Park 1988). To some observers the introduction of management policies, such as ESAs, represent an incremental rather than reforming approach to modern farming, especially as regards state farm policies (Baldock et al. 1990). Certainly the costs involved in a fundamental, rather than marginal, reform of existing measures that support an industrialized agriculture are still under debate (Council for the Protection of Rural England 1989). Also much depends on the attitudes of farmers towards conservation for the implementation of new measures at the farm level (Carr and Tait 1991). Nevertheless, a needed convergence of farm and environmental policies seems to be taking place, and this theme is developed in more detail in the next chapter.

References

Allen, R. (1980) *How to save the world: strategy for world conservation.* Kogan Page, London

Baker, O. E. (1926) Agricultural regions of North America. *Economic Geography* 2: 459–94

Baldock, D. (1984) *Wetland drainage in Europe.* International Institute for Environment and Development, London

Baldock, D., Cox, G., Lowe, P. and **Winter, M.** (1990) Environmentally Sensitive Areas: incrementalism or reform? *Journal of Rural Studies* 6: 143–62

Bayliss-Smith, T. P. (1982) *The ecology of agricultural systems.* University Press, Cambridge.

Bergthorsson, P., Bjornsson, H., Dyromundsson, O., Gudmundsson, B., Helgadottir, A. and **Jonmundsson, J. V.** (1988) The effects of climatic variations on agriculture in Iceland. In Parry, M. L., Carter, T. R. and Konijn, N. T. (eds) *The impact of climatic variations on agriculture. Vol 1 Assessments in cool temperate and cold regions.* Kluwer, Dordrecht, The Netherlands

Briggs, D. J. and **Courtenay, F. M.** (1985) *Agriculture and environment.* Longman, London.

Carr, S. and **Tait, J.** (1991) Farmers' attitudes to conservation. *Built Environment* **16**: 218–31

Carson, R. (1962) *Silent spring.* Houghton-Mifflin, Boston

Crosson, P (1986) Agricultural development – looking to the future. In Clark, W. C. and Munn, R. E. (eds) *Sustainable development of the biosphere* University Press, Cambridge, pp 104–36

Crutzen, P. J. and **Graedel, T. E.** (1986) The role of atmospheric chemistry in environment–development interactions. In Clark, W. C. and Munn, R. E. (eds) *Sustainable development of the biosphere.* University Press, Cambridge, pp 213–49

Council for the Protection of Rural England (1989) *Conserving the countryside: costing it out.* CPRE, London

Duckham, A. N. and **Masefield, G. B.** (1970) *Farming systems of the world.* Chatto and Windus, London

Eddowes, M. (1976) *Crop production in Europe.* University Press, Oxford

FAO (1978) *The state of food and agriculture, 1977.* FAO, Rome

FAO (1984) *Land, food and people.* FAO, Rome

Fantechi, R. and **Margaris, N. S.** (eds) (1986) *Desertification in Europe.* Reidel, Dordrecht

Frederick, K. and **Hanson, J. C.** (1982) *Water for western agriculture.* Resources for the Future, Washington DC

Goudie, A. (1981) *The human impact.* Blackwell, Oxford

Green, B. (1986) Agriculture and the environment: a review of major issues in the UK. *Land Use Policy* **5**: 301–13

Green, F. H. (1980) Current field drainage in northern and western Europe. *Journal of Environmental Management* **10**: 149–53

Hickey, J. J., Keith, J. A. and **Coon, F. B.** (1966) An exploration of pesticides in a Lake Michigan ecosystem. *Journal of Applied Ecology* **3** (suppl): 141–54

Hodge, I. D. (1989) Compensation for nature conservation. *Environment and Planning A* **21**: 1027–36

Huggett, R. and **Meyer, I** (1980) *Agriculture.* Harper and Row, London

Hunt, E. G. and **Bischoff, A. I.** (1960) Inimical effects on wildlife of periodic DDD application to Clear Lake. *California Fish and Game* **4**: 91–106

Jenkins, D. (1984) *Agriculture and the environment.* Institute of Terrestrial Ecology, Cambridge

Leach, G. (1976) *Energy and food production.* IPC Press, Guildford

Miller, G. T. (1982) *Living in the environment, 3rd edn.* Wadsworth, Belmont, California

Myers, N. (1979) *The sinking ark.* Pergamon, Oxford

North, R. (1983) *Wild Britain.* Century, London

O'Riordan, T. (1987) Agriculture and environmental protection. *Geography Review* **21**: 35–40

Park, J. R. (1988) *Environmental management in agriculture: European perspectives.* Belhaven, London

Parry, M. L. (1991) The changing use of land. In Johnston, R. J. and Doornkamp, J. C. (eds) *The changing geography of the United Kingdom.* Methuen, London, pp 7–34

Parry, M. L., Bruce, A. and **Harkness, C. E.** (1982) *Surveys of moorland and roughland change.* University of Birmingham Department of Geography, Birmingham

Potter, C. (1988) Environmentally Sensitive Areas in England and Wales: an experiment in countryside management. *Land Use Policy* **5**: 301–13

Sommer, A. (1976) Attempt at an assessment of the world's tropical forests. *Unasylva* **28**: 5–24

Tivy, J. (1990) *Agricultural ecology.* Longman, London

Trafford, B. D. (1970) Field drainage. *Journal of the Royal Agricultural Society of England* **131**: 129–52

United Nations Conference on Desertification (1978) *Round-up, plan of action, and resolutions.* United Nations, New York

Urban, F. and **Vollrath, T.** (1984) *Patterns and trends in world agricultural land use.* US Government Printing Office, Washington DC

Webb, N. R. and **Haskins, L. E.** (1980) An ecological survey of heathlands in the Poole Basin, Dorset, England in 1978. *Biological Conservation* **17**: 281–96

Westmacott, R. and **Worthington, T.** (1974) *New agricultural landscapes.* Countryside Commission, Cheltenham

World Commission on Environment and Development (1987) *Our common future.* University Press, Oxford

Worldwatch Institute (1985) *State of the world, 1985.* Norton, New York

9

Agriculture and the state

John Tarrant

State intervention has been identified as a major influence on contemporary agriculture and it forms one of the supporting structures of the food supply system in Fig. 1.1. Governments of all countries of the world have policies through which they intervene in agriculture and this chapter will try to show why this intervention is so widespread, some of the forms it takes, and some of its results. Although all countries enact agricultural policies, there is a spectrum of the scale and nature of such policies: from more or less complete control over all aspects of agricultural production and marketing, to interventions which are much less direct in their impacts.

An international perspective

Until the events of 1989/90, the USSR represented one end of the spectrum of state intervention in agriculture. On the vast majority of agricultural land, decisions on what was planted, when it was planted and harvested, and how the products were utilized were taken by the state (Hedlund 1984). Private production existed, however, with private incentive and decision-making ensuring that this production accounted for some 25 per cent of total Soviet output from only 3 per cent of the agricultural land. Almost all the state production and much of the private production was procured directly by government agencies: the economic costs of such a scale of state intervention were vast. As production costs rose so did procurement prices. On the other hand, consumer prices have remained more or less static after the 1950s with the ever-increasing burden of consumer subsidies added to the production costs of state agriculture. Consumer subsidies amounted to 23.2 million roubles for meat and dairy products alone in 1979 (Nove 1982). By the late 1980s, agricultural and consumer subsidies in the USSR amounted to some 45 per cent of the value of agricultural output. In addition agricultural investment, at about 30 billion roubles in 1980, represented some 27 per cent of total state investment (Johnson and Brooks 1983; Hedlund 1984).

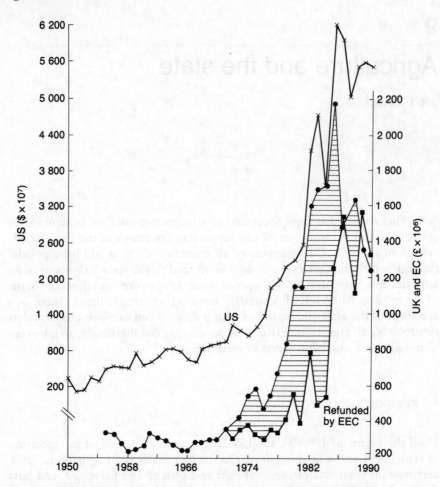

Fig. 9.1 Comparative costs of agricultural policies, UK and US. * =Receipts do not always relate to expenditure in year in which they are received (*Source: Annual review of agriculture* MAFF, various years. UK Appropriation Accounts, various years. Unpublished data supplied by National Economics Division of the Economic Research Service of the USDA)

At the other end of the spectrum we would expect to find the US, but here also the role of state intervention in agriculture is considerable and surprising. Its organization requires one of the largest departments of government, while expenditure on farm income support in the mid-1980s was estimated to be over $34bn a year (*The Economist* 1986). Although US farmers are more or less free to make their own decisions, their economies are hedged about by numerous controls and incentives. The prices they obtain for their products, while apparently the result of the operation of the free market, are manipulated by government schemes which control the volume of production, expand the size of the market, and protect the incomes of farmers. In addition, govern-

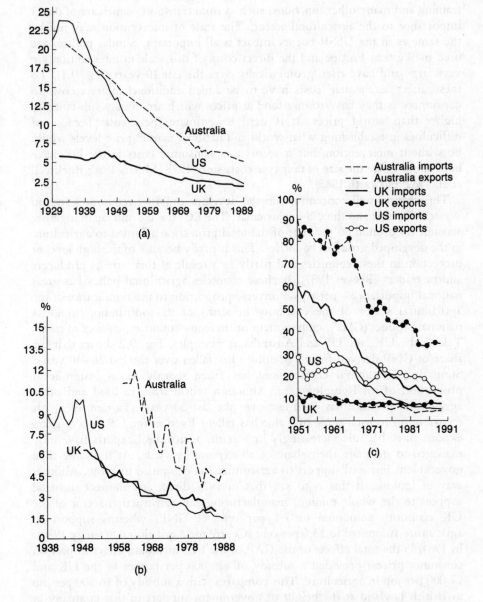

Fig. 9.2 Role of agriculture in national employment, trade and contribution to GNP. (a) agricultural employment as percentage of total employment; (b) percentage of GNP derived from agriculture; (c) agricultural trade as percentage of all trade by value (*Sources:* (a) FAO *Production yearbooks*, various years; International Labour Office *Yearbook of labour statistics* 1983; (b) *Yearbook Australia*, 1975/76; UN *Yearbook of national accounts*, 1961, 1971, 1973–88; OECD *Economic survey, Australia*, 1972, 1984; (c) FAO *Trade yearbooks*, various years)

ment spends considerable sums on agricultural research, education and training and many other functions, such as rural transport, which are of direct importance to the agricultural sector. The scale of intervention may not be the same as in the USSR but its impact is all important. Similar policies are used in Western Europe and the direct costs of this scale in intervention are very large and have risen geometrically over the last 10 years (Fig. 9.1). To these direct 'exchequer' costs have to be added additional indirect costs to consumers as they have to buy food at prices which are usually substantially higher than world prices. It is hard to estimate these costs because of difficulties in establishing what world and national market price levels would be without intervention; but it seems that consumer costs in the UK have been about twice the size of taxpayer costs since 1978 (Morris 1980; Buckwell et al. 1982; Howarth 1985).

This chapter will concentrate on the developed market economies and on Western Europe and the US in particular. The EC, the US and Japan together account for around 80 per cent of the total protection afforded to agriculture in the developed group of countries. This is partly because of the high level of protection in these countries and partly as a result of their size as producers and/or traders (Reeves 1987). In these countries agricultural policy has great national importance – perhaps in inverse proportion to the significance of the agricultural sector of the economy in terms of its contribution to gross national product (GNP), employment or, in many countries, balance of trade. Taking the UK, the US and Australia as examples, Fig. 9.2 shows that the share of GNP derived from agriculture has fallen over the last 30–40 years. Similarly, agricultural employment has fallen sharply to an insignificant proportion of total employment. Although world trade in food and other agricultural produce has risen fast over the last 30 years (Tarrant 1985), its contribution to total trade by value has fallen. Even in the US, where grain exports have become increasingly important, agricultural exports have only managed to maintain their share of all exports by value. At the same time government financial support to agriculture has continued to grow. Another way of looking at this is to say that in 1977 direct and indirect financial support to the whole mining, manufacturing and construction sector of the UK economy amounted to 7.1 per cent of GNP, whereas support to agriculture amounted to 13.6 per cent (Centre for Agricultural Strategy 1980). In 1984/85 the total effects of the CAP of the EC, including direct impact on consumer prices, provided a subsidy of £20 000 per farmer in the UK and £7 000 per job in agriculture. This compares with a subsidy of £6 500 per job in British Leyland at the height of government support to that company in 1981/82.

Reasons for state intervention

State concern with agriculture has a very long history – it is probably identifiable from the very earliest times of settled agriculture (Mumford 1961). State officials and priests had important functions in the organization of the agricultural calendar and in grain collection and storage. This central role was because of the need for security of food supplies and because of the difficulties of large numbers of farmers co-ordinating their actions to meet urban food demands without some external intervention and supervision. These reasons are still of importance today and other considerations have been added (Fig. 9.3).

Although government has been an important influence on farming in many periods of history, the origins of modern intervention in the US and in Western Europe stem from the state of agriculture during the late 1920s and

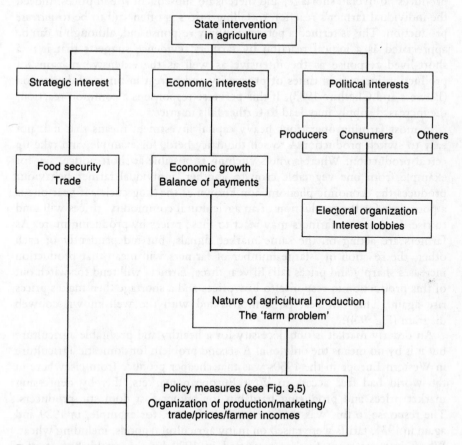

Fig. 9.3 Reasons for state intervention in agriculture

243

early 1930s. The world economic depression started, and was most severely felt, within the agricultural sector. There are many accounts of the state of agriculture at that time (Steinbeck 1962).

The problems of depressed prices and poverty within agriculture were first considered to be marketing problems. Millions of small producers, selling to a dispersed market, were seen as major obstacles to efficient and profitable farming. The government response to this view of the problem was to establish the first marketing boards (see Ch. 7). These were set up as producer organizations to enable farmers to work together, sharing economies of scale in selling their products. One special problem of agriculture had been identified. In a market economy farmers are unlike any other producers as each produces but a minute fraction of the total market demand. Communication and co-ordinated action between them are impossible without some external intervention. No one farmer can do anything to influence the nature of the market in which produce is being sold. If production is reduced, the individual farmer's share of the market is so insignificant that this action produces no overall shortage, and there is no subsequent rise in prices. Indeed the individual farmer's reaction to falling prices is often said to be to *increase* production. This is termed a perverse supply response and, although it can be appreciated as a logical reaction by farmers, evidence suggests that it is a short-lived response as the intensive as well as the extensive margins of production increase in times of prosperity and retreat in times of depression (Bowers and Cheshire 1983). If this perverse response is a common reaction, the increased supply may lead to further falls in price.

In most farming operations heavy capital investment means that it is not easy to switch production – to sell the dairy herd, for example, and take up cereal production. Where some switching is possible at short notice, say for example from one vegetable crop to another, individual farmer behaviour produces the economic phenomenon known as the 'hog cycle'. If year one is a poor year for the production of an agricultural commodity, prices will tend to rise. In year two farmers may react to these prices by producing more. As farmers are acting on the same market signals, but independently of each other, the reaction of a large number of farmers will mean that production increases sharply and prices fall. In year three, farmers will tend to switch out of this production in response to low prices and a shortage then makes prices rise again. This can best be demonstrated with the well-known cobweb diagram (Fig. 9.4).

An orderly market is one necessity for a healthy and profitable agriculture but it is by no means the only one. A second problem for domestic agriculture in Western Europe in the 1930s was that cheaper produce from elsewhere in the world had free access to West European markets, thereby depressing market prices and producing hardship for higher-cost domestic producers. The response to this was trade control. In France, for example, in 1929 and again in 1931, tariffs were raised on many agricultural goods, including wheat. Wheat imports were further restricted in 1929 by the establishment of a

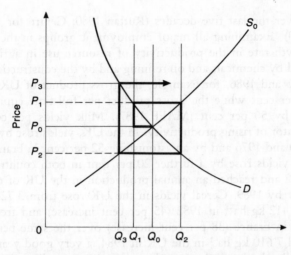

Fig. 9.4 The cobweb model. A shortage of supply reduces the quantity available from equilibrium (Q_0) to Q_1. *Prices rises to P_1.* At that increased price farmers will grow more and supply rises the following year to Q_2. Price falls to P_2. In year three, supply falls to Q_3 and price rises to P_3. Thus, provided the supply curve is steeper than the demand curve (or the supply is more elastic in its response to price changes) then the cobweb continues to grow. Such a regular development is prevented by further unexpected bad or good harvests leading to further perturbations in the supply

milling ratio which allowed the government to fix a proportion of home-produced wheat to be used in flour production. This was set at 100 per cent in 1933 as wheat production in France and French North Africa continued to increase. In the same year the government set a minimum domestic price of 115 francs per 100 k against a world price of 25–30 francs. Later, government was forced to subsidize wheat exports and to render stocks unfit for human consumption as further measures to protect domestic producers (Tracy 1982). It is in these policies, similar to those followed in Germany and Italy, that the CAP of the EC has its origins.

The root problem of agriculture, in the 1930s and still in many respects today, was farmers' incomes: more specifically, low incomes in farming in comparison to other sectors of the economy (see p. 96). The Agriculture Acts introduced in the UK and elsewhere during the 1930s made frequent reference to farm incomes in comparison to incomes in other activities and the necessity for farm incomes to be 'fair'. However, the problem of falling farm incomes did not lie fundamentally in market organization, nor in the problem of cheap imports from elsewhere, but in the changing nature of supply of, and demand for, agricultural production. Traditionally, this has been called the 'farm problem' (Josling 1974; Bowler 1979).

The basis of the 'farm problem' has been explained in earlier chapters but the main argument can be usefully summarized again. Firstly, the industrialization of agriculture (Ch. 1) has increased the productivity of all forms of agriculture over the last five decades (Ruttan 1980; Centre for Agricultural Strategy 1980). Examining all major employment groups in the UK shows that the growth rate in the productivity of resource use in agriculture was only exceeded by chemicals and oil refining and by the construction industry. Between 1973 and 1986, for example, the gross product of UK agriculture rose by 17 per cent while the gross product per full-time equivalent male worker rose by 56 per cent (MAFF 1987). Milk yields per cow provide another indicator of rising productivity. In the UK, yields rose by 24 per cent between 1960 and 1976 and by an astonishing 52 per cent in France over this period. Milk yields rose by a further 20 per cent in both countries between 1975 and 1982 and reached an annual production in the UK of nearly 5 000 litres per cow by 1989. Cereal yields in the UK rose from 3 724 kg ha^{-1} in 1969/71 to 5 412 kg ha^{-1} in 1982 (45 per cent increase) and from 3 596 to 4 959 kg ha^{-1} in France (38 per cent increase) over the same period. Wheat yields reached 7 610 kg ha^{-1} in the UK in 1984, a very good year. Although yields have fallen back a little since that year, they are still some 60 per cent higher than in 1975/77.

Table 9.1   Growth rates of population and private consumption

	Annual population growth rate (%)			Annual growth rate in private consumption (%)		
	1960–70	1970–81	1980–89	1960–70	1970–81	1980–89
UK	0.6	0.1	0.2	2.4	1.6	3.4
France	1.1	0.5	0.4	5.3	4.1	2.4
US	1.3	1.0	1.0	4.4	3.2	3.7
Australia	2.0	1.4	1.4	5.0	3.0	3.4
All developed market economies	1.1	0.7	0.7	4.3	3.0	3.1

Source: Based on data in World Bank (1991).

Secondly, the rising output from agriculture has been faced by a slower rate of increase in demand for food in developed countries (Ch. 7). The two main determinants of food demand are population and income (see Fig. 1.1). Population has continued to grow in Western Europe and in the US, but at a decreasing rate (Table 9.1). Incomes have risen faster, and with them private consumption, but unfortunately demand for food does not rise uniformly with income (Johnson 1972). People on very low incomes will spend a large proportion of any increase on food. People on high incomes may spend more on food, especially by changing their eating habits and consuming more meat and less staples such as potatoes and cereals, but only a small proportion of any extra income at that level will be spent on food. The income elasticity of food demand is not constant over all income levels. As consumer expenditure rises, therefore, a decreasing proportion of this goes towards food purchases. In 1973–75, 21.2 per cent of UK consumer expenditure was on food; this fell to 15.6 per cent by 1987. The situation is further complicated by changing eating habits. Demand for dairy products in the US has fallen and is now lower than it was in 1950. This is partly as a result of increasing concern over high-fat diets and partly the availability of alternatives such as non-dairy creams and margarines.

Thirdly, although food supply has outstripped demand, farmers have not cut back their production to create equilibrium in the market. Two arguments have been advanced to explain this behaviour. On the one hand, farmers are caught in a 'cost–price squeeze' (p. 14) with input costs rising faster than output prices (Table 9.2). Perversely, farmers increase output to combat falling profit margins. On the other hand, farmers are also caught on a 'technological treadmill' (p. 14), whereby farmers continue to invest in output-increasing (and unit-cost-reducing) farm technology. Indeed farmers press research institutions and agricultural supply firms for these inputs (Hayami and Ruttan 1971), while government grants may also be available to encourage this type of farm 'modernization' (Marris and Hess 1986).

Table 9.2   Change in prices received and paid by farmers

Price	UK	Australia (1978–88)	US	EC (10) (1973–86)
Prices received (%)	+42	+79	+23	+59
Prices paid (%)	n.a.	+99	+49	+71
Machinery (%)	+81	+119	+72	
Fertilizers (%)	+42	+88	+30	
Pesticides (%)	+53	+79	+32	
Feed (%)	+29	+52	+30	
Seed (%)	+60	+83	+45	
Energy (%)	+22	+109	+58	

Country	Prices received (%)		Prices paid (%)		Farm wages (%)	
	1975–85	1985–89	1975–85	1985–89	1975–85	1985–89
Belgium	+40	+0.7	+54	−1.4	+75	+3.1
Denmark	+63	+0.4	+74	−0.2	+91	+5.2
Greece	+191	+13.7	+174	+11.3	+220	+14.6
France	+76	+1.4	+89	0.0	+132	+4.0
Ireland	+90	+4.7	+103	−0.2	+132	+4.2
Italy	+125	+3.1	+118	+2.0	+196	+6.2
Luxemburg	+47	+3.5	+55	−1.3	n.a.	n.a.
The Netherlands	+24	+0.4	+40	−2.8	+56	+2.3
UK	+70	+2.8	+93	+2.2	+121	+5.7
West Germany	+10	−0.1	+32	−1.8	+57	+2.3

Note: As these data are in current prices, the absolute percentage changes will be partly a consequence of inflation rates. Direct inter-country comparison is possible, however, of the magnitude of the gap between prices paid and prices received, or the changes in agricultural terms of trade.

Sources: FAO (1989); Commission of the European Communities (1990).

Fourthly, the disequilibrium between food demand and supply, nevertheless, tends to depress farm output prices and thus farm incomes. Governments intervene in the market to support farm prices but thereby exacerbate the problem since farmers are not penalized in their prices for oversupplying the market. If support prices are allowed to fall, as they have in real if not in absolute terms in the EC for many products, the most rational response of producers, at least in the short term, will be to invest even more heavily in ways to reduce unit costs further. Unless support prices fall truly dramatically, such a policy may accelerate both overproduction and some of the environmentally detrimental consequences of highly mechanized and modernized agriculture (see Ch. 8). Although Bowers and Cheshire (1983) suggest that this perverse response will only exist in the short term, milk production in the UK over the last 10 years has steadily risen in the face of static or falling real prices. Dairy cattle have become concentrated into larger and more cost-efficient herds with higher levels of milking mechanization.

The 'farm problem', without any state intervention, would have led to a sharp decline in the number of farmers and the few which remained would have become larger, achieving economies of scale in production. This tendency has been very marked in New Zealand since 1984 when the government removed subsidies to agriculture (Le Heron 1988; Cloke 1989). An increasing number of bankruptcies have been seen in farming, a feature also of the US in the 1980s under a similar government policy of reducing the level of government support. Indeed the recent problems of farming have accelerated the longer-term decline in farm numbers in this century and the size of the typical farming operation has increased (Ch. 4). Governments, however, have maintained their intervention in agriculture, in part to control the negative effects of the 'farm problem' on rural communities. They have been pushed in this direction by the important political role of farmers as the most influential occupiers of large tracts of rural areas, (Johnson and Smit 1985; Senior Nello 1989). Where the political system requires representation by area rather than by population, as in, for example, the election of two senators from each state to the US Senate, then the agriculture lobby becomes politically very important. All significant agricultural legislation in the US starts life in the Senate. Also in cases where government is by coalition, minority parties, perhaps representing rural interests, can achieve political influence out of all proportion to the number of their supporters. Such was the position of the Free Democratic Party in West Germany for much of the last decade (Hendriks 1987). It drew its main support from the small farmers of Bavaria and was highly significant in German agricultural policy during the formative years of the CAP (Andrlik 1981). Although direct evidence is hard to come by (Bowler 1986a) it is also asserted that landowning members of legislatures exert considerable influence in agricultural matters. Certainly, farmer organizations and interest groups, particularly perhaps the National Farmers' Union (NFU) in the UK, play an important role in lobbying for the special interests of their members (Wilson 1977). The significance of the NFU has declined since the UK joined the EC, and a large proportion of agricultural policy is now determined in Brussels rather than in London.

Consumers have emerged as a powerful political force only since the 1960s. Governments have been concerned for much longer with aspects of food quality and safety (one of the original functions of the Milk Marketing Board in the UK) but, as we shall see in the next section, when the interests of producers and consumers can be amalgamated, powerful political associations can be forged and have great influence over policy.

Since the 1930s governments have devised a whole series of policies to protect farmers from the economic forces outlined here. There has been an emphasis on 'fair' incomes for farmers, on 'parity' with incomes in other sectors of the economy (an index of this parity is still used in US agricultural policy), and on various degrees of protection to agriculture.

Balance of payments and balance of trade may also provide government with reasons to support domestic agriculture, and a case can be advanced for net food importing and exporting countries alike. Food imports, like any other kind of import, can be reduced through protection of domestic production. This can be given added emphasis from time to time by the need to ensure secure food supplies, especially in times of war. One of the most important founding objectives of the Treaty of Rome, which established the EC, was to achieve self-sufficiency in those foods which could be produced in Europe. The rapid rise in grain prices on the world's markets between 1972 and 1975, events triggered by the surprise Soviet purchases of grain in 1972, produced a spate of self-sufficiency policies throughout the world as governments saw their supplies of food threatened by the actions of one or two giants in the food trade scene. The UK, for example, which had relied on substantial food imports for more than 100 years, produced a White Paper *Food from our own resources* (MAFF 1975). For most countries the concept of food security and self-sufficiency remains central to the formulation of agricultural policy. There may then be a strategic level to add to the economic and political forces leading to state intervention in agriculture.

Exporters may also find agricultural support crucial. Exports of agricultural goods from the US, although a slowly diminishing proportion of all exports by value (Fig. 9.2c) still remain of the utmost importance because without them the current US trade deficit would be even more serious. Of course countries relying on a large agricultural export trade have to ensure that their goods can be sold at competitive prices on world markets. This they can do either through export subsidy or through help to farmers provided in such a way that does not directly affect the price. Alternatively, as in the case of New Zealand and Australia, the need to compete on world markets may encourage a withdrawal from agricultural support (Lloyd 1987).

If we refer back to Fig. 9.3 we see that many of the factors influencing policy have become less significant than in the 1970s. Food security is less of an issue with generally low world prices and abundant food stocks world-wide. In 1987/88 world wheat stocks stood at 129 million tonnes compared with an annual world trade volume of about 102 million tonnes (Anon. 1988a). The shortages of the early 1970s seem well in the past. Although governments remain concerned with balance of payments and balance of trade issues, food exports have become less significant in terms of the value of total trade. Oil prices have fallen sharply since the 1970s and this has reduced somewhat the balance of trade pressure on oil-importing countries such as the US. The political role of producers has diminished along with their numbers, while other opposing interest groups have achieved greater importance, especially as regards environmental issues.

These changes have allowed governments in the UK, Australia, New Zealand and the US more opportunity to regard the agricultural sector of the economy in the same light as other sectors. Indeed the cost-effectiveness of agricultural policies are being more closely examined than at any time for the past 50 years. This reflects the widespread adoption in the 1980s of a political ideology which regards government intervention as expensive, inefficient and politically insupportable. In part this has been applied to agriculture, but to a much smaller extent in most countries than to manufacturing industry and to public utilities. In spite of these first signs of change, most developed market economies have a very long way to go before they can claim to have removed support from agriculture. The various mechanisms (or policy measures) available for intervention are examined in the next section.

Mechanisms of state intervention

The state attempts to counteract the outstripping of demand by supply leading, potentially, to falling farm incomes. This suggests three possible strategies for state intervention (Fig. 9.5): (1) increase demand, (2) reduce supply or (3) directly intervene to influence farm incomes. In practice, after a policy development period of over 40 years, most developed countries have agricultural policies which are combinations of all three strategies.

Policy measures for increasing demand

A government hoping to increase the demand for food has two alternatives: the state can allocate funds to all, or to sections, of the population in such a way that these funds encourage the purchase of additional food. Alternatively the state itself can purchase food and thereby create an artificial market and maintain high prices. The first is a policy extensively followed in the US through the food stamp programme: a combination of a social welfare programme and support for agriculture (Lane 1978; Spitz 1987). Although concerned with food *consumption*, this provides an example of where the

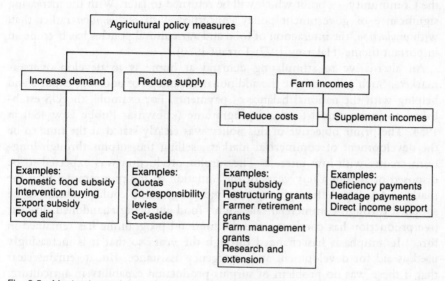

Fig. 9.5 Mechanisms of government support to agriculture

interests of producers and consumers can be harnessed together. It is no coincidence that the programme is administered by the Department of Agriculture. It has grown rapidly in importance and in 1984 accounted for 46 per cent of the total Federal spending on agriculture and 2 per cent of the total Federal budget. Families below a predetermined income level are allocated food stamps which they can exchange for food. The programme was originally devised to encourage *additional* expenditure on food. Recipients were required to purchase a dollar's worth of food before an additional dollar's worth of food stamps could be spent. This requirement was dropped in 1977 which reduced the effectiveness of the demand–stimulating side of the policy although, with an expenditure of $16 489m. in 1984, its role in agricultural support remained considerable. Unfortunately most of this subsidy benefits food-processing industries, which are responsible for such a high proportion of the value added of food, rather than farmers (see Ch. 7).

Other food subsidies have been tried from time to time in various countries. In the 1970s there were general subsidies on basic foods in the UK, for example. All people qualified for these subsidies, the purpose of which was not primarily to support the agricultural sector, although it did stimulate demand somewhat. Within the EC there are also systems for consumer subsidy: meat and dairy products from intervention storage are distributed to the poor and needy through various European charities. Even so, these schemes are motivated mainly by the need to reduce stocks of surplus commodities; the use of charities for distribution reduces EC costs considerably. Such EC subsidies have a very limited effect in stimulating additional market demand because of the other methods of market intervention used in the Community – a point which will be returned to later. With the increasing significance of government policy concerned with *consumption* rather than with *production*, the interaction of food and agricultural policies has become an important theme (Halcrow 1977; Tarrant 1980).

An alternative to stimulating demand at home is to develop overseas markets. Such policies have the additional advantage of boosting exports and helping with the national balance of payments. For example, the US established an international food aid programme (known as Public Law 480) in 1954. The prime objective of this policy was clearly stated at the time to be the development of commercial markets, selling the surplus through long-term credits with low interest. The surplus production was expected to be disposed of within a few years of the initiation of the programme; then the markets developed through the food aid programme could continue to be supplied through commercial sales, with food aid being abandoned. In fact, overproduction has continued and the food aid programme has remained in force. Its emphasis has changed through the years so that it is increasingly used as aid for development and emergency assistance. But it remains clear that if there was no problem of surplus production capability in agriculture, there would be little or no food aid. In 1972, when the USSR became a major commercial purchaser of cereals from the US, for instance, food aid shipments halved. Shipments have increased steadily again since 1975, having grown from 8.8 million tonnes in 1979/80 to 9.8 million tonnes by 1983/84; about 90 per cent of these shipments go to low-income, food-deficit countries (Singer et al. 1987).

If people cannot be persuaded to buy more food (or subsidized to do so) then the state itself can defend a predetermined price by purchasing that which the market cannot absorb. For example, in the US some 5.9 million tonnes of milk, butter and cheese, amounting to 10 per cent of total production, is purchased and stored by the government each year. In the EC, the European Agricultural Guidance and Guarantee Fund (EAGGF), operating through various national intervention boards, purchases enough of most agricultural products to ensure that their price is maintained at or above a predetermined level (Fennell 1979; Tarrant 1980). Artificially high prices within the Community require tariff control at its borders. Taxes are applied to imported foodstuffs which are in competition with EC production, raising their prices to or above EC levels. If the EC price should fall below world price levels then the reverse of this policy is applied; exports are taxed to ensure that EC producers do not take advantage of higher prices outside the Community. This situation is rare but did occur for most cereal crops between 1972 and 1975, when world prices were exceptionally high. Such intervention policy tends to be self-perpetuating: prices are generally maintained at higher levels than will ensure market clearance without intervention, so further production is encouraged; demand continues to be suppressed and further intervention buying becomes necessary as demand has remained static. In the EC, state-subsidized high prices have contributed at least as much to the production of surplus agricultural commodities as has the application of farm technology.

The produce purchased by the state has to be disposed of. In the EC some is destroyed, some rendered unfit for human consumption and used for animal feeds or for industrial purposes, and some maintained in intervention storage at very great cost. This may later be disposed of outside the Community through food aid programmes and, often heavily subsidized, sales. Thus in 1990, export refunds (subsidies) from the EC amounted to 9 358 million ECUs or to 33.8 per cent of the total EAGGF spending. This can be compared with 40.7 per cent spent on price subsidies and 19.2 per cent spent on storage. At the end of 1989, the intervention stores held about 22 000 tonnes of skimmed milk powder, 8.609 million tonnes of cereals, of which 2.264 million tonnes was of common wheat. Using these accumulated stores of food, the EC and its constituent countries have become major world food aid donors (Tarrant 1982; Singer et al. 1987).

As state-held surpluses accumulate in Western Europe and in the US, so it gets harder and harder to dispose of them without seriously disrupting the domestic market. The EC export subsidies have the effect of increasing demand in a way which, within the Community, consumer subsidies could not. Suppose West European consumers were given access to stored commodities at a subsidized price, most consumers would use this opportunity to *replace* at least some proportion of their existing market purchases. The real market demand, therefore, would fall and the intervention agencies would be required to purchase more to defend the price, thus tending to refill the storage. Efforts to stimulate demand can only be made by *increasing* the size of the total market. This explains the apparently paradoxical subsidized sales of stored commodities to Eastern Europe and the USSR, while consumer demand remains depressed in Western Europe through artificially high prices.

Policy measures for reducing supply

An alternative to increasing demand is to reduce supply. The US again provides the longest experience of policy measures in this respect (Talbot 1977). The first measures were introduced in the 1930s with the foundation of the soil and conservation reserve programmes. Initially many of these were designed to ensure that land liable to severe erosion was not used for agriculture but, from the start, some were called conservation programmes as a matter of convenience; their main objectives were production control and income redistribution for the agricultural population (Swader 1980). These conservation programmes were insufficient and surplus production continued to be a problem; consequently a land withdrawal programme was started in the early 1960s with the direct and stated objective of limiting production. Other forms of state support were not available to farmers unless they agreed to withdraw a proportion of their land from production of crops in surplus.

These and other measures were consolidated in the Food and Agriculture Act of 1965. For a brief period in the late 1960s production controls were abandoned as export demand rose, especially from India, but they were reintroduced at the end of the decade. During the world food crisis of 1972–73, when demand for US food exports rose to unprecedented levels, President Nixon was able to encourage farmers to plant 'fence-row to fence-row'. Production controls were withdrawn and the 'farm problem' seemed solved; but this was soon over as world demand fell. In fact production controls were reintroduced and had reached 30 per cent of potential output for some crops by 1984. The production control programme perhaps reached its ultimate absurdity in 1983 with the PIK schemes (payment in kind). Here the government used surplus grain, accumulated through intervention in previous years, to pay farmers for not growing grain on land idled under the programme.

Although production control continues to be an important part of the American package of support for agriculture (Spitz 1987), there are serious problems with its effectiveness. Its major difficulty is that farmers who participate naturally ensure that it is their worst land which is withdrawn, while their capital equipment and other inputs are concentrated on the remaining land (Grigg 1984). Productivity of this land rises and the effect of the land withdrawal on overall production is much reduced.

Within Western Europe, production limitation has until recently been exceptional, although the sugar-beet industry, with a fixed processing capacity, has long required that farmers should not be free to grow as much sugar-beet as they like (i.e. they produce under a quota system). The EC extended production quotas to milk in April 1984, requiring all producing countries to cut production. For example, the production quota set for the UK in 1984 was 15.7 million tonnes, representing 6.5 per cent less than 1983 production. Nationally, each dairy farmer was set a quota 9 per cent below the 1983 production level; the aim was to provide some margin which could then be redistributed to special cases of hardship. These were the small producers and those who had expanded recently, often with substantial Ministry of Agriculture encouragement. Producers who exceeded their quotas paid a tax (a superlevy) on that overproduction, which was about 22 per cent higher than the average price they received for their milk. To ease the transition, the UK government added extra incentives for dairy producers to give up milk production completely. The Ministry of Agriculture paid 13p per litre of the surrendered quota (about £650 per cow) in equal instalments over 5 years; the quota released by this means was then redistributed. Quotas have been further reduced since 1984, but milk production continues to rise; also the superlevies have proved difficult to enforce and collect in some regions of the EC (Tracy 1987). Moreover, the introduction of these levies has further distorted the agricultural economy. The quotas themselves have acquired value and are traded between farmers at prices which may exceed the value of the cows to which they refer.

More recently attention has been turned to the cereals sector in the EC. Cereal production, already in overall surplus within the Community, has continued to rise by an average of 2 per cent a year. This relentless increase in cereal production has prompted the European Commission to follow the example of the US and introduce production limitation through payments to idle farmland (Regulation 1094/88). This policy, known as 'set-aside', aims to encourage a voluntary reduction in the growing of cereals without allowing other crops already in surplus to be substituted (Buckwell 1986). From 1988/89, the UK government opted for a scheme under which, to qualify, farmers must agree to withdraw at least 20 per cent of their area of cereals to receive payments initially fixed at between £100 and £200 a hectare (Ilbery 1990). Similar payments in a number of other countries in the EC have encouraged only the marginal producers to enter the set-aside programme (Jones 1991). As this will probably concentrate the production on to land which is best suited to cereals, both on individual farms and nationally, a much larger proportion of the *area* of cereals will have to be withdrawn to bring about a certain level of reduction in *production* (Ervin 1988). Using the land so released for non-agricultural purposes (see p. 156), for farm woodland planting (MAFF 1985), or leaving it fallow under grass are options, but the need for alternative crops to replace those in surplus has stimulated a great deal of research (Centre for Agricultural Strategy 1986; Agricultural and Food Research Council 1987). In Western Europe, many of the potential new crops are sources of vegetable oil (e.g. lupins). This follows the experience of the US in the 1920s when depressed grain prices encouraged the introduction of the soyabean. Unfortunately the soyabean has now joined the list of crops in surplus production (Lockeretz 1988).

Policy measures of direct income support

If the main objective of government intervention is to protect farm incomes, then the most efficient and effective intervention could be to maintain those incomes directly without trying to modify agricultural markets. For example, together with all the other policies already outlined, a system of deficiency payments is used in the US, while an essentially similar system applied in the UK before joining the EC. Under deficiency payments, prices for the main agricultural commodities are fixed annually. In the UK this was done by the Ministry of Agriculture, in consultation with the NFU, at the annual price review. Under the system, farmers are free to sell their produce for the best available price. If the market price obtaining at sale is less than the previously agreed price then the difference, or the deficiency, is made up by direct government subsidy. In the US a 'loan rate' is fixed at the start of the year; at harvest farmers may receive a loan equal to the value of their production at this loan rate; if they are subsequently able to sell at or above this rate then the loan is repaid; if the selling price is less than the loan rate then only a part of the loan is repaid. In addition a 'target price' is fixed each year, often considerably above the loan rate. Farmers who sell below the target price have the difference made up by direct payment from the government as a deficiency payment. Production may be stored in government silos, or farmers may receive a subsidy for storing the produce on their own farms – known as the 'farm owned reserve' – in the hope of a later improvement in market conditions. Income intervention can prove expensive if the market price stays well below the target or loan price for prolonged periods. The most important difficulty, however, and the one which means that such a scheme cannot easily be applied within the EC, is that it requires direct contact between the government and every farmer. The EC method of intervention requires only an agency able to purchase from farmers or from a limited number of wholesalers – the administration of a system which required dealing individually with the incomes of millions of producers, many of whom in Western Europe are very small indeed, would be a burden well beyond the present capabilities of the European Commission.

Direct income support has been adopted in the EC, but only in specially designated 'Less Favoured Areas' (LFAs). The Directive (75/268) establishing the LFAs enables farmers in certain sparsely populated areas to receive headage payments for farm animals. These are known as hill livestock compensatory allowances (HLCAs). The subsidies are paid up to a maximum stocking density of six ewes per hectare (almost certainly too high for most areas of unimproved hill pasture, thus farmers are encouraged to improve at least part of the hill land) and 1.34 cows per hectare. Such income supplements have been widely accepted within the LFAs of the UK (Wathern et al. 1988); they are extensions of long-standing policies to support the income of farmers in difficult farming areas (Bowler 1979; Wathern et al. 1986). The ready acceptance of such payments in the UK provides interesting evidence for the debate on the efficiency and acceptability of direct income support in agriculture (Bowler 1986a).

As an alternative, or in addition to, intervention in markets farm incomes can be influenced by manipulating operating costs and/or the efficiency of agriculture. The simplest way is by subsidizing inputs such as fertilizers and machinery. Direct payments (grants) may be made available for farm modernization, the provision of water to fields, new buildings and farm amalgamation. Directive 72/159 of the EC, for example, allocated funds for farm development where specific plans had been approved. Some 230 000 such plans had been approved by 1987, 35 per cent of them applying to farms of over 50 ha. Also some 40 per cent of the development plans led to subsidized investment in cattle and cattle housing, thereby contributing directly to the increase in output of farm surplus products. The 1972 Directive was later revised by Directive 797/85 which stated that the farm investment should be directed towards the quality rather than the quantity of the produce and to the protection of the environment.

Government-sponsored research, education and extension services in agriculture can be considered as indirect payments to agriculture. The cost of the Agricultural Research Council in the UK is estimated as £121.1 m. in 1987/88 (Agricultural and Food Research Council 1988). This spending has been the subject of substantial cut-backs as government insists that 'near market' research and much of the extension service is paid for by the industry. Agricultural research and extension services funded by the Federal government in the US amounted to $977 m. in 1983/84. This excludes expenditure by state governments in support of the state university system, much of which retains an important interest in agricultural teaching and research.

A further mechanism for government support for agricultural change is through encouragement for farm amalgamation following the retirement of a number of marginal farmers. This was the main thrust of the Mansholt proposals for the reform of the CAP (Tracy 1982). Commissioner Mansholt suggested in 1968 that there should be a reduction in the number of farmers and a speeding up of the trend to larger farms. Incentives were recommended to encourage older farmers to retire and some 7 per cent of the agricultural land of Western Europe was to be taken out of production and devoted to recreation and forestry. Although these proposals were never implemented as a package, features of the plan survived through Directive 72/160 and Regulation 1096/88 for the cessation of farming. Money has been allocated to encourage the early retirement of farmers and for retraining programmes. The take-up rate for these programmes has been low; only 125 000 holdings had ceased farming by 1986, releasing 1.5 million ha to just under 200 000 other farms (Commission of the European Communities 1988). Significantly only 15 per cent of the gaining farmers under the Directive had approved farm development plans under Directive 72/159. None the less, the number of farms over 1 ha in size has fallen by 47 per cent since 1950 and the average farm size had increased from 15.3 ha of usable agricultural land in 1975 to 17.4 ha by 1985, (EC10). Great differences remain within Western Europe, from an average farm size of 68.9 ha in the UK to only 5.3 in Greece, leaving plenty of scope for further restructuring (see Table 4.1).

Any country's agricultural policy is going to be an amalgamation of many of these alternative instruments. The EC's main, and most expensive, form of intervention is through price control with market intervention and surplus disposal absorbing some 66 per cent of the Community's agricultural budget. This leaves little for all other forms of farmer support. In the US during the 1980s, approximately equal shares of expenditure on agriculture were allocated to consumer subsidies, income and price support schemes and to all other forms of agricultural spending including research, rural development and conservation.

Effects of state intervention

Trying to identify the effects of government intervention in agriculture is confounded by the difficulties in unravelling the 'government impact effects' from other processes of change in agriculture. There are three approaches. The first is to examine agriculture across an international boundary where the environmental and technological factors in agricultural production are similar. This approach has been adopted for the US/Canada border by Reitsma (1986) where policy differences provided part of the explanation for different intensities of farming. A second approach is to attempt to simulate what would happen to agriculture under different scenarios of government inter-vention including none at all (Ray and Heady 1974; Buckwell et al. 1982; Collins et al. 1990). Most simulations of withdrawal of support suggest that prices would fall – in most cases by large amounts. Many farmers would undoubtedly try to react to falling prices by producing more and would increase their investment in machinery and other inputs to the maximum extent which they could finance. Increased production would push prices even lower. In such circumstances only the strongest would survive – and farming units would become very much larger. Some simulations suggest that agriculture would then become even more specialized, with large units concentrating on a reduction of unit production costs. Other simulations suggest that the effects would be to encourage a greater degree of self-reliance within the farm; a reduction in specialization and a greater emphasis on mixed farming may result. The third approach, adopted in the following discussion, examines retrospectively the inputs and outputs of agriculture associated with various policies.

The distribution of financial assistance

State intervention almost always results in the bulk of support going to the larger farmers. If state policy is designed to mitigate the effects of economic forces of supply and demand outlined earlier, it is ironic to find that, although support may be aimed at the small farmers, much more is given to those large producers who probably need it least. The CAP of the EC spends most on supporting prices through intervention buying. As prices are raised for all producers, obviously those who produce most receive most of the subsidy. As one objective of the EC policy was to increase domestic production in order to achieve self-sufficiency, then there is some justification for a concentration of policy support on the largest producers. They are likely to be those most receptive to new ideas and new investment and from whom there will be a rapid response to the support. These are farmers Munton et al. (1988) described as 'accumulators' (see p. 23). On the other hand, there are

258

Table 9.3   Self-sufficiency in EC agricultural products (%)

	1967–71 (EC 7)	1973–74 (EC 10)	1984–85 (EC 10)	1988–89 (EC 12)
Sugar	82	100	101	124
Butter	91	98	134	n.a.
Milk fat	100	n.a.	n.a.	n.a.
Milk powder (skimmed)	n.a.	145	128	n.a.
Milk powder (whole)	n.a.	231	342	n.a.
Cheese	n.a.	103	107	n.a.
Wheat	n.a.	104	129	123
Barley	103	105	124	118
Rye	100	98	113	106
Wine	97	103	100	91
Poultrymeat	101	102	107	105
Beef	90	95	108	103
Eggs	n.a.	100	102	102
Potatoes	n.a.	101	102	101
Pigmeat	n.a.	100	102	103

Source: Based on data in Commission of the European Communities (1989).

other objectives contained within the CAP. A 'fair' income for all farmers is one such objective. At a time when the EC has reached production levels which are at or above self-sufficiency for most products (Table 9.3), there is even less justification for a system of payments which gives the largest share to the largest farmers.

Policies in the US are no more selective. Prices are maintained partly through the system of land withdrawal programmes. The larger farmer withdraws more than his smaller neighbour and in doing so receives a larger share of the support budget. Small, poor and marginal farmers receive little benefit from this massive agricultural programme. In the early 1960s farms classified by the Department of Agriculture as the smallest (annual sales of less than $5 000) accounted for 62 per cent of all farms, 24 per cent of all farm income and 26 per cent of all government payments. In contrast, farms with annual sales of over $40 000 made up only 3 per cent of farms, but 18 per cent of incomes and 15 per cent of government payments (Tarrant 1980). Incomes and support have become even more polarized as the trend to larger and larger farming units has continued. In 1984, 15 per cent of farmers received about 50 per cent of government payments and for the PIK scheme alone, 51 separate businesses each received an annual payment in excess of $1m. In spite of all the costly programmes to defend the viability of the family farm, 60 per cent of US farm income now derives from only 1 per cent of farms – the giant agribusinesses. This concentration of the benefits of a range of price support policies on the largest producers is evident in most countries (Johnson and Short 1983).

Some other forms of support may seem less liable to favour the already well-off large farmers but none the less do so. In addition to the price support functions of the EC's farm budget (EAGGF), the guidance section is designed to supply assistance to farmers to modernize their businesses and to become efficient enough to make a reasonable living. In theory it is the small, traditional farms which should benefit most from this programme. In practice those on larger farms, often better educated and more aware of the opportunities provided by the guidance expenditure schemes, are again the ones to benefit most from it (Commission of the European Communities, 1981). Subsidies on inputs obviously benefit most those who use most, and headage payments – for example on hill cattle or sheep – are worth more to those farmers who have more livestock. Such payments may be regionally specific and may apply in areas where there is a majority of small and marginal farmers – but there are few measures to prevent large payments to the few farmers who do not fit this model

In most cases deficiency payments also favour the large farmers. If farmers are paid a subsidy by government to make up the difference between the market price and some target price, then those who sell most will receive most in government payments. It is only in cases where payments take the form of direct income supplements, payable only to farmers with less than a certain level of income, that government support can be directed to the most marginal of producers. Of course, it can be argued that this support is counterproductive in the long run as it only delays the economic adjustment of the agricultural sector; such marginal farmers should be encouraged to abandon farming altogether, rather than being artificially maintained in the industry.

Policy measures which support the development of fewer, larger farmers in Western Europe were given less prominence in the 1980s (Tracy 1982). The slowdown in economic growth and environmental concerns made further transfers of population from rural to urban areas seem less desirable. Mansholt, reconsidering restructuring policies in 1979, advocated a halt to the increase in farm size (Mansholt 1979). Sweden produced new structural reform guidelines at the same time, aimed at preventing further growth of already adequately large farms (Tracy 1982). A further modification to EC structural reform, introduced by the Directive 797/85, restricted investment aid to farmers whose incomes were below the average gross wage of non-agricultural workers in the region. Such grant-aided investments were not permitted to raise farm income to more than 120 per cent of this reference income. The revised policy also paid attention to environmental concerns by enabling compensation payments to be made following agreements to use less fertilizer, pesticides and herbicides and/or limit stocking density in sensitive environments.

Attempts have begun to emerge to reduce payments to the largest farming units. In the US, for example, commodity and income support are limited to $50 000 per farm and loans to $200 000. These limits represent the realization by the administration that farm payments should be capped and it reflects the concern with the inequities in state payments as well as a need for budget savings. In the EC, in contrast, similar limitations have been contested. For example, the 'MacSharry Proposals' of 1991 suggested ways in which price support levels could be reduced with compensating increases in direct income payments restricted to smaller farms. While this type of policy reform has a history of academic support (Marsh 1987; Harvey 1989), there has been a predictable opposition from politicians and producers in those countries with large-farm structures or intensive farming systems. Their farm sectors have much to lose from any revision of the CAP, while subsidizing the economically inefficient small-farm sector, at the expense of 'efficient' farms, appears to have its basis in social and political rather than economic considerations.

This debate recalls events in 1978 when co-responsibility levies were introduced in the dairy sector. The largest producers were penalized through a levy for the overproduction of milk products within the Community. The political and farm lobby outcry from areas where there were large milk producers, such as the UK, was predictable. As the UK was not individually in surplus for milk products, unlike the EC as a whole, its milk producers felt penalized for being efficient. In this view they were being forced to prop up smaller-scale operations elsewhere in Europe. At the root of this problem is the definition of 'efficiency'. Although the UK dairy sector produces high milk yields per cow, its aggregate productivity, measured in terms of output per unit of land, labour and capital employed, is substantially less than in Denmark, The Netherlands and Belgium (Centre for Agricultural Strategy 1980). For UK farming in general, neither productivity per unit of land nor per unit of labour is the highest in Europe, nor have these measures grown fastest in the UK (Table 9.4). In the event, such levies on milk production

Table 9.4 Land and labour productivity in the EC

Country	Labour productivity			Land productivity		
	000 ECUs per man*	Annual % change		000 ECUs per ha*	Annual % change	
		1973–85	1985–88		1973–85	1985–88
Belgium	61.2	3.4	4.4	4.5	1.7	2.9
Denmark	42.8	4.6	3.8	2.4	5.6	0.9
Greece	8.2	2.9	3.2†	1.4	1.7	1.3†
France	30.6	6.2‡	4.9	1.4	1.7	1.3
Ireland	26.5	5.9	1.9	0.8	1.1	0.7
Italy	18.9	5.0§	5.9†	2.1	0.7§	2.3†
Luxemburg	31.3	5.5	1.7	1.5	0.7	0.1
The Netherlands	54.8	4.8	−2.7	7.8	4.0	1.9
UK	33.4	2.9	1.4	1.1	1.8	0.7
West Germany	27.2	4.4	4.5	2.4	4.6	1.0

* 1989 at 1985 prices.
† 1984–87.
‡ 1977–85.
§ 1980–85.

Source: Based on data in Commission of the European Communities (1989).

were insufficient to reduce production and were replaced by the 1984 quota system. Co-responsibility levies, also introduced for cereals, have done little to reduce production, have been paid for by higher consumer prices (Hubbard 1986), and are being replaced by production control policies such as set-aside.

Such conflicts are inevitable in a Community with an agriculture as diverse as in Western Europe. The very fact of introducing a common policy, with common pricing, to cover agricultural situations as diverse as in Southern Italy and the North German plain or East Anglia, was bound to produce inequalities in its impact. These inequalities are not only between farms of different sizes but also between different agricultural regions. The inequalities between farmers and between regions are mutually reinforcing. For example, EC payments are regionally selective. The origins of the CAP, as a part of a negotiated balance of advantage in Western Europe, particularly between France and Germany, ensured that the crops of northern Europe received comprehensive attention within the CAP design. Cereals, sugar-beet and dairy produce receive the lion's share of the support budget. As northern Europe also has generally larger more prosperous farmers than southern Europe, the regional crop bias compounds the farm size bias in the spatial distribution of farm support spending (Fig. 9.6).

Since 1978 rather more emphasis has been placed on the problem of agriculture in southern Europe. This redirection was given added impetus when Greece began to negotiate membership of the Community, and by an integrated plan covering the period 1985–91 for the Mediterranean, as proposed in 1983 by the Commission. A major aim of this plan was to redress the regional imbalance in CAP expenditure (Clout 1984). Considered too expensive to be implemented in full, parts of this plan subsequently found their way into regionally restricted integrated development programmes. Such elements of a structural policy for agriculture have become to a large extent a means of budgetary redistribution between member states, especially to help Italy and France meet competition first from Greece, and more recently from Spain and Portugal.

The LFA Directive of the EC (75/268) was also designed to offset the regional and crop bias in Community spending by directing payments to particular areas of agricultural and social difficulty within Western Europe (Bowler 1985). With the introduction of Spain, Portugal and Greece, the LFAs now represent 52 per cent of the usable agricultural area of the Community. The objectives were as much social as agricultural, as in the designated areas a profitable agriculture is the main means of maintaining viable communities. The regions selected for special support include areas of poorly drained soils and highlands where there has been substantial rural depopulation. Payments are available to offset high operating costs in such environments. Even in such cases, however, payments are made in accordance with the area farmed and herd sizes. Although the HLCAs have been widely adopted in the UK, only 27 per cent of farms in the LFAs in the Community as a whole receive payments. The truly marginal farmers receive little or no support.

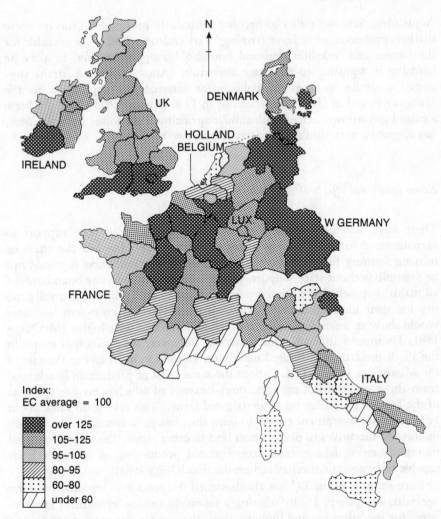

Fig. 9.6 Regional distribution of EC farm support spending, 1976–77 (*Source:* Henry 1981)

Turning now to another context for state intervention, Australian agriculture is crucially dependent on international markets: 97 per cent of the wool, 75 per cent of the sugar, 50 per cent of the beef, 70–80 per cent of the wheat and 25 per cent of the dairy produce is exported. In such circumstances the framework of policy has to be built on an agricultural sector which can survive the ups and downs of the international market. Although there are commodity pricing schemes where minimum prices are set, these are based on moving averages of past prices and are designed to smooth out short-term fluctuations in price rather than to provide long-term protection from the world market. The marginal farmers are encouraged to use the Rural

263

Adjustment Schemes either to become sufficiently profitable to survive these market conditions, or to leave farming. Various loan schemes are available for the farmer and 'rehabilitation and household support funding' to alleviate hardship in adjusting to the latter alternative (Anon. 1988a). Australia suggested a similar package of policies for international application to the Uruguay round of GATT negotiations in 1988 (Anon. 1988b), and the New Zealand government, faced with similar agricultural economic circumstances, has adopted a very similar approach (Lloyd 1987).

Some locational effects of state intervention

There have also been more general locational effects of state support to agriculture. Artificially raising prices or farm receipts has had the effect of moving farming further into the margins of production where it would not be possible without state support. In the US, for example, the boundaries of all major crop belts have moved westwards on to land which is generally too dry for their ideal production. Thus all the major US crop-growing areas would show an increase in yields if precipitation levels were higher (McQuigg 1981; Thompson 1975). Cotton growing, however, shows another trend. In the 1920s the crop was located mainly on 16 million ha in and to the east of the Mississippi valley. Today about the same level of production is achieved from about one-third of the area, three-quarters of which is now to the west of the Mississippi valley on large irrigated farms. This relocation came about as a result of government efforts to limit the growth in cotton production by quotas and incentives to plant cotton land to other crops. The efforts, coupled to rapidly rising labour costs, concentrated production on to better lands capable of large-scale mechanized production (Grigg 1984).

Price support in the EC has also hastened the process of regional and farm specialization (see p. 15) by offering a relatively riskless investment environment for specialization and higher returns than under free market conditions. Regional variations in the agricultural geography of the Community have been deepened (Bowler 1985); milk and cereal production in the UK serve as examples. On milk production, the discussion in Chapter 7 (pp. 183–4 and Fig. 7.8) has already shown the relocation of milk production within England and Wales from eastern to western counties as a result of the uniform pricing of milk. The trend has been reinforced by the interregional trading of milk quotas, while similar relocational effects of dairy quotas have been reported for milk production in Ontario (Bowler 1986b). On cereals, the area devoted to production in eastern England increased steadily between 1951 and 1966, with most of this increase being accounted for by barley production and the wheat share generally falling slowly. The situation remained fairly stable until a renewed upturn in cereal production in the mid-1970s. An increasing share of this cereal area is now devoted to wheat, a change greatly encouraged by high EC support prices (Fig. 9.7). The increase in the cereal production area is not most marked in East Anglia but in the counties of east–central England with a previously mixed farming pattern. Bedfordshire now has a larger share of its agricultural land devoted to cereals than either Norfolk or Suffolk, for example.

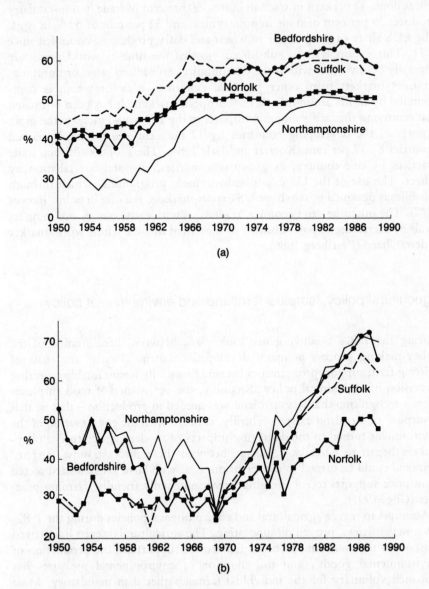

Fig. 9.7 Changes in importance of cereals and wheat in eastern England: (a) percentage of agricultural land devoted to cereals; (b) percentage of cereal area devoted to wheat (*Source:* UK agricultural census, various years)

Trade effects of agricultural policies

Finally there are the externalities of agricultural policies. EC exports, for example, represent 17 per cent of world trade in wheat, 62 per cent of trade in wheat flour, 47 per cent of trade in butter, 50 per cent of trade in non-fat dairy produce, 39 per cent of trade in poultrymeat and 52 per cent of trade in eggs. The EC's share of world trade in wheat and dairy produce has doubled since 1971. This export trade is subsidized most of the time as world prices are generally below those within the Community. Subsidized sales, or dumping, removes markets from other potential exporters, a factor which is compounded by tariffs around the West European markets. It has been estimated that removing the tariffs on grain imports to the EC would increase the grain exports of the developing countries by 12 per cent and of the developed countries by 17 per cent (Koester and Bale 1984). The adoption of such trade practices by one country, or group of countries, encourages retaliation by others. The use of the US export enhancement programme, which through subsidies is designed to win back US export markets, is a case in point (Reeves 1987). The outcome can become a 'trade war' with each country outdoing its rivals by increasing export subsidies to the extent that the whole world market is destabilized (Paarlberg 1986).

Agricultural policy, 'surplus' farmland and environmental policy

During the 1980s a convergence took place between three strands of the policy-making process in many developed countries. Firstly, the costs of existing farm support programmes became financially unsustainable requiring a revision of agricultural policy. Secondly, the persistence of food surpluses drew a recognition that less farmland was needed in production – that is, that a 'surplus' of farmland existed. Thirdly, concern with the conservation of the environment turned to the damaging impacts of modern agriculture. Policy-makers began to perceive a linkage between these three strands: 'surplus' farmland could be turned to conservation uses, with farm subsidies redirected from price supports to encouraging 'environmentally friendly' farming practices (Gilg, 1991).

Attempts to merge agricultural and environmental policies during the 1980s have, nevertheless, proved disappointing. The agricultural interest has proved resistant to the switching of funds from price supports to the production of 'environmental goods', and the adoption of environmental measures has remained voluntary for the individual farmer rather than mandatory. Most environmental measures for agriculture, therefore, appear to be 'bolted on to' existing farm policies which remain essentially unreformed (Robinson 1991). Experience in the UK is instructive.

Starting with the 1981 Wildlife and Countryside Act, farmers were success-ful in acquiring financial compensation in forgoing 'improvements' or devel-opment of their land. For example, farmers forgoing grant aid to plough up (reclaim) moorland were able to negotiate financial compensation from the Nature Conservancy Council. In 1987 the MAFF launched an ALURE (alternative land use and the rural economy) package of proposals which offered the potential for environmental protection (Cloke and McLaughlin 1989). In the event emphasis was placed on alternative uses for farmland rather than with the environment *per se*. More promising, however, have been the 1988 Farm Woodland Scheme (FWS), the 1989 Farm and Conserva-tion Grant Scheme (FCGS), and the 1990 Nitrate Sensitive Areas (NSA) programme. The FWS, for example, aims to bring about 36 000 ha of land under trees by paying a grant per hectare of land so planted. The FCGS, on the other hand, offers grants for pollution control, heather burning, bracken control, and the maintenance of hedges, stone walls, shelter belts, stiles and footbridges. The 10 NSAs so far designated allow farmers to receive payments for restricting their use of fertilizers (basic scheme) or converting from cereal to grassland production (premium scheme). At present 11 000 ha are involved at sites as far apart as Kilham in Humberside and Egford in Somerset (Fig. 9.8).

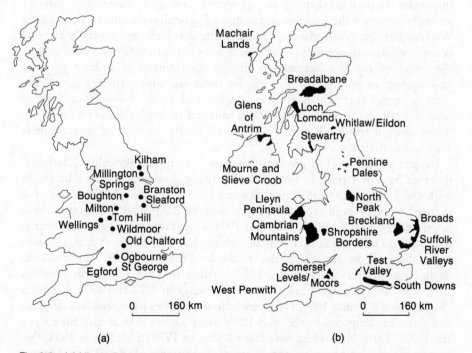

Fig. 9.8   (a) Nitrate Sensitive Areas and (b) Environmentally Sensitive Areas in the UK (*Source:* Robinson 1991: 102)

The environmental content of policies under the CAP has been even more limited. The set-aside programme, for example, has little benefit for the environment despite offering great potential. The Environmentally Sensitive Areas (ESAs) (initially under Regulation 797/85) programme has been applied with varying enthusiasm within the member states, and even within the 17 ESAs of the UK, less than 30 per cent of the eligible agricultural area was receiving grant payments by mid-1990. The grants aim to conserve the environment by paying farmers an annual premium if they follow beneficial farming practices on land under agreement within the designated areas. Once again the programme requires the voluntary collaboration of individual farmers.

At present, therefore, the convergence of agricultural and environmental policies has made early but faltering progress (Potter, 1990), but the inertia of farm policies that have been under development since the 1930s is considerable.

Conclusion

The experience of the last 50 years has shown that the state is one of the most important factors in shaping the geography of agriculture. State policies strongly influence the location and nature of agricultural industry throughout Western Europe, North America and all other developed and centrally planned economies. Because of these connections between agriculture and state policy, the effect of modern agriculture on the environment is at least partly a consequence of policy. The striving for more and more efficient agriculture (that is, agriculture which produces higher and higher yields from units of land and labour) is a process which is initiated through the operation of the 'technological treadmill' and is accelerated by the concern of governments with self-sufficiency.

In spite of this level of government effort to support agriculture, farmers' incomes have not risen in line with this rising government expenditure. In both the UK and the US, farm incomes have fallen sharply from the peak induced by very high world food prices in 1972/73. This falling trend was briefly halted in the UK in 1980 and 1984 when favourable harvests ensured that incomes of cereal farmers remained high; but over all farm types, incomes have fallen and remain low by the standards of the early 1970s. In the US the decline has been even more marked. A rising trend in total farm earnings from 1960 to 1979, accelerated by peak earnings in 1972 and 1973, has been sharply reversed since 1979. Here protectionist policies have been less effective and farm bankruptcies in the mid-1980s were higher than at any time since the 1930s. Farm borrowing rose from $50bn in 1970 to $215bn in 1984; this borrowing was more than the overseas borrowing of Brazil and Mexico together in that year.

At best, therefore, massive government expenditure on agriculture has slowed down the economic processes, the consequences of which it was designed to mitigate. In so doing it has concentrated the benefits in the hands of a small minority of large farmers. As the costs of support policies continue to rise, cut-backs in price subsidies which are being enforced will, unless the remaining support is targeted carefully, produce an acceleration in the very processes that the government policies have tried to prevent for the last 50 years as the largest farms are best able to cope with the harder economic environment (Munton et al. 1988).

One of the most abiding difficulties with agricultural policies has proved to be their inertia. Policies established in the economic situation of the 1930s have often proved all but impossible to dismantle or seriously modify in the light of different economic circumstances. The legislators of the 1930s in Western Europe and the US would recognize the fundamentals of their agricultural policies still enacted today. There are practical and political reasons for this inertia. The Department of Agriculture in the US has become one of the largest instruments of government with thousands of different programmes being administered by a host of staff in Washington and scattered all over the US. Legislation has tended to add programmes rather than modify or rethink the whole strategy – thus innovations such as the food stamp programme are added to existing structures. To radically reformulate agricultural policy in the US would be a monumental task and, with so many vested interests at stake, probably politically impossible.

Similarly in Europe, the agricultural programme so dominates the EC that major reform of procedures and programmes is made very difficult. Over 65 per cent of Community income is still spent on farm support policies compared with, for example, less than 10 per cent on regional policy and 5 per cent for overseas aid. Policies adopted at a time when security of food supplies was of the greatest importance, and directed to achieve food self-sufficiency, are very difficult to alter when that food self-sufficiency has been reached or surpassed. The introduction of milk quotas was a recognition that, once production has been increased to meet the market demand, other forms of regulation have to be applied. Indeed the level of price support has been allowed to fall in recent years with price rises at less than the rate of inflation. In addition 'stabilizers', or upper limits on the total amount of farm produce that will be subsidized at the full amount, have been introduced for the main crops and livestock. While falling levels of price support have depressed farm incomes, the impact on overproduction for the market has yet to become apparent.

Finally, agricultural policies have a number of economic effects external to the farm sector. For example, the taxpayers' costs of the CAP are not shared equally between member countries nor equally in terms of the benefits of the policy. This is because of the differing importance of agriculture to the GNP of member countries and the way the Community is financed. Revenue is collected from a levy on value added tax and from taxes on agricultural imports which are in competition with European production. Thus the UK, with a high VAT turnover, a high food import bill and an agriculture which contributes little to the GNP, is a large contributor and receives relatively little from the CAP budget. Germany is a high contributor on the basis of VAT receipts but imports less food from outside the Community. Thus inequalities are inherent in the policy between farmers, between regions and between countries.

Table 9.5   Changes in economic welfare under alternatives to the CAP, 1980 (10 million EUA)*

Country	Self-sufficiency†				Free market†			
	Producers	Consumers	Taxpayers	Total	Producers	Consumers	Taxpayers	Total
West Germany	−25.5	24.5	18.5	17.6	−65.0	90.2	27.1	52.3
France	−23.9	14.9	13.9	4.9	−52.0	53.7	20.4	22.1
Italy	−7.7	8.8	6.2	7.3	−25.4	38.6	9.0	22.2
The Netherlands	−11.3	3.2	3.4	−4.6	−22.1	11.5	5.0	−5.7
Belgium/ Luxemburg	−5.4	3.3	2.7	0.6	−11.7	10.3	3.9	2.6
UK	−12.6	15.4	9.8	12.7	−24.8	37.2	14.3	26.6
Ireland	−4.8	1.0	0.5	−3.3	−6.9	2.3	0.7	−3.9
Denmark	−6.1	1.2	1.5	−3.5	−12.5	4.6	2.2	−5.8
EEC-9	−97.3	72.3	45.4	31.7	−103.4	248.4	82.6	110.4

* EUA: European Unit of Account or European Currency Unit (ECU).
† Tabulated values for producers, consumers and taxpayers

Source: Buckwell et al. (1982).

Just how the advantages and disadvantages of the present EC agricultural policies are shared out between countries and between producers and consumers can be indicated by considering the changes which would spring from reducing protection in Western Europe (Table 9.5). The greatest gains would be shared by consumers in Germany and the UK, while the greatest losses would be felt by producers in Germany and France.

The externalities of the CAP also lead to income transfers from consumers and taxpayers in the EC to consumers in many Third World, and centrally planned, economies who benefit from the subsidized export trade that the Community is forced to adopt. Except in conditions of natural disaster, income transfers from the developed economies to the Third World would be better directed towards the rural producer rather than the urban consumer. Not only does the vast majority of EC food aid go to urban populations (Tarrant 1982), but other EC agricultural policies act against Third World producers. Trade agreements exist between the EC and many developing countries but imports of agricultural raw materials in direct competition with domestic production of the EC are subjected to tariffs and quotas (Harris 1987). By denying access to markets in Europe, income transfers take place from Third World producers to EC producers. As the EC import restrictions are applied mainly to products of temperate agriculture, they are felt particularly by countries wishing to export sugar (Grigg 1984), citrus fruits, cereals (including rice) and meat.

References

Agricultural and Food Research Council (1987) *Agricultural and Food Research Council: Annual Report 1986–87.* AFRC, London

Agricultural and Food Research Council (1988) *Agricultural and Food Research Council: Corporate Plan 1988–93.* AFRC, London

Andrlik, E. (1981) The farmers and the state: agricultural interests in West German politics, *West European Politics* **1**: 104–19

Anon (1988a) Resolving the world agricultural crisis. *World Farming Times* **5**: 9–20

Anon (1988b) World farm policy: an overview. *World Farming Times* **5**: 23–4

Bowers, J. K. and **Cheshire, P.** (1983) *Agriculture, the countryside and landuse: an economic critique.* Methuen, London

Bowler, I. R. (1979) *Government and agriculture: a spatial perspective.* Longman, London

Bowler, I. R. (1985) *Agriculture under the Common Agricultural Policy.* Manchester University Press, Manchester

Bowler, I. R. (1986a) Government agricultural policies. In Pacione, M. (ed) *Progress in agricultural geography.* Croom Helm, London, pp 124–48

Bowler, I. R. (1986b) Direct supply control in agriculture: experience in Western Europe and North America. *Journal of Rural Studies* **2**: 19–30

Buckwell, A. E. (1986) What is a setaside? *Ecos* **7**: 6–11

Buckwell, A. E., Harvey, D. R., Thompson, M. J. and **Parton, K. A.** (1982) *The costs of the common agricultural policy.* Croom Helm, London

Centre for Agricultural Strategy (1980) *The efficiency of British agriculture.* CAS report 7, University of Reading, Reading

Centre for Agricultural Strategy (1986) *Landuse alternatives for UK agriculture* CAS report 12, University of Reading, Reading

Cloke, P. J. (1989) State deregulation and New Zealand's agricultural sector. *Sociologia Ruralis* **29**: 34–48

Cloke, P. J. and **McLaughlin, B.** (1989) Politics of the alternative land use and rural economy (ALURE) proposals in the UK: crossroads or blind alley? *Land Use Policy* **6**: 235–48

Clout, H. (1984) *A rural policy for the EEC?* Methuen, London

Collins, N., Bradbury, I. K. and **Charlesworth, A.** (1990) Formulation of the European Community price review: models of change. *Journal of Rural Studies* **6**: 163–73

Commission of the European Communities (1981) *The regions of Europe:*

first periodic report on the social and economic situation of the regions of the Community. The Commission, COM(80) 816 Final, Brussels

Commission of the European Communities 1988 *The agricultural situation in the Community. 1987 report.* Office for Official Publications of the European Communities, Luxemburg

Commission of the European Communities (1990) *The agricultural situation in the Community. 1989 report.* Office for Official Publications of the European Communities, Luxemburg

The Economist (1986) Only the Russians are smiling. *The Economist* 16 Aug: 43

Ervin, D. E. (1988) Set aside programmes: using US experience to evaluate United Kingdom proposals. *Journal of Rural Studies* **4**: 181–91

FAO (1989) *Annual Production Yearbook.* FAO, Rome

Fennell, R. (1979) *The common agricultural policy of the European Community.* Granada, London

Gilg, A. W. (1991) Planning for agriculture: the growing case for a conservation component. *Geoforum* **22**: 75–9

Grigg, D. (1984) *An introduction to agricultural geography.* Hutchinson, London

Halcrow, H. G. (1977) *Food policy for America.* McGraw-Hill, New York

Harris, S. (1987) Current issues in the world sugar economy. *Food Policy* **12**: 127–45

Harvey, D. (1989) Alternatives to present price policies for the CAP. *European Review of Agricultural Economics* **16**: 83–112

Hayami, Y. and **Ruttan, V. W.** (1971) *Agricultural development: an international perspective.* Johns Hopkins University Press, Baltimore

Hedlund, S. (1984) *Crisis in Soviet agriculture.* Croom Helm, London

Hendriks, G. (1987) The politics of food: the case of FR Germany. *Food Policy* **12**: 35–45

Henry, P. (1981) *Study of the regional impact of the common agricultural policy.* Studies Collection, Regional Policy Series 21, Brussels

Howarth, R. W. (1985) *Farming for farmers.* Institute for Economic Affairs, London

Hubbard, L. J. (1986) The co-responsibility levy – a misnomer? *Food Policy* **11**: 197–201

Ilbery, B. W. (1990) Adoption of the arable set-aside scheme in England. *Geography* **75**: 69–73

Johnson, D. G. and **Brooks, K.** (1983) *Prospects for Soviet agriculture in the 1980s.* Indiana University Press, Bloomington

Johnson, G. L. (1972) Theoretical considerations. In Johnson, G. L. and Quance, C. L. (eds) *The overproduction trap in US agriculture.* Johns Hopkins University Press, Baltimore

Johnson, J. D. and **Short, S. D.** (1983) Commodity programmes, who has received the benefits? *American Journal of Agricultural Economics* **65**: 912–21

Johnson, T. and **Smit, B.** (1985) An evaluation of the rationale for farmland policy in Ontario. *Land Use Policy* **2**: 225–37

Jones, A. (1991) The impact of the EC's set-aside programme. *Land Use Policy* **8**: 108–24

Josling, T. (1974) Agricultural policies in developed countries: a review. *Journal of Agricultural Economics* **25**: 229–64

Koester, U. and **Bale, M. D.** (1984) *The common agricultural policy of the European Community.* World Bank Staff Working Paper 630, Washington DC

Lane, S. (1978) Food distribution and food stamp program effects on food consumption and nutritional achievement of low income persons in Kein County, California. *American Journal of Agricultural Economics* **60**: 108–16

Le Heron, R. (1988) State, economy and crisis in New Zealand in the 1980s: implications for land-based production of a new mode of regulation. *Applied Geography* **8**: 273–90

Lloyd, A. G. (1987) The Australia–New Zealand farm problem and the appropriate role of government. *Australian Economic Review* **3**: 3–20

Lockeretz, W. (1988) Agricultural diversification by crop introduction: the US experience with the soybean. *Food Policy* **13**: 154–66

McQuigg, J. D. (1981) Climatic variability and crop yields in high and low temperature regions. In Bach, W., Pankrath, J. and Schneider, J. H. (eds) *Food–climate interactions.* Reidel, Boston

MAFF (1975) *Food from our own resources.* HMSO, London

MAFF (1985) *Woodland as a farm crop.* HMSO, London

MAFF (1987) *Annual review of agriculture.* HMSO, London

Mansholt, S. (1979) *The Common Agricultural Policy: some new thinking.* The Soil Association, London

Marris, J. and **Hess, T. M.** (1986) Farmer uptake of agricultural land drainage benefits. *Environment and Planning A* **18**: 1649–64

Marsh, J. S. (1987) Alternative policies for agriculture in Europe. *European Review of Agricultural Economics* **14**: 11–21

Morris, C. N. (1980) The common agricultural policy. *Fiscal Studies* **1**: 17–35

Mumford, L. (1961) *The city in history; its origins, its transformations and its prospects.* Harcourt, New York

Munton, R., Eldon, J. and **Marsden, T.** (1988) Farmers' response to an uncertain policy future. In Baldock, D. (ed) *Removing land from agriculture.* Council for the Protection of Rural England, London

Nove, A. (1982) Soviet agriculture: new data. *Soviet Studies* **32**: 366–78

Paarlberg, R. L. (1986) Responding to the CAP: alternative strategies for the USA. *Food Policy* **11**: 157–73

Potter, C. (1990) Conservation under a European farm survival policy. *Journal of Rural Studies* **6**: 1–7

Ray, D. E. and **Heady, O. E.** (1974) *Simulated effects of alternative policy and economic environments on US agriculture.* Center for Agricultural and Rural Development, Report 46T, Iowa State University, Ames, Iowa

Reeves, G. W. (1987) World agricultural trade and the new GATT round. *Journal of Agricultural Economics* **38**: 393–405

Reitsma, H. A. (1986) Agricultural transboundary differences in the Okanagan region. *Journal of Rural Studies* **2**: 53–62

Robinson, G. (1991) EC agricultural policy and the environment: land use implications in the UK. *Land Use Policy* **8**: 95–107

Ruttan, V. W. (1980) Agricultural research and the future of American agriculture. In Batie, B. S. and Heady, O. E. (eds) *The future of American agriculture as a strategic resource*. The Conservation Foundation, Washington DC

Senior Nello, S. (1989) European interest groups and the CAP. *Food Policy* **14**: 101–6

Singer, H., Wood, J. and **Jennings, A.** (1987) *Food aid: the challenge and the opportunity*. Oxford University Press, Oxford

Spitz, R. G. (1987) The evolution and implications of the US Food Security Act of 1987. *Agricultural Economics* **1**: 175–90

Steinbeck, J. (1962) *The grapes of wrath*. Heinemann, London

Swader, F. N. (1980) Soil productivity and the future of American agriculture. In Batie, B. S. and Heady, O. E. (eds) *The future of American agriculture as a strategic resource*. The Conservation Foundation, Washington DC

Talbot, R. (1977) The three US food policies. *Food policy* **2**: 3–16

Tarrant, J. R. (1980) *Food policies*. Wiley, London

Tarrant, J. R. (1982) EEC food aid. *Applied Geography* **2**: 127–41

Tarrant, J. R. (1985) A review of international food trade. *Progress in Human Geography* **9**: 235–54

Thompson, L. M. (1975) Weather variability, climatic change and grain production. *Science* **188**: 535–41

Tracy, M. (1982) *Agriculture in Western Europe: challenge and response* 2nd edn. Granada, London

Tracy, M. (1987) *Structural policy under the CAP*. Arkleton Trust, Langholm, Scotland

Wathern, P., Young, S. N., Brown, I. W. and **Roberts, D. A.** (1986) Implementation of the EEC LFA policy and upland landuse in the UK. *Land Use Policy* **3**: 205–12

Wathern, P., Young, S. N., Brown, I. W. and **Roberts, D. A.** (1988) Recent upland landuse changes and agricultural policy in Clwyd, North Wales. *Applied Geography* **8**: 147–63

Wilson, G. K. (1977) *Special interests and policy making; agricultural policies and politics in Britain and the United States of America 1956–1970*. Wiley, London

World Bank (1991) *World development report*. Oxford University Press, Oxford

10

Farming at the urban fringe

Christopher Bryant

In this final chapter attention is turned to one particular context for farming, namely those urban fringe zones that surround large urban concentrations (also called metropolitan regions). The aim is to bring together a number of themes that have been introduced in earlier chapters, and to discuss the ways in which they interact in one farming context. Urban fringe areas have been chosen because processes of urbanization are increasingly dominant in developed countries, while the effects of urbanization are most clearly demonstrated at the urban fringe. Even so, we can anticipate the influence of urbanization continuing to extend further from our metropolitan areas with the passage of time.

Urbanization, associated with the process of economic development, has been one of the most general trends in society in the last half-century. It influences many facets of modern life and is accompanied by changes that have had far-reaching effects upon agriculture and the agricultural production system, including its support services and industries. This simple stimulus–response interpretation, of course, should not be taken as acceptance of the view that agricultural change is only reactive in relation to changes in the broader socio-economic system. Certain types of agricultural change originate from within the agricultural sector. For example, the development of novel types of farm enterprises can create markets and are not always a reaction to them – for instance, the development of kiwi fruit production in New Zealand in the 1960s and 1970s – and the development of new technologies of production. Furthermore, in some respects agricultural activity may influence certain aspects of the urbanization process; for example, particular farm production systems associated with individual types of landholding structure (intensive market gardening) may influence the form and spatial structure of the urban development process. Consequently, we must recognize that there is a two-way interaction between agriculture and the non-farm sectors of the socio-economic system.

This is complicated further, even within the Western capitalist context, because there are important differences in the socio-economic organization of production that originate in variations in historical development and in the relationships with the urban–industrial complex. Thus, for example, the agricultural production system in one region may differ from another in terms of the relative importance of capitalistic farm units, family farm-type units and units with vestiges of ties to a more subsistence-oriented past. To the extent that these socio-economic modes of production differ in terms of their technical, financial, economic and social characteristics and the driving force at the individual decision-making level, then the nature of the interaction between urbanization and agriculture may differ. Furthermore, at the individual farm level within a given socio-economic mode of production, many factors may influence one farmer to evaluate opportunities and resources, and choose strategies, differently from another. Indeed, in Chapter 6 an emphasis was placed on the relative importance of economic versus non-economic goals, financial resources, managerial abilities and commitment to farming as potential influences on decision-making that may vary between farmers.

The urbanized world

Urbanization has been one of the most all-pervasive trends in the twentieth century in Western society. It is a social and economic process with significant political ramifications. One of its principal manifestations has been an increasing proportion of a country's population living in urban areas and, more generally, urban-centred regions or metropolitan zones of influence (Bryant et al. 1982). It is much more, however, than the mere outward expansion of cities and the spread of non-farm land-use activities and functions into the surrounding countryside. Intertwined with the general process of economic development, it gathered momentum with the transformation of economies following the Industrial Revolution, first from a primary base towards secondary activities and, subsequently, towards an increasing emphasis on tertiary and quaternary activities. This transformation of the organization of production has been characterized both by growth in non-farm sectors and by reorganization of primary, especially agricultural, production.

The principal demographic manifestation was the rural to urban exodus, a phenomenon that has decimated the populations of many agricultural regions since the middle of the nineteenth century in Europe. Even in more recently settled areas, for example in many parts of North America including the Canadian Prairies and eastern Canada and in Australia, this process has taken its toll. In the Maritime provinces in eastern Canada, almost as soon as agricultural settlement reached its maximum extent in the early 1900s, the farm population began a precipitous decline. Underlying the massive movements involved were perceived and real differences in life-style, standards of living and income-earning opportunities that appeared to favour urban areas.

More recently, since the Second World War in most countries, other life-style considerations have led many people to leave the built-up urban environment with its congestion and lack of natural open space to live in the countryside or in the small towns and villages surrounding our major urban areas. It is not, however, that 'urbanization' has stopped, for the structural transformation of production continues. It is simply that the geographic manifestation of the process has changed. The evolution of transportation and communications technology – the automobile, the phone, the mass media, computer-based data transfer – have permitted some people to maintain all the important ties with the 'urban' environment while enjoying the qualities of small town or country living. Extending this argument, it is a simple step to see how few areas in the developed world lie beyond the influence of the life-styles and values that have come to be associated with urbanization. Urbanization of society can no longer be conceived in terms of the built-up core and the immediate urban fringe. Even the more extensive geographic area of the regional city with its built-up core and surrounding countryside, towns and villages tied together into a functioning whole by flows such as commuting, recreational travel, messages and goods, is a narrow expression of the broad influences of urbanization.

What then are the urbanization influences that impinge upon farming? One framework sees the urbanization process as creating a set of stimuli that may elicit a response from the agricultural sector (e.g. Bryant et al. 1982). Specifically, the urbanization process can be associated with:

1. An increase in the demand for labour, because of the link with growth in employment opportunities in the urban–industrial complex;
2. An increase in the demand for land for non-agricultural functions, as urban areas and regions have expanded;
3. A concentration in the demand for agricultural products and services, because of the population-concentrating effects of the urbanization process at the regional scale.

In Fig. 10.1 these three stimuli are labelled 'metropolitan forces' because they are particularly associated with the growth of the larger metropolitan areas. In this context, other forces that influence agriculture are labelled 'non-metropolitan forces' and have been discussed in other chapters of this book. As is readily apparent, however, some of the latter have been strongly interrelated with the general processes of economic development and urbanization. For instance, changes – and specifically increases – in living standards are also part of the mechanism that has fuelled the movement of labour from agriculture to the non-farm sectors and urban regions.

Indeed, it is important to recognize that both metropolitan and non-metropolitan forces impinge upon agriculture simultaneously, and they can combine to produce very different situations for farming and agricultural development. In addition, the particular configuration of different regional environments, in terms of their biophysical, economic and cultural dimensions, provide different contexts within which the different forces play themselves out. At the micro-level, farm-level factors influence the ways in

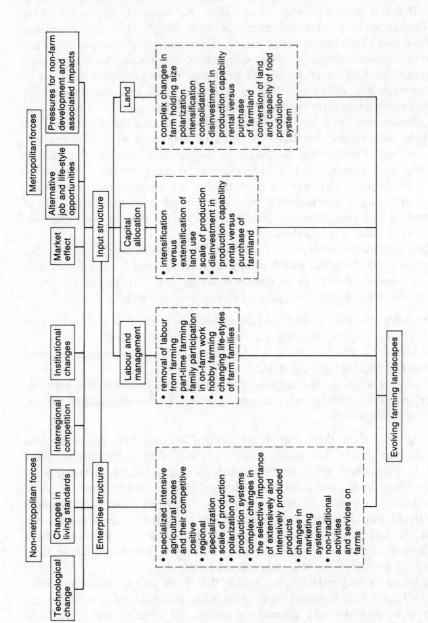

Fig. 10.1 External forces of change and selected farming responses

which individual decision–makers are able to respond to different situations (Fig. 10.2).

The importance of the regional environment is multifaceted. First, different physical environments influence the range of agricultural production possibilities; given some regional specialization, agricultural structures around cities, as elsewhere, respond to enterprise-specific forces (e.g. broad market changes). Second, various agricultural technologies have been adapted to certain enterprises and biophysical environments. In Canada, in the agricultural environments of south-west Ontario and the Prairies, the pull of agricultural labour into non-farm employment, together with technological change, gave rise to farm consolidation and mechanization; but in the Maritimes, with generally a much poorer biophysical environment for agriculture, the effects have often been the abandonment, rather than a restructuring, of agriculture.

From the perspective of agricultural development and progress, the different forces may either be complementary or conflict to produce an environment fraught with difficulties for farming, or combine to produce particular forms of agriculture which are neither unambiguously negative or positive, for

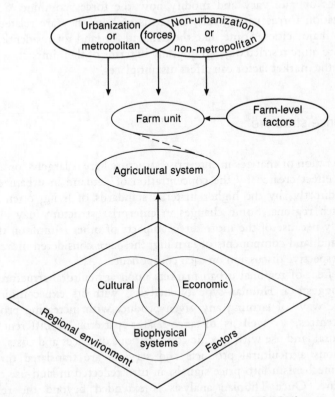

Fig. 10.2 Factors influencing farm change (*Source:* After Bryant 1984: 311)

example hobby farming (Bryant et al. 1984). Clearly, in terms of evaluating the response of farming to these various situations, it is not simply a question of looking for growth or decline in agriculture, but also for different types of structural transformation of the farming system.

Farming's response

Some farming responses to the range of forces influencing agriculture are summarized in Fig. 10.1. Earlier chapters of the book have dealt with farming responses to non-metropolitan forces in terms of modification to four major components of the farm system, namely enterprise structure (Chapter 6), labour, management and capital (Chapter 3), and the land resource (Chapter 8). In this chapter attention is given to farming decisions affecting these components related to the metropolitan forces influencing agriculture. It is appropriate to emphasize again that the various urbanization-generated stimuli do not play themselves out in the same geographic arena, although there is overlap. First, as noted above, the regional environments in which the processes operate vary and modify how the forces combine. Second, the urbanization forces vary greatly in terms of how they are related distance-wise to major cities. Some, like the demand for land for residential development, are quite restricted geographically in terms of their impact, while others such as the market factor can affect distant places.

Enterprise structure

Consideration of changes in enterprise structure here is largely confined to the market effect created by the concentration of demand in urbanized regions and, indirectly, by the higher material standards of living often associated with such regions. Some changes in enterprise structure may also be felt indirectly because of the more direct impacts of other stimuli on the labour, capital and land components of farming; these are considered more explicitly in the respective discussion on each component.

The effect of regional urban markets upon agricultural structure immediately suggests a Thünian-type of analysis, with its expectation of more intensive ways of farming any given product with increasing proximity to urban centres, as well as distance-related patterns of different types of agricultural land use with specific combinations of prices and costs. Demands for various agricultural products and services are translated through the market mechanism into price signals, in turn reflected in land-use changes in agriculture. Once Thünian analysis is extended beyond the relationship between distance to market and the comparative advantage this gives to different locations in the production of different commodities, to include geographic variations in productivity relationships and interregional trade, it is clear that it is not only those agricultural areas close to major urban markets that benefit from urban demands (Gregor 1982).

Many specialized intensive agricultural areas, for example vegetable, fruit and milk production zones, and even vineyard areas, developed at an early date around cities to serve the urban market. Examples include the orchard areas west and north of Paris (France) (AREEAR 1976), market gardening around Adelaide (Australia) (Smith 1972), the varied intensive agriculture around Auckland (New Zealand) (Moran 1979) and dairying near Los Angeles (USA) (Gregor 1982). However, the relative spatial monopoly, once conferred on these agricultural zones, has been diminished by improvements in transportation and food-handling and processing technology. One of the problems facing the Niagara Fruit Belt in southern Ontario, for instance, has been competition from producing areas in the US, better placed in terms of production costs; and intensive vegetable production in the Lower Fraser Valley in British Columbia has had to contend with stiff competition from Mexican produce. In Western Europe, even before the creation and development of the European Community, some specialized agricultural areas near large cities had experienced severe competition following developments in rail and, subsequently, truck transportation. Vegetable areas near Paris suffered early on from competition from Mediterranean areas. As an indication of the change involved, the Paris region supplied 81 per cent of its own vegetable and fruit requirements by tonnage in 1895 but only 46 per cent by 1950 (Bryant 1984). These types of problems have become exacerbated with competition from other European countries as the EC has grown larger.

Another related consequence of declining transportation costs has been the tendency, noted by several observers (Morgan and Munton 1971; Grigg 1984), for the location of agricultural production to reflect physical and economic productivity patterns more closely. Increasing regional specialization has thus been suggested (see p. 15). Wong (1983), for instance, demonstrates how this process has influenced changing agricultural patterns in Canada from 1941 to 1976 in areas both adjacent to and remote from urban areas alike. Thus, the difficulties encountered by some urban fringe agricultural areas may be accounted for, partially at least, by changing patterns of competition at the broad regional scale, and attempts by local producer groups to develop customer loyalty through promoting the quality and freshness of local produce must be seen in this context.

None the less, there are still advantages for certain types of farm enterprise in a location close to the market. For instance, Lawrence's (1988) study of 15 US metropolitan counties shows that horticultural speciality crops and vegetables increased substantially in terms of sales value from 1949 to 1982 (Table 10.1). Similarly, in southern California, the most heavily urbanized areas have continued to experience increases in the intensity (expressed as sales value per unit area) and farm size (in terms of sales values and farm area)

Table 10.1   Changing agricultural production chracteristics in 15 US metropolitan counties*

Farm enterprise	1949–82 % change in:			Levels of intensity in metropolitan countries compared to all US agriculture	
	Area or animals on census farms	Sales value†	Intensity‡	1949	1982
Horticultural specialities	4	138	196	1.50	1.63
Field crops	−38	58§	n.a.	n.a.	n.a.
Other crops	n.a.	n.a.	234¶	1.35	1.37
Vegetables	−59	22	233	1.65	1.23
Fruits and nuts	−67	−42	152	1.41	0.95
Poultry	−73	−66	153‖	1.88	1.27
Dairy	−86	−73	176	2.23	1.09
Total land in farms	−50	n.a.	n.a.	n.a.	n.a.
Total harvested cropland	−45	n.a.	n.a.	n.a.	n.a.
All crops	n.a.	n.a.	243	3.06	3.08

* The metropolitan counties (and their respective metropolitan areas) are: Queens, Nassau, Suffolk (New York); Los Angeles (Los Angeles); Cook, DuPage, Kane, Lake (Chicago); Dallas, Tarrant (Dallas – Fort Worth); Baltimore City and County (Baltimore); Multnomak, Clackamas, Washington (Portland); Polk (Des Moines).
† All sales are converted to 1967 dollars using the Producer Price Index for Farm Products.
‡ Calculated as the 1967 dollar sales per unit area or per producing animal.
§ Corn, wheat and hay.
¶ All crops less horticulture, vegetables and fruits and nuts.
‖ Calculated for each year as the total poultry sales divided by number of chickens.

Source: Adapted and compiled from Lawrence (1988: Tables 1–4).

(Gregor 1988) (Table 10.2). Furthermore, the seasonality of agricultural production still means that locally produced products may maintain a certain monopoly on local and regional markets, despite higher production costs. Good examples are the production of peaches in the Niagara Fruit Belt in southern Ontario and the continued, though declining, fruit areas in the Paris region. Developments in storage technology help adjust the flow of production to the market to optimize such advantages, but can equally, however, be used to the disadvantage of local production.

Proximity to large urban markets, despite general decreases in transportation costs, also remains important for some farming enterprises in other ways. The size of many metropolitan markets has unquestionably encouraged the development of large-scale, specialized and land-intensive agricultural operations in proximity to the markets, for example broiler operations, intensive pig production, beef feed-lot operations and dry-lot dairying in many metropolitan regions. Such land-intensive, industrialized operations are also encouraged by higher land prices to some extent, although such operations are not confined to the urban fringe as Gregor (1982) has demonstrated for the US. But the sheer volume of production of the larger operations makes a market location attractive because it reinforces economies of large-scale

Table 10.2 Changing patterns of agricultural structure and intensity, southern California, 1950–80

(a)

Counties	% urban population 1980	% change 1950 to 1980 in:		
		Population	Value of land and buildings/ area*	Value of crops sold/unit area harvested
Los Angeles	98.9	80.1	72.2	469.8
Orange	99.7	793.8	118.5	1 039.2
San Diego	93.2	234.4	711.1	699.1
San Bernardino	90.1	217.8	−15.7	72.1
Riverside	82.4	290.0	730.1	237.9
Ventura	94.6	361.6	267.5	341.8
Santa Barbara	90.8	204.1	441.1	308.1
California	91.3	123.6	n.a.	204.8

(b)

Counties	% change 1950 to 1980 in:		
	Value produced per farm†	Share of farms >260 acres‡	Share of farms selling >$250 000
Los Angeles	363.7	97.8	3.8
Orange	1 373.2	220.5	3.7
San Diego	468.5	−40.5	6.7
San Bernardino	1 133.5	81.4	3.7
Riverside	569.7	−4.5	4.1
Ventura	256.5	−20.2	3.4
Santa Barbara	153.8	−4.0	11.2
California	449.5	27.3	6.8

* Constant 1967 dollars.
† Value of products produced on farm minus value of products bought in but produced on other farms.
‡ Excluding farms of < 10 acres.

Source: Compiled from Gregor (1988; Tables 1–3, 12 and 15).

production. This undoubtedly underlies Wong's (1983) observations, for the regions centred on Canada's 23 census metropolitan areas, that the scale of operations of pig, poultry and dairy farms was greater in those regions, in practically all provinces and census years from 1914 to 1976, than for the farming areas outside those metropolitan regions.

Similarly, products which are very bulky, low in value and potentially perishable remain sensitive to transport costs. One of the best examples in North America is sod, or turf, farming which is carried on around cities in central Canada and the US. This activity is largely oriented to the urban development market and the demands for an 'instant environment' by people buying new homes in suburbia, especially in the 1960s and 1970s, reflecting the high material standards of living in these urban areas. The product is bulky, low in value and cannot be stored for long without deterioration; thus, it is rare for transportation of turf to exceed 40 or 50 km.

Market proximity confers advantages for some enterprises in terms of providing opportunities for short-circuiting 'normal' marketing channels through the development of certain types of direct sale commercialization (Moran 1979; Johnson et al. 1987). These were dealt with more fully in Chapter 7 and include large-volume sales direct to supermarkets, factories, schools and other 'collective' customers, as well as direct sales from the farmer to the individual consumer, such as in farm markets, on-farm sales (e.g. farmshops) and the tremendous variety of different types of pick-your-own sales (Thomson 1981; Laureau 1983; Johnson et al. 1987; Smith 1987).

Finally, while not involving agricultural produce directly, other opportunities are presented by the urban market and are very much linked in one way or another to the recreational demand of the essentially urban-based or oriented population. First, certain types of non-traditional agricultural services (discussed as non-farming diversification on p. 156) can be performed by farmers, for example the boarding of horses and even the development of riding stables ('horsiculture'). Clearly demands are placed by such activities on agricultural production and land use in terms of feed and pasture requirements. Of course, it is not only the farmer who develops these enterprises; just as frequently, and probably much more so, it is from among the heterogeneous group of 'exurbanites' that the owners and managers of riding stables are found. Second, there are opportunities for farmers to perform non-agricultural services, for example running a 'vacation farm', offering farmhouse accommodation, or renting out an agriculturally obsolete barn for indoor parking of recreational vehicles (for example, snowmobiles, camper vans and boats in the North American context) and long-term storage generally. Evidence of the magnitude of participation by farmers in offering these services near cities has been provided by Ilbery (1991) in the context of Birmingham's urban fringe. Earlier, in a survey of 1983–84 near Paris, a similar wide range of adjustments on farms were noted – the development of garden plots for rental to nearby urban residents, the rental of indoor barn space for vehicle storage, the conversion of a disused chicken shed to provide locked storage space and the creation of a tennis club, to name but a few (Bryant 1984). All these activities involve an explicit integration, or increasing contact, between farming and the non-farm population at the farm level. Furthermore, the opportunities from an entrepreneurial perspective may provide important sources of supplementary income to the farm family, even though such developments are often frowned upon in areas where conservation of the agricultural land resource is identified with maintaining a separation between traditional agricultural and non-agricultural activities.

Overall, then, the stimulus of the large urban market created by the urbanization process has presented many opportunities for farm production. The sheer size of many market concentrations, coupled with the accompanying higher incomes, has an inevitable influence on the enterprise structure of surrounding agricultural regions. Certain advantages of the near-urban location have been lost, however, following developments in transportation technology. Thus, the effects of a growing and changing urban market can be felt in many different regions. At the same time, certain advantages remain and these have been bolstered by increasing involvement of the urban population in making recreational weekend and day trips to the countryside and combining these with purchases of farm produce. The same phenomenon can affect vacation areas or areas attractive to retired people. Finally, it is important to note that within the areas surrounding cities, all this is taking place within a general regional agricultural environment which is oriented to markets much broader than the regional urban market. The picture that thus emerges from the perspective of enterprise structure is a complex, layered one, with some sectors developing and responding to national and even international forces, while others are more closely dependent upon the regional urban market.

Labour and management

Urbanization stimuli influence labour and management in agriculture in at least three ways. First, growth in employment opportunities in the non-farm sector may provide incentives for full-time or part-time movement of farm operators, family and hired labour out of agriculture when the gap between relative earnings and non-farm employment – that is, the opportunity cost of remaining in agriculture – becomes too high. This is reinforced by considerations of income stability and life-style as well as by the general erosion of farm profit margins through the cost–price squeeze. Second, the demand for exurban or countryside living encompasses part of the phenomenon of part-time and hobby farming, combining residential, leisure-time and agricultural production objectives. This contributes, therefore, to a change in the organization of agricultural production in an area. These first two processes underlie the rapid shift in the proportion of farm and non-farm populations in areas near cities, a good example of which is shown in Fig. 10.3. Third, life-style considerations may lead to changes in farm production systems as some segments of the farming population develop 'urban' needs, such as for vacations and reduced working hours. Potential impacts include both further mechanization and reduction in labour-intensive enterprises.

A permanent reduction in the farm labour force, including the pool of farm entrepreneurs, has been a widespread feature of agricultural systems throughout the developed world and has been discussed in some detail in Chapter 3. Generally though, it is possible to see in the overall process of substitution of capital for labour, including farm enlargement and farm labour reduction, a complementary situation between urbanization and agricultural development.

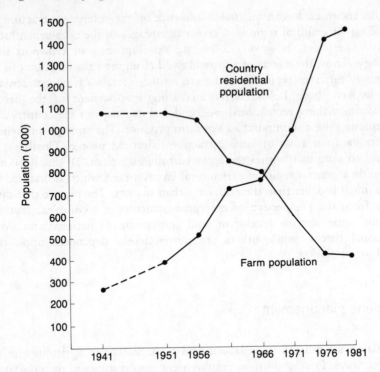

Fig. 10.3   Country residential and farm population, 1941–81, for Canadian regional cities over 40 000 in 1976 (50 km radius) (*Source:* Based on Bryant and Russwurm 1982)

Urbanization stimuli combine with other forces, for example technological change in agriculture, to produce results that up until recently have been regarded as beneficial from the perspective of agricultural development and progress.

Patterns of decline in the farm labour force and in the number of farms (and therefore in the number of farm entrepreneurs) are complex, because regional environments differ so much and because the geographic 'reach' of the pull of non-farm employment has varied substantially over time. Thus, while areas of new agricultural development exist, for example in parts of the Canadian Prairies, few established agricultural areas have been immune from the decline of farm labour and farm numbers (McCuaig and Manning 1982). Many factors influence the patterns, including farm income levels and business size opportunities for agricultural intensification, as well as non-farm employment opportunities. In rural Victoria, Australia, for example, Williams and Mac-Aulay (1971) suggested that decreases in the rural labour force were least, and even some gains were made, in areas where there were irrigation possibilities and land development schemes.

Absolute farm income levels are not the only trigger in the decline, however. The presence of alternative job opportunities is also important (Williams and MacAulay 1971), and relative incomes in comparison to locally or regionally available non-farm employment opportunities are of more interest than absolute income levels in farming. In a study of temporal patterns of farm labour force change by region in Canada, 1946–73, Smit (1979) found that farmers tended to leave agriculture when non-farm employment opportunities and wages were favourable, and similar results were found for unpaid family labour and hired labour. Such impacts are not, of course, confined to the areas surrounding cities, for the mass media have speeded up the diffusion of knowledge of opportunities and life-styles in other regions. None the less, for the early stages of urbanization and industrialization in France, Pautard (1965) argued that the rate of agricultural depopulation was greatest in close proximity to the urban–industrial complex, for example the Paris region, despite the higher absolute farm wages and returns there. This encouraged rapid transformation of the agricultural structure in this region and brought higher levels of mechanization there initially. Wong (1983) also reports that the number of farm machinery types per farm was greater in Canadian metropolitan regions than in the hinterland areas, though the differences were small.

At the individual farm level, too, it is also not necessarily the case that the people to leave the industry first, when agricultural conditions deteriorate, are those with the lowest absolute returns from farming. For example, the farmer who feels the cost of 'opportunities forgone' may be the young, competent entrepreneur with a wide range of options still in front of him/her, just as much as the farmer barely able to make ends meet.

The addition of another activity off the farm can also be a response to either relatively low farm incomes or the availability of alternative employment. It leads to part-time farming or what is now termed 'pluriactivity' by the farm family (Fuller 1990). This combination of farming with another off-farm (farm or non-farm) activity is, however, not a modern phenomenon, and in temperate latitudes has long been encouraged by the seasonality of production demands on the agricultural labour supply. This is one of the principal reasons why in defining part-time farming, some lower limit on the amount and/or share of the entrepreneur's time spent on an off-farm job, and/or on the income received from it, has usually been found necessary (Gasson 1988). In the face of difficult conditions for farming in some areas and increasing non-farm job opportunities, however, the phenomenon has increased to become a significant part of the agricultural structure (Mage 1982; Tweeten 1983 and Table 10.3). This has often resulted in significant concentrations of part-time farming close to major urban areas, for example in south-western Ontario, south-east England and the north-east states in the US. In Canada, while the proportion of farmers reporting off-farm work (Wong 1983) in metropolitan regions by province is sometimes greater than in their respective hinterland areas, the scale of part-time farming (measured by the days of off-farm work per farm reporting off-farm work) is usually greater in the metropolitan

Table 10.3 Proportion of farm households with regular* off-farm work (OFW): the study areas

EC study areas ranked from highest % regular pluriactivity to lowest	% households: with any member with OFW	% households: farmer with regular OFW	% households: spouse (not farmer) with regular OFW	% households: other family members only with regular OFW	% households: with >0.5 AWU† on-farm non-agricultural work
Freyung–Grafenau (FRG)	81	56	4	21	1
Agueda (Portugal)	68	29	22	17	2
Euskirchen (FRG)	65	41	4	20	1
Udine (Italy)	53	20	9	24	29
S. Lazio (Italy)	50	30	6	14	27
Asturias (Spain)	48	14	13	21	2
Languedoc (France)	48	36	8	21	2
Buckingham (UK)	44	25	14	5	14
Maas–Waal (The Netherlands)	44	27	7	10	5
Ireland (East)	43	18	6	19	3
Devon (UK)	43	20	13	10	9
Calabria (Italy)	43	30	2	11	17
Grampian (UK)	40	20	12	8	8
Fthiotis (Greece)	40	26	4	10	4
Catalunya (Spain)	39	13	8	18	4
Korinthia (Greece)	36	22	3	11	7
Andalucia (Spain)	33	18	4	11	1
Ireland (West)	33	14	5	14	4
Savoie (France)	33	15	6	12	7
Picardie (France)	27	8	13	6	1
Average (%) of EC study areas	58	24	8	13	7
Non-EC study areas					
Austria (West)	70	23	15	33	34
Austria (SE)	69	36	12	22	19
West Bothnia (Sweden)	72	57	11	4	3
Le Chablais (Switzerland)	44	31	6	7	8
Average (%) of non-EC study areas	64	36	11	17	17

* Regular off-farm work = full-time and part-time regular employment
† AWU = annual work unit.
(*Source:* Fuller, 1990)

regions than elsewhere. It is therefore not surprising that part–time farming is frequently thought of as an urban–oriented phenomenon.

An interesting set of questions concerns whether part–time farming represents a means of exit from or entry to farming on a full-time basis (Hodge and Whitby 1981; Mage 1982). The first question raises the issue of whether part-time farming is really only a transitional or temporary phenomenon. Clearly, while part-time farming has become a relatively permanent feature of modern agricultural structure (see Table 10.3), it may only represent a transitional stage for the individual farm and farm family. However, Canadian evidence suggests that part-time farming is not dominantly a 'way out' of agriculture, because less than 5 per cent of the full-time farming operators in Ontario in 1966 who stopped farming by 1976 were in the part-time category in 1971 (Mage 1982). Regarding the second question, it is clear that some people also move into part-time farming with a reduced capital outlay, hoping eventually to move into farming full time. This can be illustrated again with Canadian evidence; over 20 per cent of the part-time operators in Ontario in 1966 had achieved full-time farming status by 1976 (Mage 1982).

Part-time farming has other dimensions to it that accentuate the urban-oriented locational structure noted above. For instance, some people have developed a farming enterprise in order to supplement income from their principal activity and/or to provide an income supplement upon retirement – the 'worker-peasant' smallholdings around many of the urban–industrial areas of Western Europe are good examples of this. Finally, another type of part-time farmer is the hobby farmer. Unlike the other types of part-time farmer discussed above, the hobby farmer has added a farming enterprise to his or her existing range of non-farm activities because of the pleasure to be derived from farming. A major distinguishing characteristic of hobby farmers is that they are generally much less dependent upon farming income for their livelihood – though this does not mean the farming income is unimportant absolutely. Hobby farmers are particularly concentrated in urban regions because of their other jobs (Troughton 1976), and in some regions are associated with horse-raising (Thomson 1981). However, they are found on a wide range of farm types although there is a natural tendency to emphasize enterprise with lower labour demands.

There is, therefore, a complex pattern of different types of part-time farming, and this makes analysis of many of the questions posed regarding part-time farming very difficult. Not only do the origins of such people differ widely, but part-time farmers can be found across a wide range of farm business sizes, even though much of the research emphasizes the small-scale part-time farm operation in urban regions (Vail 1987). Part-time, especially hobby, farmers have often been regarded as relatively inefficient farmers. Certainly, labour supply constraints on such operations are important and encourage less labour-intensive forms of farming systems, for example cash cropping or beef production (Troughton 1976). But for operations of similar scale and enterprise mix, the evidence on whether part-time farmers are any less or any more efficient than their full-time counterparts is ambiguous. In Munton's (1983) analysis of land maintenance standards (admittedly not a variable directly related to agricultural productivity) for a sample of farms in London's metropolitan green belt, it was found that part-time farmers had higher maintenance standards, while hobby farmers had lower maintenance standards than full-time farmers.

Where part-time farming becomes a permanent feature for the farmer, other effects may develop. It may act as a brake on farm consolidation and amalgamation, contributing to a fossilization of the farm structure, at least from the land base perspective. This thus creates another 'layering' effect in the farm structure with part-time and hobby farmers as part of the agricultural mosaic around cities. Furthermore, at the lower end of the business-size scale, it is not likely that the farm operation will be able to support the farm family, which reinforces the need for the off-farm activity. This situation has been frowned on in many countries because of the implied view that farming should receive the farmer's undivided attention, and that part-time farming represents a dilution of farming which will lead to its demise. Against this, it should be noted that other forms of agricultural adjustments can take place within the fixed land base; for example, custom and shared field operations provide one way of coping with the high threshold requirements of much farm machinery (e.g. Vail 1987). Thus, while the total farm income generated may be inadequate to support a family, the enterprise itself may none the less be economically viable.

Part-time farming centres upon the involvement of the farm entrepreneur in one or more off-farm activities. However, another important response to income pressures and opportunities is the participation of farm family members in off-farm activities (pluriactivity). This provides a major method of coping with surplus family labour and frequently involves an important tie between the farm and the urban environment. These types of linkages, together with interaction between farm and non-farm population that develops through frequenting the same churches, social organizations and schools, are also partly responsible for other adjustments on the farm because of life-style influences. One of the reasons for a decline in activities, such as dairying and cattle farming in some metropolitan regions, for instance the Paris region, is the combination of scarcity of reliable labour to operate such activities and the desire of many farmers and their families to have more leisure time available. These influences partly account too for some of the 'extras' found increasingly on farm equipment, such as air-conditioned cabins and stereos on major equipment. Data to support such explanations must be derived from direct farm surveys. In the absence of much empirical evidence, the above comments are simply offered as interesting avenues for future research.

Capital allocation

Capital, and access to sources of capital, has become a major preoccupation of modern farming, as economic and technological development forces have dictated more and more non-farm inputs be used in the agricultural production process (see Chapter 3). Capital may be placed into the acquisition of farmland, into other fixed assets (buildings, fixed installations and equipment – though the latter is potentially more flexible because of more rapid resale possibilities) or into working capital. The discussion on enterprise structure and labour has already touched on important ramifications for capital allocation. Thus, changing opportunities for different enterprises clearly have implications for capital requirements and allocation. Furthermore, changing patterns of labour supply and cost, influenced by labour demand in the non-farm sectors, are linked with increasing use of labour-saving technology on the farm, farm enlargement (because of the higher threshold land base requirements of much of the new technology) and reduced labour input.

Other relationships between farming and its urbanized environment influence capital allocation. Close to cities, many observers have noted with dismay the inflation of farmland prices. Farmland, seen as a possible source of development land, has frequently reached prices far in excess of its agricultural use value; and general inflationary conditions, as existed in many developed countries in the 1970s, contributed to this as non-farm interests saw investment in land as a 'safe' investment (Munton 1976). The financial implications of this are twofold, in terms of capital investment in the land resources by farmers and in the context of pressures for farm amalgamation and consolidation in the urban fringe.

First, higher land prices may be sufficient to prevent some farmers from expanding their land base, given that many farmers do prefer to own land. A fossilization effect in terms of the land base is thereby encouraged, although the higher land costs can contribute to quite different forms and conditions of agriculture, for instance, the smaller part-time and hobby farm on the one hand or intensification on the other.

Second, farmland rental may be encouraged. Although this has become an increasingly frequent way of expanding a farm's land base throughout the agricultural system, simply because of the higher demands on farm capital for machinery and equipment, it is reinforced around cities where land prices increase (Bryant et al. 1982). It can lead to permanent improvement in farm structure if the renting environment is regarded as stable, and in this case provides an excellent example of stress (higher land prices) ultimately leading to a 'better' agricultural production environment. It may, however, represent a precarious situation where the land is owned by developers and/or speculators, which then has the potential to contribute to poor husbandry practices and disinvestment in agriculture (Munton 1983).

Disinvestment in farming in urbanizing environments has been, indeed, a major area of interest to geographers (Lucas and van Oort 1991). Disinvestment possibilities can be linked to many of the factors already discussed which might influence the returns from farming – changing labour costs, machinery and equipment costs and changing market opportunities, especially when opportunities for substitution of inputs and enterprises are also limited. Here, we shall confine ourselves to those factors related to urbanization stimuli

291

linked to non-farm demand for land and the relationships between ongoing farm and non-farm land-use activities. These represent some of the indirect impacts, therefore, of non-farm land-use development on the productivity of agricultural investment (Bryant et al. 1982). In any event, these two opposing processes can be present simultaneously, and this partly underlies the observation of polarization in farm size and enterprise structure in some metropolitan regions.

The most widely discussed aspect of disinvestment, and the only one with any significant theoretical framework, is the effect of the anticipation (sometimes discussed in terms of uncertainty) of urban development upon farm investment in areas with rapid and extensive urban growth. The first significant formalization of the ideas was in the Thünian-type analysis put forward by Sinclair (1967). He suggested that where expectations of urban development were high, the value of the land for agricultural investment would be low (Fig. 10.4). The closer to the edge of urban expansion, the lower the value of land for agricultural investment. In extreme situations, disinvestment could take place and even abandonment of the land might occur. Bryant (1974) developed the framework further, focusing on the mechanism involved. High expectations of urban development effectively created shorter planning horizons for farm investment, so that capital investment involving long amortization periods would be most likely to be affected. In this way, continued intensification of certain types of enterprise was still possible (those with few fixed costs) in areas with a high potential for urban development, other intensive forms of production could be expected to decline, and certain forms of more extensive agricultural production could be expected to increase, such as field crops (see Table 10.1). The land value curve for agriculture based on this mechanism, in addition to distance costs with respect to the urban market, first rises and then falls. The curve represents, in a very graphical way, both the conflicts and the opportunities faced by some farm enterprises close to major urban areas (Fig. 10.4) (Boal 1970).

The evidence for the magnitude and significance of this mechanism is ambiguous, partly because the process does not operate in isolation from other forces which may counteract it, and partly because farm operators do not react to the same stimuli in the same way (Moran 1979; Bryant 1981). Thomson's (1981) analysis of agricultural land-use structure and change around the seven metropolitan counties of England, from 1967 to 1974, is typical of aggregate analyses in the urban fringes; the study found evidence both of extensive land uses in the urban fringe (even including some idling of land) as well as the intensification of other types of land use. Moran's (1979) research emphasized the mitigating role of other influences; he observed that adjacent to Auckland's western suburbs, intensive vineyard and orchard areas continued to function normally, partly because of the large amount of capital investment in farming that had already been made. Furthermore, and in a more general vein, adaptive strategies can be developed sometimes to cope with an uncertain future. For example, fixed investment can be more mobile

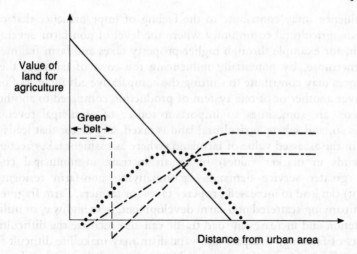

Fig. 10.4 Hypothetical relationships between distance from urban area and value of land for agriculture: (——) Thünian model; (–·–) Sinclair model; (. . .) one example from the Bryant model; (– – –) green belt (*Source:* Bryant and Russwurm. 1982)

than outward appearances suggest, such as easily dismantled greenhouses and barns. Finally, one has to weigh the importance of cost savings through disinvestment against the forgone revenue from a potential decrease in production (due to the disinvestment) and the capital gains from actual sale of land for development. Where the expected development does not materialize, this will only reinforce tendencies not to disinvest.

Because of the lack of specific data on the disinvestment process, it is not possible to evaluate the magnitude of the influence of potential urban development pressures. It would appear reasonable to expect there to be a direct relationship between the level of urban development pressures and the impact of urban development anticipation. The impacts could be hypothesized as related to distance from urban development areas and, depending upon the specific geographic configuration of development pressures, would take on various combinations of concentric and sectoral patterns. Detecting such relationships, however, would require extensive farm-level surveys. These would have to be conducted in a variety of regional environments since different types of agriculture can be expected to yield different responses, giving rise to variation both between and within regions (Lucas and van Oort 1991). Furthermore, another complicating factor is the variability of urban growth pressures over time. In many urban regions in the developed world in the late 1970s and early 1980s, there was a significant slowing of growth, for example the Paris agglomeration and many cities in south-western Ontario; however, in other regions, growth pressures remained very high in the late 1980s, for example around London (UK), or returned to very high levels, for example the Toronto metropolitan region in south-western Ontario.

Other influences may contribute to the feeling of impermanence that can develop in an agricultural community where the level of non-farm development is high, for example through higher property taxes and farm fragmentation. Furthermore, by potentially influencing the costs of farming, these other influences may contribute to shifting the comparative advantages of one enterprise over another or of one system of production compared to another. Property taxes are sometimes an important source of municipal revenue; where this is so, and where agricultural land is taxed, any factor that leads to an increase in the assessed value of farmland (where assessment takes account of local trends in market values) or to an increase in municipal costs (potentially greater service demands accompanying non-farm residential development) can lead to increased property taxes for farmers. Farm fragmentation accompanying scattered non-farm development, new highway or utility line construction and increases in road traffic can also increase the difficulties and the costs of farming. Trespass and vandalism may make life difficult for some types of farmers, such as orchard growers and market gardeners. Finally, other incompatibilities are linked to the adjacency of farm and non-farm land uses. For instance, farm odours can give rise to complaints which may ultimately restrict some forms of agricultural operation such as pig farming. Individually, many of these stimuli or pressures do not appear capable of leading to agricultural decline (Bryant 1981). But when they all combine in an area, many types of farming become beset with difficulties and the agricultural structure may degenerate; for example, dairy farming can be plagued by trespass and carelessness when field gates are left open, and can be faced with severe problems of labour shortages. Wage differentials and differences between urban and rural life-styles can be contributory factors in local disputes over farming practices. The point regarding the reinforcing effect of several pressures is developed in the synthesis later in this chapter.

Land

All of the discussion so far could equally be related to the agricultural land resource – what is produced on it, how it is used and how it is combined with other inputs into a productive farming system. This only serves to underline the interrelatedness of the components of the farm system. One link between farming and urbanization that has not been discussed yet, however, is the actual use and conversion of land for non-farm land uses and activities.

A limited number of non-farm activities can be carried on while the land remains in agricultural use. They mainly involve recreational types of activities, such as cross-country skiing and hiking and more passive leisure-time pursuits such as countryside viewing. Frequently, these activities are carried on without the permission of the farmer; where a public footpath or trail system exist, this is not necessary in any case. These activities can be carried out without interfering with the farm operating system. Yet, they are also a potential source of concern where the farmland is not respected: crops can be trampled down, gates left open and theft of farm produce occur. Evidence on the extent of these problems is very limited, but that they exist is undeniable (Bryant 1981). Even so, in many areas the problems are limited to a small minority of the population. As we shall see later, such recreational activities on farmland have been increasingly fostered in some countries, and finding ways of integrating farmland and farm systems with these non-farm activities in some form of multi-purpose use scheme is a significant challenge.

The actual conversion of farmland to urban use is undoubtedly the most obvious agricultural impact of urbanization. It stimulated a widespread concern during the 1970s regarding the extent of farmland loss and its importance for the capacity of the food production system. Several reviews of the work undertaken on this topic were subsequently published, for example: Best (1981), Bryant (1986a), Bryant et al. (1982). The most important points from the perspective of agricultural production are:

1. Much of the debate centred on the measurement of the rates of land lost from agriculture. The ambiguity of earlier measurements based on official statistics, such as agricultural census returns, is notorious (Best 1981). Rates of presumed loss to urban development were often grossly exaggerated because of the inclusion of losses due to farmland abandonment and afforestation. Increasing use of more direct measurement, such as from air photo analysis, has given a much more realistic picture of actual conversion rates (Ilbery and Evans 1989).
2. The quality of farmland converted was of major concern in many countries. Many cities developed in areas with a good agricultural resource base (Furuseth and Pierce 1982). Not surprisingly, much of the land converted was of good quality; these tendencies in themselves, therefore, do not constitute evidence of development 'prefering' good-quality farmland because of cheaper servicing costs, for example. In the context of accretionary urban development – subdivisions, industrial parks, major institutional developments, such as hospitals and universities – the possibilities for moulding development to avoid better-quality land are limited because of needed scale economies in development and servicing constraints. In the context of development that takes place in the surrounding countryside, or contiguous to smaller settlement nodes, the possibilities are wider. Indeed, some types of non-farm development, such as rural estate development, may well be attracted to wooded areas and rolling topography (Bryant et al. 1982).
3. The debate over farmland loss was overtaken in the 1980s by a concern with 'surplus' farmland, as demonstrated by the structural oversupply of markets in developed countries. The debate has shifted to the problem of what to use 'surplus' farmland for, rather than the possible shortage of farmland.

From the perspective of the individual farm entrepreneur, non-farm land-use development has certainly not always been viewed negatively. The farmer often wears two hats: as farm manager-cum-entrepreneur, who must make a livelihood from farming, and as landowner, who sees in non-farm development the opportunity to realize capital gains.

1. Even in the farming role, the farmer may see an advantage in selling a piece of land for a non-farm residence and utilizing the capital realized for improvements to the operating farm unit. The opportunities for doing this are much greater close to cities (in the absence of tight planning controls to the contrary). In Canada, for example, this is implied in the dramatic increase in the rural non-farm (ex-urban) population in the city's country-side during the 1960s and 1970s (see Fig. 10.3). These opportunities also vary regionally, first because the process of regional city development was initiated much earlier in some regions than in others, and second, because growth pressures are themselves unevenly distributed. Excellent examples of such regional differences can be found in many countries; for example, in Canada, contrast the mushrooming growth of Toronto over several decades and the almost lethargic evolution of Regina by comparison.

2. Farmers age like all of us, and many become interested in the capital value of their farm as a means of financing their retirement. Selling off parts of the farm's land base is just one of the possible links between farm investment decisions and stages in the life cycle of the farm family (Moran 1988).

3. Finally, some farmers have been quick to see that sale of their land at prices far in excess of agricultural use values can permit them to sell and purchase a farm or farms elsewhere, and still leave capital to spare for other improvements. The relocation process implied here, often forced but sometimes consciously planned for, is not a rare phenomenon; the reconstitution of agricultural areas, in the face of urban expansion, through relocation has been observed around many cities, for example Paris (Bryant 1984) and Adelaide (Smith 1972).

Evolving farming landscapes

A picture thus emerges of farming undergoing a complex set of changes in relation to the stimuli produced by the urbanization process, as well as responding to other forces and processes. Most agricultural regions appear to be influenced to some degree, though the various stimuli–response relationships differ significantly in terms of geographic extent within a given region, as well as between regions (see above). It is in the areas close to major cities that farming is perhaps most complex, because all of the urbanization forces as well as a variety of other forces are at work there.

From the discussion so far, it is evident that at the scale of the urban region there is a complex 'layering effect'. Different processes, all involving urbanization stimuli to varying degrees, have created various polarizing tendencies. On the one hand, some processes favour farm enlargement, mechanization and labour reduction; yet other processes favour the creation of small-scale, often part-time farms. On the other hand, some processes may encourage disinvestment while others encourage intensification and specialization. There is no contradiction in these statements – urbanization is a complex process, with both negative and positive ramifications for agriculture. Neither individual farmers, nor the various agricultural sectors, can be expected to respond in a completely homogeneous manner.

Hence, it is common to observe polarization tendencies in agricultural change around major cities. For example, Lawrence (1988) highlights both the increase in the importance of certain intensively produced horticultural enterprises in the metropolitan counties he studied, at the same time as a substantial increase in the sales values of field crops (see Table 10.1). Field crops dominated the innermost counties in terms of area cultivated, but horticultural specialties dominated in terms of value. Gregor's (1988) results also show how, in southern California, the two most urbanized counties (Los Angeles and Orange) in his study area experienced the greatest increase in the share of farms greater than 260 acres (105 ha) in size (see Table 10.2). Finally, Wong (1983) suggests a polarization in terms of farm size in the urban regions of southern Ontario, reflecting both extensification and intensification processes, as well as the development of part-time and hobby farms in metropolitan regions. Thus, any aggregate picture of agriculture in urban regions (see e.g. Tables 10.1 and 10.2) reflects the operation of all the various forces simultaneously, and it is difficult to tease out the relationships without undertaking detailed disaggregated analyses.

In an attempt to summarize the complexities involved, a threefold categorization of evolving farming landscapes can be suggested for agricultural areas around cities (Bryant 1984). Based on the contemporary processes at work, rather than on the purely morphological components of the landscape, it incorporates a broad range of influences, both positive and negative. The framework emphasizes the role of the individual farmer, and also deals with both agricultural productivity and landscape aspects. It, therefore, serves as a useful stepping-stone to the final discussion on land-use planning and management.

Thus, landscapes of (1) agricultural degeneration, (2) agricultural adaptation and (3) agricultural development are distinguished:

1. In *landscapes of agricultural degeneration* it is suggested that a whole range of negative urbanization stimuli combine to produce very difficult conditions for agriculture – excessive fragmentation, very high land prices, serious problems with theft and vandalism, and so forth. In addition, negative non-urbanization forces and regional environment factors are often

associated with such zones of agricultural decline, for example severe competition from other producing areas and a relatively poor physical resource base. Where conditions are extreme, it is even possible for the land to become abandoned and certainly to experience disinvestment.

2. In *landscapes of agricultural adaptation* there are still negative urbanization stimuli, but they are outweighed by positive urbanization stimuli, non-urbanization forces and regional environment influences. Farmers respond to the opportunities present (e.g. enhanced farm enlargement, direct sales) and adapt to the negative stimuli (e.g. fencing of intensively farmed areas).

3. Finally, in *landscapes of agricultural development* negative stimuli related to non-farm development and land use are miminal. So-called 'normal processes' of agricultural development dominate – farm business enlargement, mechanization, labour reduction.

No systematic attempt has yet been made to evaluate the extent of these different categories of changing farm landscapes. However, some evidence from the Paris region and southern Ontario helps to place them into perspective. In the Paris region, one of the corner-stones of land-use planning in the urban fringes has been the 'green belt'; it has been incorporated into the proposed 1980 regional Master Plan and used as a guide in public decision-making since the early 1980s (Bryant 1986b). The 'green belt' includes about 94 000 ha of green and open space lying roughly between 10 and 30 km from the centre of the Paris agglomeration. It includes about 31 000 ha of relatively sound agricultural areas (landscapes of agricultural adaptation and development) and 6 000 ha of fragile agricultural land deemed in need of special protection (landscapes of agricultural degeneration). This compares with about 10 000 ha of agricultural land identified as 'interstitial agriculture' in the mid-1970s (see Bryant 1986b for an account of the research on agriculture in this region). The green belt's fragile agricultural areas, therefore, account for a little over half of this, and the remainder – except where conversion to other uses has occurred – is in other parts of the green belt as somewhat less fragile agricultural areas. Beyond the 'green belt', the agricultural zones can be classed primarily as 'landscapes of agricultural development', with some significant areas of 'agricultural adaptation' closer to the agglomeration. The fragile areas in the 'green belt' are fragile partly because of urban development pressures and partly because of various combinations of poorly adapted farm structures (e.g. highly fragmented and small fields), strong competition (e.g. in tree fruit production), and poor physical environment (e.g. in some of the narrower valley bottoms).

Johnston and Bryant (1987) have described the results of an analysis of 214 farmers undertaken in 1985 in the area between the west edge of Toronto, Hamilton and Guelph. The area contains a great variety of physical environments as well as levels of urbanization. The latter include the city of Brampton, one of the most rapidly developing urbanizing areas in southern Ontario (its population increased from 71 000 to 149 000 between 1971 and

1981). In the interviews, very few unambiguous indicators of agricultural degeneration were found, except right on the edge of urban development zones in Brampton and Oakville. On the one hand, in the most urbanized municipalities, one group of farmers had developed large-scale, relatively extensive cash-cropping operations involving large amounts of rented land, while another group had developed much more intensive operations on smaller land bases, using various forms of direct selling. Interestingly enough, it was in the municipalities intermediate between the rural hinterland and the expanding cities that the proportion of part-time farm operators was highest.

The threefold categorization of farming landscapes presented above is directly related to changes in agricultural productivity and structural change. It can also be linked to landscape changes. Landscapes of 'agricultural degeneration' can be associated with changes viewed negatively from the perspective of landscape amenity – poorly tended fields and buildings, areas fragmented by roads and housing, and even abandoned fields. However, in 'landscapes of agricultural adaptation', some of the positive responses from the agricultural performance perspective, such as greenhouses and intensive market gardening with its roadside piles of packing cases at harvest, are seen as eyesores by the amenity-conscious. And modern farm technology, quite apart from growing criticism from the land and wildlife conservation movements, generates complaints from the amenity-conscious group too – for example, in response to cultivation of former pastureland (the 'idyllic' landscape), silos in proximity to valued architectural sites, and modern metal buildings. These represent a relatively new set of pressures for farming to deal with and, in a sense, represent yet another peculiarity of our urbanized world, with the development of another 'urban' demand on agricultural land – support of amenity. How we cope with the variety of demands on the agricultural land resource, and indirectly on the farm system, is another story in itself.

Land-use planning and management

In terms of the societal or collective significance of many of the relationships discussed above, it has been farming's responses to the urban demand for land for various functions that have attracted the most widespread attention (see Furuseth and Pierce 1982). The negative impacts of these demands on agricultural productivity and performance have been singled out partly because anything that threatens agriculture potentially threatens our very existence. Even so, the significance of many of these impacts noted above have not been satisfactorily evaluated (Bryant 1986a). Given the widespread surplus agricultural production problems in the Western world, the concern has been increasingly expressed in terms of long-term production capability or potential rather than current food production volumes. Put in the context of the different landscapes introduced earlier, the concern can be expressed as how to prevent 'landscapes of agricultural degeneration' from forming and how to upgrade existing ones.

A wide range of land-use planning approaches has been developed to combat the presumed problems (see reviews in Bryant and Russwurm 1982; Furuseth and Pierce 1982; Pacione 1984). With the focus of concern on the land resource, most approaches have relied heavily upon physical land-use planning – including the allocation of land to different uses, the control of property fragmentation processes, control of the form of development and particularly attempts at the proscription of non-farm development in designated agricultural reserves. These approaches have been applied in different forms in a variety of administrative contexts, ranging from relatively centralized systems, like the UK, to very decentralized ones, like the US and Canada. They have ranged from the imposition of controls from relatively centralized authority (e.g. the British Columbia Land Commission Act) to much more voluntary, bottom-up approaches (such as the California Land Conservation Act). They have met with varying degrees of success, which in itself partly reflects the range of demands on the land resources that have to be met somewhere. Furthermore, the priorities that society places on different functions of the land resource have not been constant, either geographically or temporally.

The limitations encountered by planning have been of three types. First, underlying most landscapes of 'agricultural degeneration' are other forces in addition to the urban/non-farm demand for land, for instance alternative and more attractive job opportunities, competition from other producing areas, or farm structures poorly suited to farm modernization. Traditional, physical land-use planning has been almost powerless in the face of such forces which generally operate at a much broader geographic scale. Second, and linked to the first point, there has been a lack of ability to move beyond the traditional scope of land-use planning to incorporate concerted efforts with structural planning (remodelling farm structure at the local level) and to fine-tune the whole management effort to different situations. Third, our ability to influence the quality of the agricultural land converted to urban functions is limited severely in the context of accretionary urban growth (as opposed to more dispersed development in the countryside), because scale economies of development and servicing constraints are powerful imperatives.

These issues are complicated even further by the existence of other collective functions for agricultural land. Landscape or amenity-related functions for agricultural land are many – including the maintenance of a pleasant environment, providing contrast with the built environment (a regional landscape composition function), maintaining the cultural and historic heritage embedded in the landscape, helping shape urban form and providing outdoor recreational opportunities of a relatively informal nature. Several of these have been especially influential in Western Europe; in France, for instance, these concerns for agricultural land in planning near Paris are much more clearly articulated than the concerns for the productive land resource (Bryant 1986b).

Clearly, agricultural land conservation and landscape conservation, or providing access to rural areas for urban populations, are not always easy to reconcile. This point has been underlined by Munton's (1983) work in London's metropolitan green belt. Planning and management for such functions can become viewed by the farming community as an additional set of constraints and pressures to deal with. Once more, the issues involved – the reconciliation of agricultural land conservation, including compatible agricultural development, and landscape conservation – lie beyond the scope that has developed for land-use planning in most countries. It is in such a context that the more persuasive management approaches, founded upon interaction with farmers and property owners on an individual basis, in order to lay out opportunities and strategies, and exemplified in the urban fringe management projects of the UK Countryside Commission and the green belt and natural open space zones (*zones naturelles d'équilibre*) of the Paris region, hold so much interest (Bryant 1984). In terms of the 'landscapes of agricultural degeneration' discussed earlier, the persuasive management approach becomes even more significant when it is realized that many of the agricultural zones in such landscapes are relatively small in geographic extent, and are managed by a relatively small number of farm operators and their families. Thus, the role of individual farm family circumstances assumes greater significance in the fortunes of such zones; they must be thoroughly understood and taken into account in developing effective public strategies for such areas.

Conclusion

The pace of change in farming and the non-farm sectors has quickened in the second half of the twentieth century. The process of urbanization, and the associated ones of industrialization and economic development, have strengthened the ties between the two. Any process of change in a decision-making context has the potential of creating stress, and the ramifications of the urbanization process for farming are no exception. However, opportunities may also be presented and farm entrepreneurs may be able to cope with and manage the stress positively. The schematic threefold categorization of evolving farming landscapes underlines the fact that there are many areas within the orbit of cities in which farming is still alive and well, although inevitably changing.

From the planning perspective, it is important to realize that society's values in agricultural land vary both geographically and through time. Any evaluation of the planning response to the negative impacts of urbanization upon farming must be seen in terms of the broader set of societal goals and objectives. In a similar vein, it is important for students of farming in our urbanized world to appreciate the diversity of farming responses to the various stimuli presented. In a sense, this represents one of the strengths of farming within the orbit of our major cities – it is complex with different strata responding to different mixes of goals and objectives, and it is dynamic. We do need, however, to be concerned about the areas of agricultural degeneration, either because of agricultural production or landscape conservation concerns or both. By taking a more comprehensive view of the farm as an operating system and its relationship to its encompassing environment, we gain in understanding and in appreciating the limitations and opportunities for our management of the processes involved.

References

AREEAR (Atelier Régional d'Etudes Economiques et d'Aménagement Rural) (1976) *L'agriculture spécialisée en Ile de France*. Ministry of Agriculture, Paris

Best, R. H. (1981) *Land use and living space*. Methuen, London

Boal, F. W. (1970) Urban growth and land value patterns. *Professional Geographer* **22**: 79–82

Bryant, C. R. (1974) The anticipation of urban expansion: some implications for agricultural land use practices and land use zoning. *Geografia Polonica* **28**: 93–115

Bryant, C. R. (1981) Agriculture in an urbanizing environment: a case study from the Paris region, 1968 to 1976. *Canadian Geographer* **21**: 27–45

Bryant, C. R. (1984) The recent evolution of farming landscapes in urban-centred regions. *Landscape Planning* **11**: 307–26

Bryant, C. R. (1986a) Agriculture and urban development. In Pacione, M. (ed) *Progress in agricultural geography*. Croom Helm, Beckenham, pp 167–94

Bryant, C. R. (1986b) Farmland conservation and farming landscapes in urban-centred regions: the case of the Ile-de-France region. *Land and Urban Planning* **13**: 251–76

Bryant, C. R. and **Russwurm, L. H.** (1982) North American farmland protection strategies in retrospect. *GeoJournal* **6**: 501–11

Bryant, C. R., Russwurm, L. H. and **McLellan, A. G.** (1982) *The city's countryside: land and its management in the rural–urban fringe*. Longman, London

Bryant, C. R., Russwurm, L. H. and **Wong, S. Y.** (1984) Agriculture in the urban field: an appreciation. In Bunce, M. F. and Troughton, M. J. (eds) *The pressures of change in rural Canada*. Monograph 14, Department of Geography, Atkinson College, York University, Toronto, pp 12–33

Fuller, A. (1990) From part-time farming to pluriactivity: a decade of change in rural Europe. *Journal of Rural Studies* **6**: 361–71

Furuseth, O. J. and **Pierce, J. T.** (1982) *Agricultural land in an urban society*. Association of American Geographers, Washington

Gasson, R. (1988) *The economics of part-time farming*. Longman, London

Gregor, H. F. (1982) *Industrialization of US agriculture: an interpretive atlas*. Westview Press, Boulder, Colorado

Gregor, H. F. (1988) Urbanization and agricultural industrialization in Southern California. Paper presented to the *IGU Commission on Changing Rural Systems*. Auckland, New Zealand

Grigg, D. (1984) *An introduction to agricultural geography*. Hutchinson, London

Hodge, I. and **Whitby, M.** (1981) *Rural employment: trends, options, choices*. Methuen, London

Ilbery, B. W. (1991) Farm diversification as an adjustment strategy on the urban fringe of the West Midlands. *Journal of Rural Studies* **7**:

Ilbery, B. W. and **Evans, N. J.** (1989) Estimating land loss on the urban fringe: a comparison of the agricultural census and aerial photograph/map evidence. *Geography* **74**: 214–21

Johnson, T. G, **Marshall, J. P.** and **O'Dell, C. R.** (1987) A proposed urban agricultural enterprise. In Lockeretz, W. (ed) *Sustaining agriculture near cities*. Soil and Water Conservation Society, Ankeny, Iowa, pp 37–47

Johnston, T. R. and **Bryant, C. R.** (1987) Agricultural adaptation: the prospects for sustaining agriculture near cities. In Lockeretz, W. (ed) *Sustaining agriculture near cities*. Soil and Water Conservation Society, Ankeny, Iowa, pp 9–21

Laureau, X. (1983) Agriculture péri-urbaine: des entreprises pour demain. *L'Agriculture d'Enterprise* **171**: 3–42

Lawrence, H. W. (1988) Changes in agricultural production in metropolitan areas. *Professional Geographer* **40**: 159–75

Lucas, P. and **van Oort, G.** (1991) Response of farmers to the loss of land caused by urban pressure. In van Oort, G., van den Berg, L., Groenendijk, J. and Kempers, A. (eds) *Limits to rural land use*. Pudoc, Wageningen, pp 96–104

McCuaig, J. D. and **Manning, E. W.** (1982) *The evolution of agricultural land use in Canada: process and consequences*. Lands Directorate, Environment Canada, Ottawa.

Mage, J. (1982) The geography of part-time farming. *GeoJournal* **6**: 301–12

Moran, W. (1979) Spatial patterns of agriculture on the urban periphery: the Auckland case. *Tijdschrift voor Economische en Sociale Geografie* **70**: 164–76

Moran W (1988) The farm equity cycle and enterprise choice. *Geografiska Annaler* **20**: 84–91

Morgan, W. B. and **Munton, R. J.** (1971) *Agricultural geography*. Methuen, London

Munton, R. J. (1976) An analysis of price trends in the agricultural land market of England and Wales. *Tijdschrift voor Economische en Sociale Geografie* **67**: 202–12

Munton, R. J. (1983) *London's green belt: containment in practice*. John Wiley and Sons, London

Pacione, M. (1984) *Rural geography*. Harper and Row, London

Pautard, J. (1965) *Les disparités régionales dans la croissance de l'agriculture française*. Gauthier-Villars, Paris

Sinclair, R. J. (1967) Von Thünen and urban sprawl. *Annals of the Association of American Geographers* **57**: 72–87

Smit, B. (1979) Regional employment changes in Canadian agriculture. *Canadian Geographer* **23**: 1–17

Smith, D. L. (1972) The growth and stagnation of an urban fringe market gardening region – Virginia, South Australia. *Australian Geographer* **12**: 35–48

Smith, S. N. (1987) Farming near cities in a bimodal agriculture. In Lockeretz, W. (ed) *Sustaining agriculture near cities*. Soil and Water Conservation Society, Ankeny, Iowa, pp 77–90

Thomson, K. J. (1981) *Farming in the fringe: an analysis of agricultural census data drawn from parishes around the six metropolitan counties and London*. Countryside Commission, Cheltenham

Troughton, M. J. (1976) *Landholding in a rural–urban fringe environment: the case of London, Ontario*. Occasional Paper 11, Lands Directorate, Environment Canada, Ottawa

Tweeten, L. (1983) The economics of small farms. *Science* **219**: 1037–41

Vail, D. (1987) Suburbanization of the countryside. In Lockeretz, W. (ed) *Sustaining agriculture near cities*. Soil and Water Conservation Society, Ankeny, Iowa, pp 23–36

Williams, D. B. and **MacAulay, T. G.** (1971) Changes in rural population and work force in Victoria, 1961–66. *Australian Geographical Studies* **9**: 161–71

Wong, S. Y. (1983) Agricultural change in Canada: 1941–1976. Unpublished PhD thesis, Department of Geography, University of Waterloo, Canada

Conclusion

An attempt has been made in this book to integrate the 'traditional themes' of agricultural geography with newly emerging interpretations and research issues. For instance, to the traditional concern with agricultural regions, farming types, agricultural resources and farmer decision-making (Bowler 1984) has been added the emerging literature on agricultural industrialization, the urban impact on farming, agriculture–environment relationships and political economy interpretations of agricultural restructuring. The concept of the 'food supply system' (see Fig. 1.1), also termed the agro-food system, has been used to integrate this broader understanding of the processes that impinge on contemporary agriculture, although no attempt has been made to cover every possible topic. Moreover, the farm or production sector has been retained as the integrating focus of the book.

The approach adopted has been to identify the main 'structural' or macro-economic, political and social processes acting upon farming as a land-based productive activity. From this platform, an exploration has been conducted of those processes and structures in agriculture that have resulted in the spatially differentiated or uneven transformation of the farm sector. As agriculture has become more dependent on external industrial and finance capitals, both upstream and downstream of the farm sector, so different regional agricultural systems have been transformed in varied ways and at different times. More formally, new territorial divisions of capital and labour have emerged, each farming region displaying geographical and historical specificity in the out-come (Marsden et al. 1986). Regional farm sizes, farm types, land tenures and natural resource endowments appear to be the main mediating structures in the uneven development of agriculture, with external capitals and state intervention forming two major processes of transformation. But the uneven development of agriculture is also evident at the level of individual farms. Contemporary agriculture is not characterized by unilinear trends in its transformation; rather, individual farm businesses have scope for manoeuvre under, and resistance to, the influence of external capitals as they seek to transform the food supply system. Diversity, not uniformity, is evident in trends between farms as well as farming regions. In this respect, attention has

305

been given to the diverse internal relations of the farm business as well as their external relations with non-farm capitals and the state (see Table 4.9). Even so, compared with the industrial sector, agriculture shows a degree of inertia in its regional and local farming systems.

The industrialization of farming, as described in this book, has taken place during a 'productionist' phase in the development of agriculture. The 1980s, nevertheless, have witnessed the development of an 'international farm crisis' which probably signals the end of that phase and the entry, in the 1990s, into a post-productionist era. Faced by the escalating costs of state-financed farm price support programmes, the oversupply of domestic markets, rising farm indebtedness, increasingly unacceptable environmental impacts of modern farming practices, and the internationalization of agro-industrial capital, a new economic and political agenda for agriculture is emerging. Three trends seem particularly influential: further developments in the biotechnology associated with food production; the deeper integration of the farm population into regional rural economies; and the production of 'environmental goods' as a valid use of farmland. Each trend has implications for the development of individual farm businesses, farming regions and national agricultural sectors.

On *biotechnology* (a fourth agricultural revolution?) most commentators identify the increasing yields available in the future to both crop and livestock production from developments in genetic engineering. New crop varieties, for example, are likely to be more efficient in transforming inorganic fertilizers into useful food, more drought-resistant, more resistant to particular pests and diseases, and able to be harvested within a shorter growing season. Powerful economic interests support the new biotechnology: it complements the existing industrial model of agriculture and offers scope for a new cycle of agro-industrial accumulation (Commission of the European Communities 1989). Food processors, for example, view developments in biotechnology as an assurance of the continuation of the supply of their cheap raw materials, while competition between countries, regions and individual farms is likely to ensure the adoption of such developments. These two aspects are linked: food processors are now organized as 'food transnationals' and seek out their raw materials competitively between producers at a global scale (Bijman et al. 1986). Biotechnology, therefore, is likely to perpetuate the development of large, agro-industrial farm businesses and the intensively farmed agricultural regions in which they are located.

Turning now to *agriculture and the rural economy*, a trend is already under way for the farm population to be integrated into the wider rural economy by either taking off-farm employment (pluriactivity) or diversifying their farm businesses through non-farm enterprises (see p. 156). Evidence has been presented already that both developments are selective by farm business and farming region; non-farming diversification, for example, is most prevalent in urban fringe and tourist regions, with an emphasis on farm accommodation, equine enterprises and direct farm sales. Pluriactivity and diversification are often viewed as twin processes in the development of an 'alternative agriculture', in which farm production itself plays a much reduced role in providing an income for the farm family.

The increased validity of *producing 'environmental goods'* is the third likely trend for agriculture in the 1990s (OECD 1989). Already farmers in the EC are being financed to manage their land in environmentally sensitive ways, whether through Environmentally Sensitive Areas (ESAs) (Potter 1988), planting more farm woodland, reducing the application of nitrogen, or providing access to and through farmland for recreational purposes. In the UK, financial compensation for not destroying wildlife habitats, countryside stewardship payments and farm pollution control grants are already available; Potter (1986) and Robinson (1991) have described these measures in more detail. The 1990s are likely to see the state becoming more involved in financing and developing programmes which encourage the production of environmental goods, for example by ensuring that set-aside land (see p. 268) produces greater environmental benefits.

Just as there has been no unitary pattern in the contemporary transformation (restructuring) of agriculture, so the competitive and atomistic structure of the farm sector will ensure a varied response to these three projected trends. Adopting the typology of farm business structure introduced on p. 23, six 'paths of farm business development' can be anticipated:

1. Extension of the industrial model of farm business development based on traditional products and services (agro-industry);
2. Redeployment of farm resources (including human capital) into new agricultural products or services (farming diversification);
3. Redeployment of farm resources (including human capital) into new non-farm products or services (non-farming diversification);
4. Redeployment of human capital into an off-farm occupation (pluriactivity);
5. Maintenance of traditional farm production and services with either reduced inputs and/or reduced income (marginalized businesses);
6. Hobby or semi-retired farming.

While an individual farm business may combine two or more of these 'paths', for example by combining agro-industrial farming (1) with farming diversification (2), for most farms the typology reduces to three strategies for future business development. These are:

(a) maintaining a full-time, profitable food production to the farm business (paths 1 and 2);
(b) diversifying the income base of the farm business by restructuring resources into non-farm enterprises and occupations (paths 3 and 4);
(c) surviving as a marginalized farm business at a low level of income, perhaps supported by investment income, pensions or other direct state payments.

Individual farm businesses are likely to make a transition from strategies (a) to (b) and to (c) under the joint effects of competitive market capitalism and the farm-family life cycle. More difficulty is likely to be encountered in making the reverse transition owing to the need for scale economies and/or large capital investments to re-enter profitable full-time farming or diversified enterprises. Farm businesses adopting strategy (a) (pathways 1 and 2) will have to asborb the consequences of developments in biotechnology, including the production of generic food components for the food-processing industries (e.g. starch, glucose, vegetable protein), with a structure of relatively few but very large production units able to supply most food requirements for the population.

Following the general argument developed in this book, farm business strategies, when taken in aggregate, are likely to translate into regional variations in the dominant type of farm business development. Urban fringe areas, for example, may well be dominated by paths 3 and 4, and upland/ mountain farming regions by paths 3 and 5. This theme has been explored for the EC by Conrad (1987: 240); he speculates that three ideal–typical 'agro-structural types of region' will result:

Type A. Regions in which intensive agriculture is practised, especially to produce food. These areas generally have good quality soils and are agriculturally competitive. [paths 1 and 2; strategy (a)]

Type B. Regions constituting mainly agrarian–touristic peripheries (with) skilful coordination and completion of production processes and services in the primary, secondary and tertiary sectors, the production of local specialities, the exploitation of market niches and the supply of regional markets. [paths 3 and 4; strategy (b)]

Type C. Less favoured areas with hardly any positive prospects for future development and a marked tendency towards depopulation (under forestry, nature protection and extensive agriculture). [paths 5 and 6; strategy (c)]

Finally, an issue is beginning to emerge in North America that transcends the debate on 'paths of farm business development': it concerns the very *sustainability of agriculture* in the medium to long term, including the relationship between agricultural production and the prospective changes in global climatic regimes (Parry 1990). The debate surrounding 'sustainable agriculture' has been introduced into the geographical literature by Brklacich et al. (1990), Pierce (1990, 297–313) and Troughton (1991), although in origin the ideas can be traced to the late 1960s and early 1970s when societal concern with the over-exploitation of natural resources first emerged (Merril 1976). On the definition of 'sustainability', there is no single objective function, there being at least three criteria to be met. Summarizing, these are:

(a) *environmental* – to maintain or enhance the agricultural resource base;
(b) *socio-economic* – to provide equitable economic rewards to individual farms and rural communities in the production sector;
(c) *productionist* – to produce a sufficient food supply.

Keeney (1988: 102) puts these criteria together in the following definition of 'sustainability': 'agricultural systems that are environmentally sound, profitable, and productive and that maintain the social fabric of the rural community'. To these goals of 'sustainability', however, should be added:

(d) *budgetary* – to absorb an acceptable proportion of state (public) expenditure;
(e) *political* – to maintain the political support of society.

In Western Europe these last two goals are particularly important. While there are few prospects of state assistance to farming being withdrawn altogether, the escalating budgetary costs of the CAP will have to be brought within 'sustainable' limits. Moreover, political support for the farm sector, both among politicians and voters, is necessary to sustain farm support programmes. At present, such political support is ebbing, if successive attitude surveys amongst the general public are a guide.

Looking at the application of 'sustainability' to farming systems, to date the debate has been conducted largely in macro-conceptual or theoretical terms (e.g. Hill 1985) or else in the context of agro-ecosystems (e.g. Jackson 1988; Troughton 1991). Although some writers emphasize that 'sustainable agriculture' is a goal rather than a set of prescriptions for exactly how agricultural systems should operate, checklists of compatible farming practices are widely available. These include: diversified land use, integration of crop and livestock farming, crop rotations, organic manures, nutrient recycling, low energy inputs, low inputs of agrichemicals, and biological disease control. When applied to farming, the checklists produce a continuum of increased application of ecological principles to the farm business:

1. *Diversified farming*: the introduction of a variety of crops (including timber) and/or livestock into the farm business, including crop rotations and 'environmentally friendly' farming practices (environmental goods). Other farming practices associated with agro-industry need not be affected.
2. *Low input–output farming*: a reduction of purchased inputs per hectare of farmland leading to reduced levels of production. Farming practices associated with agro-industry can be continued at a lower level of intensity, compensated by 'environmentally friendly' farming practices.
3. *Organic farming*: removal of agrichemicals and inorganic fertilizers from farm inputs. Ecological farming practices are required to compensate for the loss of 'industrial' inputs.

It can be argued that all three types of 'sustainable' agriculture are already recognized in the EC under the CAP: permissive legislation to stimulate diversification, extensification (low input–output) and conversion (organic) is certainly enacted (Regulations 797/85, 1094/88 and 1272/88 respectively). However, with the exception of organic farming – for which, at the time of writing, a detailed EC programme of support is still being devised – the existing programmes do not meet the challenge of significantly reducing the energy and agrichemical dependencies of modern farming, nor lowering the level of price support needed to maintain their economic viability. Organic farming appears to meet most of the criteria for a 'sustainable' agriculture in the medium term, and could be adopted by farm businesses as a type 2 pathway of development. At present, however, there are too many vested interests in maintaining the agro-industrial system of food production, while economists have questioned the conversion costs to organic farming, its capacity to feed the population, and its economic viability as far as large farms are concerned (Bateman and Lampkin 1986). The geography of 'sustainable agriculture' has yet to be explored but, together with the other potential agricultural trends outlined in this Conclusion, it will probably form a central theme in the next generation of books which seek to examine the geography of agriculture in developed market economies.

References

Bateman, D. and **Lampkin, N.** (1986) Economic implications of a shift to organic agriculture in Britain. *Agricultural Administration* **22**: 89–104

Bijman, J., van der Doel, K. and **Junne, G.** (1986) *The international dimension of biotechnology in agriculture.* European Foundation for the Improvement of Living and Working Conditions, Dublin

Bowler, I. R. (1984) Agricultural geography. *Progress in Human Geography* **8**: 255–62

Brklacich, M., Bryant, C. R. and **Smit, B.** (1990) Review and appraisal of the concept of sustainable food production systems. *Environmental Management* **15**: 1–14

Commission of the European Communities (1989) *The impact of biotechnology on agriculture in the European Community to the year 2005.* The Commission, Brussels

Conrad, J. (1987) Alternative land use options in the European Community. *Land Use Policy* **4**: 229–42

Hill, S. B. (1985) Redesigning the food system for sustainability. *Alternatives* **12**: 32–6

Jackson, W. (1988) Ecosystem agriculture: the marriage of ecology and agriculture. In Allen, P. and van Dusen, D. (eds) *Global perspectives on agroecology and sustainable agricultural systems.* University of California, Santa Cruz, pp 15–19

Keeney, D. R. (1989) Toward a sustainable agriculture: need for clarification of concepts and terminology. *Journal of Alternative Agriculture* **4**: 101–5

Marsden, T., Munton, R., Whatmore, S. and **Little, S.** (1986) Towards a political economy of capitalist agriculture: a British perspective. *International Journal of Urban and Regional Research* **10**: 489–521

Merril, R. (ed) (1976) *Radical agriculture.* Harper and Row, New York

OECD (Organization for Economic Co-operation and Development)

(1989) *Agriculture and environmental policies: opportunities for integration.* OECD, Paris

Parry, M. (1990) *Climate change and world agriculture.* Earthscan, London

Pierce, J. T. (1990) *The food resource.* Longman, London

Potter, C. (1986) The environmental effects of CAP reform. *Countryside Planning Yearbook* **7**: 76–88

Potter, C. (1988) Environmentally Sensitive Areas in England and Wales: an experiment in countryside management. *Land Use Policy* **5**: 301–13

Robinson, G. M. (1991) EC agricultural policy and the environment: land use implications in the UK. *Land Use Policy* **8**: 95–107

Troughton, M. J. (1991) Ecological assessment of modern agriculture. In van Dort, G., van den Berg, L., Groenendijk, J., and Kempers, A. (eds) *Limits to rural land.* Pudoc, Wageningen, pp 141–202

Index